Lecture Notes in Computational Science and Engineering

10

Editors
M. Griebel, Bonn
D. E. Keyes, Norfolk
R. M. Nieminen, Espoo
D. Roose, Leuven
T. Schlick, New York

Springer
Berlin
Heidelberg
New York
Barcelona
Hong Kong
London
Milan
Paris
Singapore
Tokyo

Hans Petter Langtangen
Are Magnus Bruaset
Ewald Quak
Editors

Advances in Software Tools for Scientific Computing

With 138 Figures and 32 Tables

Editors

Hans Petter Langtangen
Department of Informatics
University of Oslo
P.O. Box 1080
0316 Oslo, Norway
e-mail: hpl@ifi.uio.no

Are Magnus Bruaset
Numerical Objects A.S.
P.O. Box 124 Blindern
0314 Oslo, Norway
e-mail: amb@nobjects.com

Ewald Quak
SINTEF
Applied Mathematics
P.O. Box 124 Blindern
0314 Oslo, Norway
e-mail: Ewald.Quak@math.sintef.no

Library of Congress Cataloging-in-Publication Data applied for

Die Deutsche Bibliothek - CIP-Einheitsaufnahme
Advances in software tools for scientific computing / Hans Petter Langtangen ... ed. - Berlin ; Heidelberg ; New York ; Barcelona ; Hong Kong ; London ; Milan ; Paris ; Singapore ; Tokyo : Springer, 2000
(Lecture notes in computational science and engineering ; 10)
ISBN 3-540-66557-9

Mathematics Subject Classification (1991): 65-02, 65F01, 65F50, 65G10, 65L05, 65M60, 65P05, 65Y05, 65Y25, 68-02, 68N15, 73G20, 73V15, 76M10

ISBN 3-540-66557-9 Springer-Verlag Berlin Heidelberg New York

Cover design: Friedhelm Steinen-Broo, Estudio Calamar, Spain
Cover production: *design & production* GmbH, Heidelberg
Typeset by the authors using a Springer TEX macro package

Printed on acid-free paper SPIN 10723676 46/3143 - 5 4 3 2 1 0

Preface

To make full use of the ever increasing hardware capabilities of modern computers, it is necessary to speedily enhance the performance and reliability of the software as well, and often without having a suitable mathematical theory readily available. In the handling of more and more complex real-life numerical problems in all sorts of applications, a modern object-oriented design and implementation of *software tools* has become a crucial component. The considerable challenges posed by the demand for efficient object-oriented software in all areas of scientific computing make it necessary to exchange ideas and experiences from as many different sources as possible.

Motivated by the success of the first meeting of this kind in Norway in 1996, we decided to organize another *International Workshop on Modern Software Tools for Scientific Computing*, often referred to as *SciTools'98*. This workshop took place in Oslo, Norway, September 14–16, 1998. The objective was again to provide an open forum for exchange and discussion of modern, state-of-the-art software techniques applied to challenging numerical problems. The organization was undertaken jointly by the research institute SINTEF Applied Mathematics, the Departments of Mathematics and Informatics at the University of Oslo, and the company Numerical Objects AS. In total, 80 participants from 11 different countries took part in the workshop, featuring one introductory and five invited addresses along with 29 contributed talks. Based on this program, 11 research papers were selected, which - after undergoing a careful refereeing process - form the contributions to this book. It is a pleasure for the editors to thank all authors and all other speakers and participants at SciTools'98 for their part in making this workshop a success.

Reflecting the goals of the workshop, the contributions in this volume cover a wide spectrum of important research topics in object-oriented programming: the numerical treatment of both ordinary and partial differential equations, software libraries for numerical linear algebra, and the evaluation and quality testing of existing libraries. Application areas include structural analysis, computational fluid dynamics (CFD), material technology, and geographical information systems (GIS).

The previous SciTools'96 meeting also resulted in a book, *Modern Software Tools for Scientific Computing*, edited by E. Arge, A. M. Bruaset and H. P. Langtangen (Birkhäuser, 1997). Both that book and the present one provide numerous examples on how to utilize modern software tools in various areas of scientific computing. The books also report experiences with the development and application of such tools. This type of information is particularly valuable as long as the amount of literature on scientific computing software is still sparse. In the present book, we have focused more on the *ideas* underlying software design and abstractions. Recently available technologies, such as templates in C++ and the Java language, are focused. Furthermore, several new topics are included: sparse grids, direct solvers, validated methods, parallel ODE solvers and GIS. The very advanced applications in Zabaras' chapter should be of particular interest to engineers. All chapters provide lots of ideas and experience with numerical software design that should be accessible to a reader even if the particular application area is outside the reader's special area of interest.

Siek and Lumsdaine present the design and functionality of a STL-like linear algebra library in C++, seeking to combine the convenience of template programming with high performance. Veldhuizen's work presents Blitz++, an array library in C++, which is based on expression templates and template metaprogramming in order to achieve high performance. Dobrian, Kumfert, and Pothen provide a first attempt at designing an object-oriented package for general sparse system solvers. Zumbusch reviews the basic concepts of sparse grids for the efficient numerical solution of higher-dimensional PDEs and shows how the corresponding techniques can be implemented in a modular and flexible software framework. The chapter by Arthur offers an examination of the current advantages and disadvantages of using Java, comparing it to C, C++, F90, Perl, Python and Tcl. Nedialkov and Jackson give an overview of the theory and the implementation of interval arithmetic as a tool for increasing the reliability of numerical simulations, with special focus on ordinary differential equations. The contribution of Barnes and Hopkins presents the results of a study on the evolution of software quality of the widely used LAPACK library from netlib. The chapter by Zabaras deals with very complicated mathematical models from material technology, involving heat transfer, viscous flow and large deformations during metal forming as well as optimization and control of such forming processes. He demonstrates how the resulting very complex simulation software can be designed cleanly using object-oriented techniques. Okstad and Kvamsdal present adaptive finite element methods based on field recovery and error estimation, using an object-oriented software framework, with applications to structural analysis and CFD. Petcu and Drăgan outline the requirements for flexible software for the solution of systems of ODEs, including an example of such software. Finally, in the chapter of Halbwachs and Hjelle, it is described how object-oriented programming techniques are used in software for geographical information systems.

As mentioned before, all 11 chapters in this volume underwent a thorough peer review, and the editors would like to thank all the referees who gave freely of their time to judge the submitted manuscripts. Their comments were a great and indispensable help in order to secure the highest possible quality of this book.

The editors also take this opportunity to thank the Norwegian Research Council for their financial support by way of the strategic research program *Numerical Computations in Applied Mathematics* (grant no. 110673/420).

The longstanding support granted by the council was indispensable in creating and maintaining the scientific environment which enabled the organization of the SciTools meetings and the preparation of this volume.

Many thanks go to Roger Hansen for his struggle with putting all the contributions together in book form according to the publisher's style requirements. The assistance from the staff at Springer-Verlag in Heidelberg, especially Martin Peters, Leonie Kunz and Thanh-Ha Le Thi, is also greatly acknowledged.

Hans Petter Langtangen
Are Magnus Bruaset
Ewald Quak

Oslo, August 1999

Table of Contents

A Modern Framework for Portable High-Performance Numerical Linear Algebra

Jeremy Siek and Andrew Lumsdaine

Laboratory for Scientific Computing, University of Notre Dame, USA
Email: {jsiek,lums}@lsc.nd.edu

Abstract. In this chapter, we present a generic programming methodology for expressing data structures and algorithms for numerical linear algebra. We also present a high-performance implementation of this approach, the Matrix Template Library (MTL). As with the Standard Template Library, our approach is five-fold, consisting of generic functions, containers, iterators, adapters, and function objects, all developed specifically for high-performance numerical linear algebra. Within our framework, we provide generic functions corresponding to the mathematical operations that define linear algebra. Similarly, the containers, adapters, and iterators are used to represent and to manipulate concrete linear algebra objects such as matrices and vectors. To many scientific computing users, however, the advantages of an elegant programming interface are secondary to issues of performance. There are two aspects to how we achieved high performance in the MTL. The first is the use of *static polymorphism* (template functions) with modern optimizing compilers, enabling extreme flexibility with no loss in performance. The second is the application of abstraction to the optimization process itself. The Basic Linear Algebra Instruction Set (BLAIS) is presented as an abstract interface to several important performance optimizations. Our experimental results show that MTL with the BLAIS achieves performance that is as good as, or better than, vendor-tuned libraries, even though MTL and the BLAIS are written completely in C++. We therefore conclude that the use of abstraction is not a barrier to performance, contrary to conventional wisdom, and that certain abstractions can in fact facilitate optimization. In addition, MTL requires orders of magnitude fewer lines of code for its implementation, with the concomitant savings in development and maintenance effort.

1 Introduction

Software construction for scientific computing is a difficult task. Scientific codes are often large and complex, representing vast amounts of domain knowledge. They also process large data sets so there is an additional requirement for efficiency and high performance. Considerable knowledge of modern computer architectures and compilers is required to make the necessary optimizations, which is a time-intensive task and further complicates the code.

The last decade has seen significant advances in the area of software engineering. New techniques have been created for managing software complexity and building abstractions. Underneath the layers of new terminology (object-oriented, generic [33], aspect-oriented [22], generative [9], metaprogramming [36]) there is a core of solid work that points the way for constructing better software for scientific computing: software that is portable, maintainable and achieves high performance at a lower price.

One important key to better software is better abstractions. With the right abstractions each aspect of the software (domain specific, performance optimization, parallel communication, data-structures etc.) can be cleanly separated, then handled on an individual basis. The proper abstractions reduce the code complexity and help to achieve high-quality and high-performance software.

The first generation of abstractions for scientific computing came in the form of subroutine libraries such as the Basic Linear Algebra Subroutines (BLAS) [11,12,19], LINPACK [10], EISPACK [32], and LAPACK [1]. This was a good first step, but the first generation libraries were inflexible and difficult to use, which reduced their applicability. Moreover the construction of such libraries was a complex and expensive task. Many software engineering techniques (then in their infancy) could not be applied to scientific computing because of their interference with performance.

In the last few years significant improvements have been made in the tools used for expressing abstractions, primarily in the maturation of the C++ language and its compilers. The old enmity between abstraction and performance can now be put aside. In fact, abstractions can be used to aid performance portability by making the necessary optimizations easier to apply. With the intelligent use of modern software engineering techniques it is now possible to create extremely flexible scientific libraries that are portable, easy to use, highly efficient, and which can be constructed in far fewer lines of code than has previously been possible.

We present the Matrix Template Library (MTL) [23], a package for high-performance numerical linear algebra. In the construction of MTL, we have focused on the use of generic and metaprogramming techniques to deliver a concise implementation in C++ of BLAS-level functionality for a wide variety of matrix formats and arithmetic types.

In the first sections we focus on how generic programming idioms guided the design of the MTL, while in the later sections we discuss performance issues such as the ability of modern C++ compilers to optimize abstractions and how template metaprogramming techniques can be used to express loop optimizations. We do not explicitly discuss the "object-orientedness" of the MTL, but it should be clear that this is an underlying theme.

The Matrix Template Library is in its second generation. The first version focused on abstractions at the vector level [20,21]. The current version of MTL has been redesigned using generic and meta-programming techniques.

2 Generic Programming and the Standard Template Library

First we give a short description of generic programming with a few examples from the Standard Template Library (STL). For a more complete description of STL refer to [34]. For an introduction to STL and generic programming refer to [2,25,35]. For the rest of the chapter we will assume the reader has a basic knowledge of STL and generic programming.

Generic programming has recently entered the spotlight with the introduction of the Standard Template Library (STL) into the C++ standard [13]. The principal idea behind generic programming is that many algorithms can be abstracted away from the particular data structures on which they operate. Algorithms typically need the functionality of *traversing* through a data structure and *accessing* its elements. If data structures provide a standard interface for these operations, generic algorithms can be freely mixed and matched with data structures (called *containers* in STL).

The main facilitator in the separation of algorithms and containers in STL is the *iterator* (sometimes called a "generalized pointer"). Iterators provide a mechanism for traversing containers and accessing their elements. The interface between an algorithm and a container is specified by the types of iterators exported by the container. Generic algorithms are written solely in terms of iterators and never rely upon specifics of a particular container. Iterators are classified into broad categories, some of which are: `InputIterator`, `ForwardIterator`, and `RandomAccessIterator`. Figure 1 depicts the relationship between containers, algorithms, and iterators.

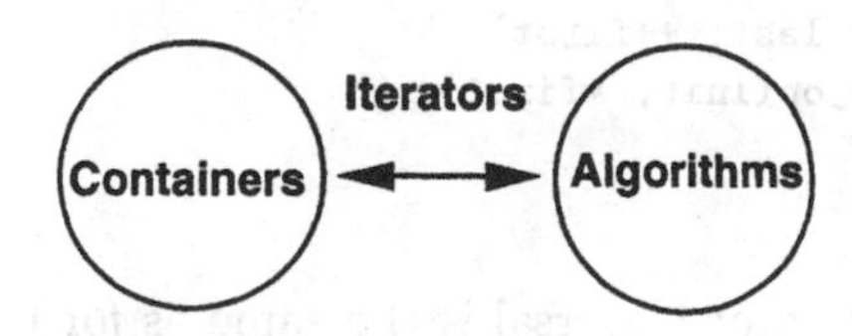

Fig. 1. Separation of containers and algorithms using iterators.

The STL defines a set of requirements for each class of iterators. The requirements are in the form of which operations (functions) are defined for each iterator, and what the meaning of the operation is. As an example of how these requirements are defined, an excerpt from the requirements for the STL `RandomAccessIterator` is listed in Table 2. In the table, `X` is the iterator type, `T` is the element type pointed to by the iterator, `(a,b,r,s)` are iterator objects, and `n` is an integral type.

For a concrete example of generic programming we will look at the algorithm `accumulate()`, which successively applies a binary operator to each element of a container. A typical use of `accumulate()` would be to sum

expression	return type	note
a == b	bool	*a == *b
a < b	bool	b - a > 0
*a	T&	dereference a
a->m	U&	(*a).m
++r	X&	r == s $\rightarrow$ ++r == ++s
--r	X&	r == s $\rightarrow$ --r == --s
r+=n	X&	same as n of ++r
a + n	X	{ tmp = a; return tmp += n; }
b - a	Distance	(a<b) ? distance(a,b) : -distance(b,a)
a[n]	convertible to T	*(a + n)

Table 1. Excerpt from the STL random-access iterator requirements.

the elements of a container, using the addition operator. The code below shows how one could implement the accumulate() algorithm in C++. The first and last arguments are iterators that mark the beginning and the end of the container. All of the arguments to the function have template types, so that the algorithm can be used with any container that exports the InputIterator interface.

```
template <class InputIterator, class T, class BinaryOperator>
T accumulate(InputIterator first, InputIterator last,
             T init, BinaryOperator binary_op)
{
  for (; first != last; ++first)
    init = binary_op(init, *first);
  return init;
}
```

The syntax for iterator traversal is the same as for pointers, specifically operator++() increments to the next position. There are several other ways to move iterators (especially random access iterators), as listed in Table 2. To access the container element pointed to by the iterator, one uses the dereference operator*(). One can also use the subscript operator[]() to access at an offset from the iterator. The code below is an example of how one could use the accumulate() template function with a vector and with a linked list (both from the STL).

```
// using accumulate with a vector
std::vector<double> x(10,1.0);
double sum1 = accumulate(x.begin(),x.end(),0.0,plus<double>());
// using accumulate with a linked list
std::list<double> y;
// copy vector's values into the list
```

```
std::copy(x.begin(), x.end(), std::back_inserter(y));
double sum2 = accumulate(y.begin(),y.end(),0.0,plus<double>());
assert(sum1 == sum2); // they should be equal
```

Each STL container provides its own iterators that fulfill the standard interface requirements, but that are implemented in terms of the concrete data structure. In C++ this is facilitated by the ability to nest type definitions within classes. Consider the example below of a vector and list class. The `Vector` class implements its iterator by just using a pointer. The `List`'s iterator is implemented in terms of the `node` class. Note that in each class, the iterator type has the same name, i.e., `iterator`. The common naming scheme gives a uniform way to access the iterator types within a container.

```
template <class T> class Vector {
  typedef T value_type;
  typedef T* iterator;          // Vector::iterator is a pointer
  ...
};
template <class T> class List {
  typedef T value_type;
  struct node { T data; node* next; };
  class iterator {              // List::iterator is a nested class
    node* curr;
    iterator& operator++() { curr = curr->next; return *this; }
  };
  ...
};
```

When dealing with some container class, we can access the correct type of iterator using the double-colon scope operator, as is demonstrated below in the function `foo()`.

```
template <class ContainerX, class ContainerY>
void foo(ContainerX& x, ContainerY& y) {
  typename ContainerX::iterator xi;
  typename ContainerY::iterator yi;
  ...
}
```

3 Generic Programming for Linear Algebra

The advantages of generic programming coincide with the library construction problems of numerical linear algebra. The traditional approach for developing numerical linear algebra libraries is combinatorial in its required development effort. Individual subroutines must be written to support every desired combination of algorithm, basic numerical type, and matrix storage format. For a library to provide a rich set of functions and data types, one might need to code literally hundreds of versions of the same routine. As an example, to provide basic functionality for selected sparse matrix types, the NIST implementation of the Sparse BLAS contains over 10,000 routines and a custom code generation system [29].

The combinatorial explosion in implementation effort arises because, with most programming languages, algorithms and data structures are more tightly coupled than is conceptually necessary. That is, one cannot express an algorithm as a subroutine independently from the type of data that is being operated on. As a result, providing a comprehensive linear algebra library — much less one that also offers high performance — would seem to be an overwhelming task.

Fortunately, certain modern programming languages, such as Ada and C++, support generic programming by providing mechanisms for expressing algorithms independent of the specific data structure to which they are applied. A single function can then work with many different data structures, drastically reducing the size of the code. As a result, maintenance, testing, and optimization become much easier.

If we are to create the generic algorithms for numerical linear algebra, there must be a common interface, a common way to access and traverse the vectors and matrices of different types. The STL has already provided a model for traversing through vectors and other one-dimensional containers by using iterators. In addition, the STL defines several numerical algorithms such as the `accumulate()` algorithm presented in the last section. Thus creating generic algorithms to encompass the rest of the Level-1 BLAS functionality [19] is relatively straightforward.

Matrix operations are slightly more complex, since the elements are arranged in a two-dimensional format. The MTL algorithms process matrices *as if* they are containers of containers (the matrices are not necessarily implemented this way). The matrix algorithms are coded in terms of *iterators* and *two-dimensional iterators*, as depicted in Figure 2. An algorithm can choose which row or column of a matrix to process using the two-dimensional iterator. The iterator can then be dereferenced to produce the row or column vector, which is a first class STL-style container. The one-dimensional iterators of the row vector can then be used to traverse along the row and access individual elements.

Figure 3 shows how one can write a generic matrix-vector product using iterators. The `matvec_mult` function is templated on the matrix type and the iterator types (which give the starting points of two vectors). The first two lines of the algorithm declare variables for the TwoD and OneD iterators that will be used to traverse the matrix. The `Matrix::row_2Diterator` expression extracts the `row_2Diterator` type from the `Matrix` type to declare the variable `i`. Similarly, the `Matrix::RowVector::iterator` expression extracts the `iterator` type from the `RowVector` of the matrix to declare variable `j`.

We set the `row_2Diterator` `i` to the beginning of the matrix with `i = A.begin_rows()`. The outer loop repeats until `i` reaches `A.end_rows()`. For

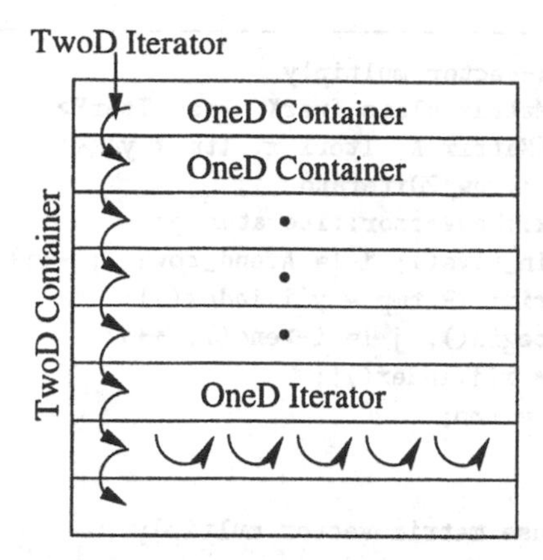

Fig. 2. The TwoD iterator concept.

the inner loop we set the `RowVector::iterator j` to the beginning of the row pointed to by `i` with `j = i->begin()`. The inner loop repeats until `j` reaches the end of the row `i->end()`. The computation for the matrix-vector product consists of multiplying an element from the matrix `*j` by the appropriate element in vector `x`. The index into `x` is the current column, which is the position of iterator `j`, given by `j.index()`. The result is accumulated into `tmp` and stored into vector `y` according to the position of iterator `i`.

The generic matrix-vector algorithm in Figure 3 is extremely flexible, and can be used with a wide variety of dense, sparse, and banded matrix types. For purposes of comparison we list the traditional approaches for coding matrix-vector products for sparse and dense matrices. Note how the indexing in the MTL routine has been abstracted away. The traversal across a row goes from `begin()` to `end()`, instead of using explicit indices. Also, the indices used to access the `x` and `y` vectors are abstracted through the use of the `index()` method of the iterator. The `index()` provides a uniform way to access index information regardless of whether the matrix is dense, sparse, or banded. A more detailed discussion of banded matrices is given in Section 5.6.

4 MTL Algorithms

The Matrix Template Library provides a rich set of basic linear algebra operations, roughly equivalent to the Level-1, Level-2 and Level-3 BLAS, though the MTL operates over a much wider set of datatypes. The Matrix Template Library is unique among linear algebra libraries because each algorithm (for the most part) is implemented with just one template function. From a software maintenance standpoint, the reuse of code gives MTL a significant advantage over the BLAS [11,12,19] or even other object-oriented libraries like TNT [28] (which still has different subroutines for different matrix formats).

```
// generic matrix-vector multiply
template <class Matrix,class IterX,class IterY>
void matvec_mult(Matrix A, IterX x, IterY y) {
  typename Matrix::row_2Diterator i;
  typename Matrix::RowVector::iterator j;
  for (i = A.begin_rows(); i != A.end_rows(); ++i) {
    typename Matrix::PR tmp = y[i.index()];
    for (j = i->begin(); j != i->end(); ++j)
      tmp += *j * x[j.index()];
    y[i.index()] = tmp;
  }
}
// BLAS-style dense matrix-vector multiply
for (int i = 0; i < m; ++i) {
  double t = y[i];
  for (int j = 0; j < n; ++j)
    t += a[i*lda+j] * x[j];
  y[i] = t;
}
// SPARSPAK-style sparse matrix-vector multiply
for (int i = 0; i < n; ++i) {
  double t = y[i];
  for (int k = ia[i]; k < ia[i+1]; ++k)
    t += a[k] * x[ja[k]];
  y[i] = t;
}
```

Fig. 3. Simplified example of a generic matrix-vector product, with a comparison to the traditional approach to writing dense and sparse matrix-vector products.

Because of the code reuse provided by generic programming, MTL has an order of magnitude fewer lines of code than the Netlib Fortran BLAS [26], while providing much greater functionality and achieving significantly better performance. The MTL implementation is 8,284 words for algorithms and 6,900 words for dense containers. The Netlib BLAS total 154,495 words and high-performance versions of the BLAS (with which MTL is competitive) are even more verbose.

In addition, the MTL has been designed to be easier to use than the BLAS. Data encapsulation has been applied to the matrix and vector information, which makes the MTL interface simpler because input and output is in terms of matrix and vector objects, instead of integers, floating point numbers, and pointers. It also provides the right level of abstraction to the user — operations are in terms of linear algebra objects instead of low-level programming constructs such as pointers.

Function Name	Operation
Vector Algorithms	
`set(x,alpha)`	$x_i \leftarrow \alpha$
`scale(x,alpha)`	$x \leftarrow \alpha x$
`s = sum(x)`	$s \leftarrow \sum_i x_i$
`s = one_norm(x)`	$s \leftarrow \sum_i \mid x_i \mid$
`s = two_norm(x)`	$s \leftarrow (\sum_i x_i^2)^{\frac{1}{2}}$
`s = infinity_norm(x)`	$s \leftarrow \max \mid x_i \mid$
`i = max_index(x)`	$i \leftarrow$ index of max $\mid x_i \mid$
`s = max(x)`	$s \leftarrow$max x_i
`s = min(x)`	$s \leftarrow$min x_i
Vector Vector Algorithms	
`copy(x,y)`	$y \leftarrow x$
`swap(x,y)`	$y \leftrightarrow x$
`ele_mult(x,y,z)`	$z \leftarrow y \otimes x$
`ele_div(x,y,z)`	$z \leftarrow y \oslash x$
`add(x,y)`	$y \leftarrow x + y$
`s = dot(x,y)`	$s \leftarrow x^T \cdot y$
`s = dot_conj(x,y)`	$s \leftarrow x^T \cdot \bar{y}$
Matrix Algorithms	
`set(A, alpha)`	$A \leftarrow \alpha$
`scale(A,alpha)`	$A \leftarrow \alpha A$
`set_diagonal(A,alpha)`	$A_{ii} \leftarrow \alpha$
`s = one_norm(A)`	$s \leftarrow max_i(\sum_j \mid a_{ij} \mid)$
`s = infinity_norm(A)`	$s \leftarrow max_j(\sum_i \mid a_{ij} \mid)$
`transpose(A)`	$A \leftarrow A^T$
Matrix Vector Algorithms	
`mult(A,x,y)`	$y \leftarrow Ax$
`mult(A,x,y,z)`	$z \leftarrow Ax + y$
`tri_solve(T,x)`	$x \leftarrow T^{-1}x$
`rank_one_update(A,x,y)`	$A \leftarrow A + xy^T$
`rank_two_update(A,x,y)`	$A \leftarrow A + xy^T + yx^T$
Matrix Matrix Algorithms	
`copy(A,B)`	$B \leftarrow A$
`swap(A,B)`	$B \leftrightarrow A$
`add(A,C)`	$C \leftarrow A + C$
`ele_mult(A,B,C)`	$C \leftarrow B \otimes A$
`mult(A,B,C)`	$C \leftarrow AB$
`mult(A,B,C,E)`	$E \leftarrow AB + C$
`tri_solve(T,B,alpha)`	$B \leftarrow \alpha T^{-1} B$

Table 2. MTL generic linear algebra algorithms.

Adapter Class	Helper Function
`scaled1D`	`scaled(x)`
`scaled2D`	`scaled(A)`
`strided1D`	`strided(x)`
`row/column`	`trans(A)`

Table 3. MTL adapter classes and helper functions for creating algorithm permutations.

Function Invocation	Operation
`add(x,y)`	$y \leftarrow x + y$
`add(scaled(x,alpha),y)`	$y \leftarrow \alpha x + y$
`add(x,scaled(y,beta))`	$y \leftarrow x + \beta y$
`add(scaled(x,alpha),scaled(y,beta))`	$y \leftarrow \alpha x + \beta y$

Table 4. Permutations of the `add()` operation made possible with the use of the `scaled()` adapter helper function.

Table 2 lists the principal operations implemented in the MTL. One would expect to see many more variations on the operations to take into account transpose and scaling permutations of the argument matrices and vectors — or at least one would expect a "fat" interface that contains extra parameters to specify such combinations. The MTL introduces a new approach to creating such permutations. Instead of using extra parameters, the MTL provides matrix and vector *adapter classes*. An adapter object wraps up the argument and modifies its behavior in the algorithm.

Table 3 gives a list of the MTL adapter classes and their helper functions. A helper function provides a convenient way to create adapted objects. For instance, the `scaled()` helper function wraps a vector in a `scaled1D` adapter. The adapter causes the elements of the vector to be multiplied by a scalar inside of the MTL algorithm. There are two other helper functions in MTL, `strided()` and `trans()`. The `strided()` function adapts a vector so that its iterators move a constant stride with each call to `operator++`. The `trans()` function switches the orientation of a matrix (this happens at compile time) so that the algorithm "sees" the transpose of the matrix. Table 4 shows how one can create all the permutations of scaling for a `daxpy()`-like operation.

The example below shows how the matrix-vector multiply algorithm (generically written to compute $y \leftarrow A \times x$) can also compute $y \leftarrow A^T \times \alpha x$ with the use of adapters to transpose `A` and to scale `x` by `alpha`. Note that the adapters cause the appropriate changes to occur within the algorithm; they are not evaluated before the call to `matvec::mult()` (which would hurt performance). The adapter technique drastically reduces the amount of code that must be written for each algorithm. Section 5.6 discusses the details of how these adapter classes are implemented.

```
// y <- A' * alpha x
matvec::mult(trans(A), scaled(x, alpha), y);
// equivalent operation using BLAS
dgemv('T', M, N, alpha, A_ptr, A_ld, x_ptr, 1, 1.0, y_ptr, 1);
```

To demonstrate the MTL algorithm interface in use, we present an extended example of LU factorization, implementing both the pointwise algorithm as well as the blocked algorithm using the MTL functions.

4.1 Pointwise LU Factorization Example

The algorithm for LU factorization is given in Figure 4 and the graphical representation of the algorithm is given in Figure 5, as it would look part way through the computation. The black square represents the current pivot element. The horizontal shaded rectangle is a row from the upper triangle of the matrix. The vertical shaded rectangle is a column from the lower triangle of the matrix. The L and U labeled regions are the portions of the matrix that have already been updated by the algorithm. The algorithm has not yet reached the region labeled A'.

for $i = 1 \ldots N$
  find maximum element in column(i) (starting below the diagonal)
  swap the row of maximum element with row(i)
  scale column(i) by $1/A(i,i)$
  let $A' = A(i:N, i:N)$
  $A' \leftarrow A' + xy^T$ (rank one update)

Fig. 4. LU factorization pseudo-code.

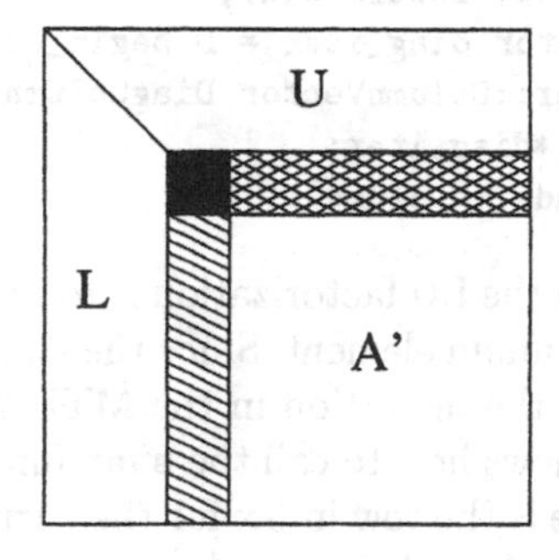

Fig. 5. Diagram for LU factorization.

We create an upper and lower triangle view of the original matrix A. We use typedefs to create the specific triangular matrix type from the MTL class triangle. We then create the U and L matrix objects and pass the matrix A to their constructors. The L and U objects are just handles, so their creation is inexpensive (constant time complexity).

```
typedef triangle<Matrix, unit_upper> Unit_Upper;
typedef triangle<Matrix, unit_lower> Unit_Lower;
Unit_Upper U(A);
Unit_Lower L(A);
```

We can access the rows and columns from L and U through their TwoD iterators, which are created with the begin_rows() and begin_columns() methods. Dereferencing the TwoD iterators gives a row or column.

```
typedef typename Unit_Upper::RowVector UpperRow;
typedef typename Unit_Lower::ColumnVector LowerColumn;
Unit_Upper::row_2Diterator row_iter = U.begin_rows();
Unit_Lower::column_2Diterator column_iter = L.begin_columns();
// ...
UpperRow row = *row_iter;
LowerColumn column = *column_iter;
```

The first operation in the LU factorization is to find the maximum element in the column. The operation operates on a single vector (a column) so it would be in the vec namespace of the MTL. The MTL algorithms are divided into several namespaces depending on the type of arguments they apply to, similar to the Level-1,2,3 division of the BLAS. The namespaces are vec, vecvec, mat, matvec, and matmat.

When we search for the maximum element, we want to include the element in the pivot row, so we cannot use the L triangle object defined above since it is unit diagonal. Instead we create another lower triangle D. We can then simply apply the vec::max_index() function to the dcolumn object.

```
typedef triangle<Matrix, lower> D(A);
Lower::column_2Diterator diag_iter = D.begin_columns();
typedef typename Lower::ColumnVector DiagColumn;
DiagColumn dcolumn = *diag_iter;
int max = vec::max_index(dcolumn);
```

The next operation in the LU factorization is to swap the current row with the row that has the maximum element. Since the swapping operation acts on two vectors, we will find the operation in the MTL vecvec namespace. The following code snippet shows how to call the swap function to interchange the rows. The pivot variable is the row index for the current pivot element. Note how rows from the matrix A are being used as arguments in the vector-vector swap algorithm. In MTL, rows and columns are first class vector objects (with some caveats for sparse rows and columns). Similarly, submatrices in

MTL are first class matrix objects, and can be used in the same way as a matrix.

```
if (max != pivot)
  vecvec::swap(A.row_vector(pivot), A.row_vector(max));
```

The third operation in the LU factorization is to scale the column under the pivot by $1/A(i,i)$, which can be performed with the `scale()` MTL algorithm, which we apply to the `column` variable defined earlier (from the lower triangle). An exception is thrown if the pivot is zero. The MTL uses the C++ exception handling mechanisms to guard against such divide by zero situations, and also to ensure proper dimensions of vector and matrix arguments.

```
PR s = diag[0];
if (s == PR(0)) throw zero_pivot();
s = PR(1) / s;
vec::scale(column, s);
```

The final operation in the LU factorization is to update the trailing submatrix according to $A' = A' + xy^T$. This operation acts on a matrix and some vectors, so it can be found in the MTL `matvec` namespace. We first create the submatrix A' from matrix A using the `sub_matrix()` method. The algorithm `rank_one_update()` can then be applied to the submatrix, using the `row` and `column` variables from the upper and lower triangle. The complete LU factorization implementation is given in Figure 6.

```
Matrix subA = A.sub_matrix(next, next,
                           A.nrows() - next, A.ncols() - next);
matvec::rank_one_update(subA, scaled(column, PR(-1)), row);
```

4.2 Blocked LU Factorization Example

The execution time of many linear algebra operations on modern computer architectures can be decreased dramatically through blocking to increase cache utilization [5,6]. In algorithms where repeated matrix-vector operations are done (`rank_one_update()` for instance), it is beneficial to convert the algorithm to use matrix-matrix operations to introduce more opportunities for blocking.

The LU factorization algorithm can be reformulated to use more matrix-matrix operations [15]. First we split matrix A into four submatrices, using a blocking factor of r (i.e., A_{11} is $r \times r$).

$$A = \begin{bmatrix} A_{11} & A_{12} \\ A_{21} & A_{22} \end{bmatrix}$$

```
template <class Matrix, class Pvector>
void lu_factorize(Matrix& A, Pvector& ipvt) throw(zero_pivot)
{
  typedef typename Matrix::PR PR; // numeric precision type
  // several other typedefs ...
  Lower D(A);  Unit_Upper U(A);  Unit_Lower L(A);

  Lower::column_2Diterator diag_iter = D.begin_columns();
  Unit_Upper::row_2Diterator row_iter = U.begin_rows();
  Unit_Lower::column_2Diterator column_iter = L.begin_columns();

  while (diag_iter.index() < std::min(A.nrows(), A.ncols())) {
    DiagColumn diag = *diag_iter;
    UpperRow row = *row_iter;
    LowerColumn column = *column_iter;
    // find pivot
    int max = vec::max_index(diag);
    int pivot = diag_iter.index();
    int next = pivot + 1;
    ipvt[pivot] = max + 1;
    // swap the rows
    if (max != pivot)
      vecvec::swap(A.row_vector(pivot), A.row_vector(max));
    // make sure pivot is not zero
    PR s = diag[0];
    if (s == PR(0)) throw zero_pivot();
    s = PR(1) / s;
    // update the column under the pivot
    vec::scale(column, s);
    // update the submatrix
    Matrix subA = A.sub_matrix(next, next, A.nrows() - next,
                               A.ncols() - next);
    matvec::rank_one_update(subA, scaled(column, PR(-1)), row);
    ++diag_iter; ++row_iter; ++column_iter;
  }
  ipvt[diag_iter.index()] = diag_iter.index() + 1;
  DiagColumn diag = *diag_iter;
  PR s = diag[0];
  if (s == PR(0)) throw zero_pivot();
  s = PR(1) / s;
}
```

Fig. 6. Complete MTL version of pointwise LU factorization.

Then we formulate $A = LU$ in terms of the blocks.

$$\begin{bmatrix} A_{11} & A_{12} \\ A_{21} & A_{22} \end{bmatrix} = \begin{bmatrix} L_{11} & 0 \\ L_{21} & L_{22} \end{bmatrix} \begin{bmatrix} U_{11} & U_{12} \\ 0 & U_{22} \end{bmatrix}$$

From this we can derive the following equations for the submatrices of A. The matrix on the right shows the values that should occupy A after one step of the blocked LU factorization.

$$\begin{aligned} A_{11} &= L_{11}U_{11} \\ A_{12} &= L_{11}U_{12} \\ A_{21} &= L_{21}U_{11} \\ A_{22} &= L_{21}U_{12} + L_{22}U_{22} \end{aligned} \qquad \begin{bmatrix} L_{11} \setminus U_{11} & U_{12} \\ L_{21} & \tilde{A}_{22} \end{bmatrix}$$

We find L_{11}, U_{11}, and L_{21} by applying the pointwise LU factorization to the combined region of A_{11} and A_{21}. We then calculate U_{12} with a triangular solve applied to A_{12}. Finally we calculate $\tilde{A}_{22}$ with a matrix product of L_{21} and U_{12}. The algorithm is then applied recursively to $\tilde{A}_{22}$.

In the implementation of block LU factorization MTL can be used to create a partitioning of the matrix into submatrices. The use of the submatrix objects throughout the algorithm removes the redundant indexing that a programmer would typically have to do to specify the regions for each submatrix for each operation. The `partition()` method of the MTL matrix cuts the original matrix along particular rows and columns. The partitioning results in a new matrix whose entries are the submatrices. In the code below we create the partitioning that corresponds to Figure 7 with the matrix object `Ap`. The partitioning for Figure 8 is created with matrix object `As`.

```
index_list row_sep2 = j, j + jb;
index_list col_sep2 = j;
Matrix Ap = A.partition(row_sep2, col_sep2);

index_list row_sep1 = j, j + jb;
index_list col_sep1 = j, j + jb;
Matrix As = A.partition(row_sep1, col_sep1);
triangle<Matrix, unit_lower> L_11(As(1,1));
```

Figure 7 depicts the block factorization part way through the computation. The matrix is divided up for the pointwise factorization step. The region including A_{11} and A_{21} is labeled A_1. Since there is pivoting involved, the rows in the regions labeled A_0 and A_2 must be swapped according to the pivots used in A_1.

The implementation of this step in MTL is very concise. The `Ap(1,1)` submatrix object is passed to the `lu_factorize()` algorithm. We then use the `multi_swap()` function on `Ap(1,0)` and `Ap(1,2)` to pivot their rows to match `Ap(1,1)`.

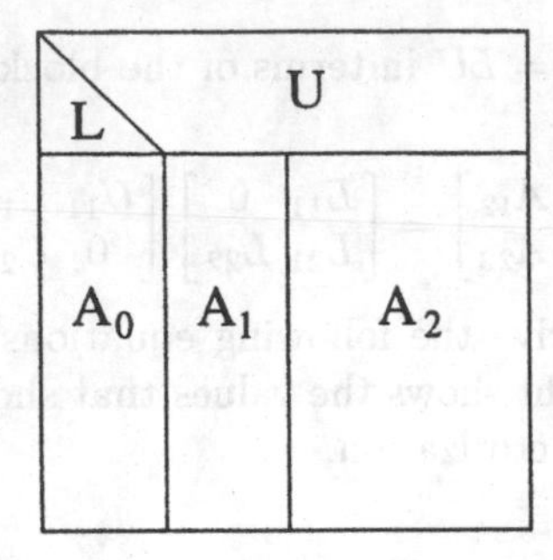

Fig. 7. Pointwise step in block LU factorization.

```
Pvector pivots(jb);
lu_factorize(Ap(1,1), pivots);
multi_swap(Ap(1,0), pivots);
multi_swap(Ap(1,2), pivots);
```

Once A_1 has been factored, the A_{12} and A_{22} regions must be updated. The submatrices involved are depicted in Figure 8. The A_{12} region needs to be updated with a triangular solve. This version of triangular solve can be found in the `matmat` namespace of the MTL.

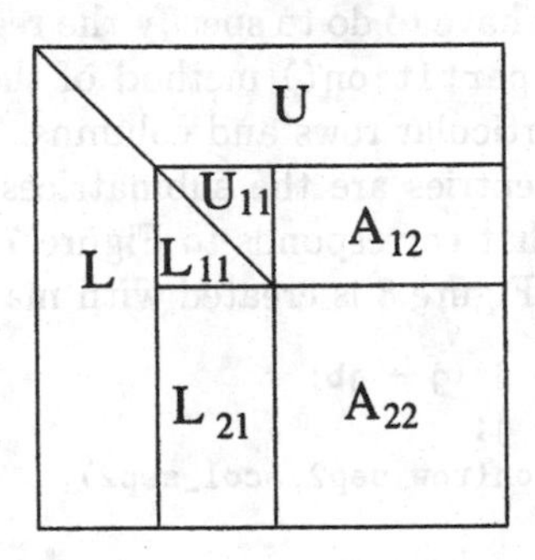

Fig. 8. Update steps in block LU factorization.

To apply the `tri_solve()` algorithm to L_{11} and A_{12}, we merely call the MTL routine and pass in the `L_11` and `As(1,2)` matrix objects.

```
matmat::tri_solve(L_11, As(1,2), matmat::left_side());
```

The last step is to calculate $\tilde{A}_{22}$ with a matrix-matrix multiply according to $\tilde{A}_{22} \leftarrow A_{22} - L_{21}U_{12}$. To implement this operation we use the `As(1,2)` and `As(2,1)` matrix objects, which have been overwritten in the previous steps with U_{12} and L_{21}.

```
matmat::mult(scaled(As(2,1), PR(-1)), As(1,2), As(2,2));
```

```
template <class Matrix, class Pvector>
void block_lu(Matrix& A, Pvector& ipvt) throw(zero_pivot)
{
  typedef typename Matrix::PR PR; // numeric precision type
  // more typedefs ...
  const int BF = LU_BF; // blocking factor
  const int M = A.nrows();
  const int N = A.ncols();
  Pvector pivots(BF);

  if (min(M, N) <= BF || BF == 1)
    lu_factorize(A, ipvt);

  for (int j = 0; j < min(M, N); j += BF) {
    int jb = min(min(M, N) - j, BF);

    // set up the submatrices
    index_list row_sep2 = j, j + jb;
    index_list col_sep2 = j;
    Matrix Ap = A.partition(row_sep2, col_sep2);

    index_list row_sep1 = j, j + jb;
    index_list col_sep1 = j, j + jb;
    Matrix As = A.partition(row_sep1, col_sep1);

    triangle<Matrix, unit_lower> L_11(As(1,1));

    // Factor block A_1
    lu_factorize(Ap(1,1), pivots);
    // record the pivot indices in ipvt
    for (int i = j; i < min(M, j + jb); ++i)
      ipvt[i] = pivots[i-j] + j;
    // interchanges the rows in A_0
    if (j > 0)
      multi_swap(Ap(1,0), pivots);
    if (j + jb < M) {
      // interchange the rows in A_2
      multi_swap(Ap(1,2), pivots);
      // update block row A_12
      matmat::tri_solve(L_11, As(1,2), left_side());
      // update A_22
      if (j + jb < M)
        matmat::mult(scaled(As(2,1), PR(-1)), As(1,2), As(2,2));
    }
  }
}
```

Fig. 9. MTL version of block LU factorization.

The complete version of the MTL block LU factorization algorithm is given in Figure 9.

5 MTL Components

The Matrix Template Library defines a set of data structures and other components for representing linear algebra objects. An MTL matrix is constructed with layers of components. Each layer is a collection of classes that are templated on the lower layer. The bottom-most layer consists of the numerical types (`float`, `double`, `interval` etc.). The next layers consist of OneD containers followed by TwoD containers. The TwoD containers are wrapped up with an *orientation*, which in turn is wrapped with a *shape*. The code below shows the general form of the layering, as well as a couple of concrete examples. Note that some TwoD containers also subsume the OneD type, such as the contiguous `dense2D` container.

```
// general form
shape < orientation < twod < oned < num_type > > > >
// triangular matrix
typedef triangle<column<array2D<dense1D<double > > >, upper>
   UpperTriangle;
// dense contiguous matrix, extended precision
typedef row< dense2D< doubledouble > > MatrixEP;
```

The MTL matrix type specifications may seem intimidating but typically the user would use `typedef`s as in the above example to create nicknames for the particular matrix types he or she uses. In addition, a simplified matrix specification interface is planned for the next version of MTL.

Each layer in the MTL matrix plays a specific role. The numeric type encapsulates the representation of the matrix element, in addition to defining the meaning of the basic operations over the elements such as addition and multiplication. The storage layer of the MTL matrix (the OneD and TwoD components) specify where in memory each row or column can be found, and how one can traverse through the elements. The role of the orientation and shape layers of MTL is to associate the appropriate row and column indices with each element that is stored.

To illustrate the roles of each layer, we will step through all the layers of a dense row-major and a dense column-major contiguous matrix in the implementation of the element access operation `operator()(int i, int j)` (where `i` is the row index and `j` is the column index of the desired element). The role of the MTL storage layer is to specify where in memory an element is to be found. Therefore the `operator()(i,j)` at the storage level (for a rectangular dense matrix) looks like this:

```
reference operator()(int major_index, int minor_index) {
  return *(major_index * ld + minor_index);
}
```

There is no reference to row or column (i or j) at the storage level of the operation. The storage level concepts consist only of the TwoD (major axis) and OneD (minor axis) containers. Here the major_index refers to which OneD container the element can be found in. The minor_index picks out the particular element within the OneD container. A dense matrix is typically represented with a contiguous piece of memory. The leading dimension (ld) is the length of the stride from the beginning of a OneD container to the beginning of the next OneD container. For a row-major matrix ld equals the number of columns. For a submatrix, the ld is not the number of columns of the submatrix, but the number of columns in the original matrix.

There are other storage formats for dense matrices, such as the banded and packed formats of the BLAS. The following code shows how one might implement the operator()(i,j) for the banded storage format. ndiag is the number of diagonals in the band, and bandwidth is a pair that consists of the sub (the number of diagonals below the main diagonal) and the super (the number of diagonals above the main diagonal).

```
// banded matrix storage
reference operator()(int major_index, int minor_index) {
  return *(i * ndiag + max(0, sub(bandwidth) - i) + j);
}
```

Though *orientation* is the next layer in the matrix template interface, in terms of the implementation of operator()(i,j), the shape adapters come next. For a banded matrix, only the elements in the band are stored. For instance, the first element of each OneD segment is not at minor index 0, but at index $max(majorindex - sub, 0)$. Therefore one must calculate this starting index for each OneD segment and subtract that from the minor_index. The implementation is as follows:

```
// banded shape adapter
reference operator()(int major_index, int minor_index) {
  return twod(major_index, minor_index - max(major_index
                                      - sub(bandwidth), 0));
}
```

The diagonal shape adapter is much different. In this matrix format, the diagonals of the matrix are stored contiguously in memory. To access an element in the TwoD storage, one must calculate which diagonal the element is in, and then how far down the diagonal the element lies. The implementation is as follows:

```
// diagonal shape adapter
reference operator()(int major_index, int minor_index) {
  size_type diag = major_index - minor_index + super(bandwidth);
  if (major_index >= minor_index) return twod(diag, minor_index);
  else return twod(diag, major_index);
}
```

The final level in the implementation of the operator()(i,j) is the orientation adapters, which map row and column indices to major and minor indices. Here is the code for the row-major and column-major adapters.

```
// row-major adapter
reference operator()(int i, int j) { return shape(i,j); }
// column-major adapter
reference operator()(int i, int j) { return shape(j,i); }
```

The concrete classes in each of the template layers will be discussed in the following sections, though we will first take some space to discuss the MTL object model.

5.1 MTL Object Memory Model

The MTL object memory model is handle based. This differs from the Standard Template Library. Object copies and assignment are *shallow*. This means that when one vector object is assigned to another vector object it becomes a second handle to the same vector. The same applies to matrices. The example below demonstrates this. The vector y reflects the change made to vector x, since they are both handles to the same vector. Vector z, however, does not reflect the change in x, since it is a different vector.

```
mtl::dense1D<double> x(5, 1.0);
print_vector(x);
> 1 1 1 1 1
mtl::dense1D<double> y = x;
mtl::dense1D<double> z(N);
vecvec::copy(x, z);
x[2] = 3;
print_vector(y);
> 1 1 3 1 1
print_vector(z);
> 1 1 1 1 1
```

The main reason that the handle-based object model was used for MTL was that MTL makes heavy use of adapter helper functions to modify the arguments to MTL algorithms (see Section 5.6). The adapter functions return temporary objects that are typically passed as arguments directly to MTL algorithms. If the MTL algorithms use pass-by-reference for all their arguments then the C++ compiler will emit a warning. Using const pass-by-reference solves the problem for **in** parameters but not for **out**. MTL solves this problem by instead passing the matrix and vector arguments by value, and by making the matrix and vector objects handles to the underlying data structures.

The MTL containers make considerable use of STL containers (though we use our own higher-performance version of the vector container). The underlying STL objects within the MTL OneD containers are reference counted to make the memory management easier for the user. This is especially helpful when the MTL matrix and vector object are used in the construction of larger object-oriented software systems.

In C++ the reference counting can be non-intrusive to the classes that must be reference counted. This is especially important if one wishes to use containers supplied by other libraries (such as STL). The implementation of reference counting smart pointers in Figure 10 derives from the `Handle` class in [35].

5.2 OneD Iterators

The OneD iterators used in the Matrix Template Library are a slight variation of the STL iterators. For the most part the iterators used in MTL containers fall into the `random_access_iterator_tag` category, though many of the MTL algorithms only require `input_iterators`. The only operation the MTL adds to the requirements is for an `index()` method, which returns the current position of the iterator. For an iterator traversing a row of a matrix, the `index()` method would return the current column number. As mentioned in Section 3, the `index()` method allows algorithms for sparse and dense matrices to be unified. Table 5 gives the requirement list for the MTL OneD iterators. There are a couple performance issues that come up in the design of iterators. We discuss these in detail in Section 6.1.

expression	return type	note
`a.index()`	`size_type`	position of the iterator

Table 5. MTL OneD iterator requirements (in addition to the STL bidirectional or random access iterator requirements).

5.3 OneD Containers

The one-dimensional containers in the MTL serve two purposes. They are used as vector objects and as the building blocks for TwoDimensional containers. The interface for the OneD containers is modeled closely after the STL container interface. In fact, the MTL OneD container requirements add only a half dozen requirements to the STL reversible container requirements. They are listed in Table 6. The only other difference, as mentioned in Section 5.1, is that MTL OneD containers use handle semantics for copy and assignment.

```
template <class Object>
class refcnt_ptr {
  typedef refcnt_ptr<Object> self;
public:
  refcnt_ptr() : object(0) { }
  refcnt_ptr(Object* c)
    : object(c), count(new int(1)) { }
  refcnt_ptr(const self& x) : object(x.object), count(x.count) {
    this->inc();
  }
  ~refcnt_ptr() { this->dec(); }
  self& operator=(Object* c) {
    if (object) this->dec();
    object = c;
    count = new int(1);
  }
  self& operator=(const self& x) {
    if (object) this->dec();
    object = x.object;
    count = x.count;
    this->inc();
    return *this;
  }
  Object& operator*() { return *object; }
  const Object& operator*() const { return *object; }
  Object* operator->() { return object; }
  const Object* operator->() const { return object; }
  void inc() { (*count)++; }
  void dec() {
    (*count)--;
    if (*count <= 0) {
      delete object;
      delete count;
    }
  }
protected:
  Object* object;
  int* count;
};
```

Fig. 10. Reference counting pointer implementation.

expression	return type	note
X::PR	T	precision type
X::sparsity	sparsity tag	sparse or dense?
X::scaled_type	scaled1D<X>	used by scaled()
a.nnz()	size_type	number of non-zeroes
a.resize(n)	void	post: a.size() == n
a[n]	reference const_reference for constant a	return element with index n

Table 6. OneD container requirements (in addition to the STL reversible container requirements).

Class dense1D. The dense1D container is implemented using our high performance version of the STL vector class. The dense1D container adds reference counting to vector, and wraps the vector's iterators to match the MTL requirements. There are several ways to construct dense1D containers. The dense1D container will allocate and manage its own memory if created with the normal constructor dense1D(size_type n). Also, one can create a dense1D that is aliased to an existing buffer in memory with dense1D(T* d, size_type n). In this case the dense1D assumes no responsibility for the memory. The aliasing constructor is useful for interoperability with other libraries and languages, such as Fortran. Below is an example of creating dense1D containers using both methods.

```
const int N = 3;
double dx[] = { 1, 2, 3};
mtl::dense1D<double> x(dx, N), y(N);
y[0] = 3; y[1] = 0; y[2] = -1;
if (vecvec::dot(x,y) == 0)
  cout << "Vectors x and y are orthogonal." << endl;
```

Class sparse1D. The sparse1D class adapts an STL-style container into a sparse vector. The adapted container must have elements that are index-value pairs. The vector, list, and set STL containers all work well with the sparse1D adapter. The time and space complexity of the various operations on a sparse1D container depend on the adapted container. For instance, random insertion into a set is $O(\log n)$ while it is $O(n)$ for a vector. The main purpose of the sparse1D container in the MTL is to allow for flexible construction of many sparse matrix types through composing different types of sparse vectors. The array2D class is used to compose vectors, and will be discussed in Section 5.5.

The code below is a short example of creating a sparse1D with a set, inserting a few elements, and then accessing the values and indices of the resulting vector. The normal iterators of the sparse1D, returned by begin(), give access to the values of the elements (through dereference) and also the indices (through the index() method on the iterator). If one wishes to view the indices only (the non-zero structure of the vector), one can use the nz_struct() method to obtain a container consisting of the indices of the elements in the sparse vector.

```
typedef mtl::sparse1D<std::set< pair<int, double> > > SparseVec;
SparseVec x;
for (int i = 0; i < 5; ++i)
  x[i*2] = i;
ostream_iterator couti(cout);
std::copy(x.begin(), x.end(), couti);
SparseVec::IndexArray ix = x.nz_struct();
std::copy(ix.begin(), ix.end(), couti);
> 01234
> 02468
```

Class compressed1D. The compressed1D container uses the *array of values* and *array of indices* storage format. Again, the main purpose of this class is to be used in the compositional construction of sparse matrices.

5.4 TwoD Iterator

A two-dimensional iterator is very similar to a one-dimensional iterator. The same set of operations are applicable: dereference, increment, etc. The main difference is that when a TwoD iterator is dereferenced, the return type is a OneD container instead of a single element. The TwoD iterator *type* can be retrieved through the row_2Diterator or column_2Diterator type definitions inside the matrix classes, which in turn are mapped to the major_iterator or minor_iterator definition within the TwoD container of a matrix. The major_iterator corresponds to the "normal" traversal of the two-dimensional container. For instance, the major_iterator of a row-major matrix is the row_2Diterator. The minor_iterator would correspond to the column_2Diterator, which strides across the OneD containers.

A TwoD iterator *object* can be obtained from a matrix object with a call to the begin_rows() or begin_columns() method.

There are many difference instances and implementations of TwoD iterators in the MTL, but the user does not need to know anything about the details of each one. All TwoD iterators have the same interface and behave in a similar fashion. Table 7 gives the interface requirements for the MTL TwoD iterators.

5.5 TwoD Containers

The MTL TwoD container interface is an orientation-free view into the storage format of the matrix. The interface is formulated with respect to *major*

expression	return type	note
X::PR	T	numerical precision type
X::value_type	OneD	
X::reference	OneD&	
X::pointer	OneD*	
a.index()	size_type	position of iterator
*a	OneD& const OneD& for const a	

Table 7. TwoD iterator requirements (in addition to the STL bidirectional or random access iterator requirements).

and *minor* instead of row and column. This results in some significant code reuse since the same TwoD container can be used for both row and column oriented matrices, when combined with the appropriate orientation adapter (see Section 5.6).

The main role of the TwoD container is to provide TwoD iterators that traverse across the major and minor axis vectors of a matrix. A large class of the standard matrix formats allows for efficient traversal of this kind, and it is a fairly straightforward exercise to implement the TwoD interface on top of these matrix formats. The Matrix Template Library defines a set of requirements for TwoD containers and provides implementations of many of the common matrix formats conformant to this set of requirements. A class X satisfies the requirements of a TwoD container if the expressions in Table 8 are valid for class X.

Class dense2D. The dense2D container handles the standard dense matrix format where all the elements are stored contiguously in memory. Similar to the dense1D container, the dense2D container can manage its own memory, or can be aliased to an existing memory region. The example below shows the use of a dense2D container to create a matrix, and perform a rank-one-update ($A \leftarrow A + \alpha x y^T$). Note that the dense2D container must be wrapped in an orientation to be used with the MTL operations.

```
using namespace mtl;
const int M = 4, N = 3;
column< dense2D<double> > A(M, N);
dense1D<double> x(M), y(N);
double alpha = 2.5;
// fill A, x, and y
matvec::rank_one_update(A, scaled(x, alpha), y);
```

Class array2D. The array2D container composes OneD containers in a generic fashion to create a two-dimensional container. Any class that fulfills the OneD

expression	return type	note
X::PR	T	precision type
X::value_type	T	T is Assignable
X::reference	lvalue of T	
X::const_reference	const lvalue of T	
X::pointer	pointer to X::reference	
X::difference_type	signed integral type	
X::size_type	unsigned integral	
X::MajorVector	OneD	
X::MinorVector	OneD	
X::MajorVectorRef	OneD&	
X::MinorVectorRef	OneD&	
X::ConstMajorVectorRef	const OneD&	
X::ConstMinorVectorRef	const OneD&	
X::major_iterator	TwoD iterator	also const and reverse versions
X::minor_iterator	TwoD iterator	
X::sparsity	sparsity tag	sparse or dense?
X::scaled_type	scaled2D<X>	used by scaled()
X u;		default constructor
X();		default constructor
X u(a);		copy constructor
X u = a;		assignment
X u(size_type m, size_type n)		create an $m \times n$ container
(&a)->~X();		destructor
a.begin_major();	major_iterator	
a.end_major();	major_iterator	
a.begin_minor();	minor_iterator	
a.end_minor();	minor_iterator	
a(i,j)	reference const_reference for a const a	element access
a.major_vector(i)	MajorVectorRef	
a.minor_vector(j)	MinorVectorRef	
a.major()	size_type	major dimension
a.minor()	size_type	minor dimension
a.sub_matrix(i,j,m,n)	X	create sub matrix

Table 8. TwoD container requirements.

container requirements can be combined with the **array2D** to form a TwoD container. This allows for a wide variety of matrix types to be created. It also allows for more flexibility than is available in a contiguous matrix format, though operations on large **array2D** matrices can be less efficient due to poor cache behavior. Below is an example of creating a sparse row-oriented matrix out of the **compressed1D** and **array2D** containers. The system of equations $Ax = b$ is then solved using the QMR iterative solver from the Iterative Template Library, an MTL based iterative solver [31].

```
using namespace mtl; using namespace itl;
typedef row< array2D < compressed1D <double> > > Matrix;
Matrix A(5, 5);
A(0, 0) = 1.0;  // fill A ...
dense1D<double> x(A.nrows(), 0.0), b(A.ncols());
vec::set(b, 1.);
// create ilu preconditioner
ILU<matrix> precond(A);
//create iteration control object
int max_iter = 50;
basic_iteration<double> iter(b, max_iter, 1.e-6);
//QMR iterative solver algorithm
QMR(A, x, b, precond.left(), precond.right(), iter);
//verify the result
dense1D<double> b1(A.ncols());
matvec::mult(A, x, b1);
vecvec::add(b1, scaled(b, -1.), b1);
assert( vec::two_norm(b1) < 1.e-6);
```

Class compressed2D. The **compressed2D** container provides the MTL TwoD container interface for the traditional compressed row/column sparse matrix storage format. This format consists of three arrays. The first array consists of indices that indicate the start of each *major* vector. The second array gives the *minor* indices for each element in the matrix, and the third array consists of the matrix element values. To allow for interoperability, there is an aliasing constructor for the **compressed2D** that allows the user to create a MTL sparse matrix from preexisting memory that is in the **compressed2D** format. An example of constructing matrices with the **compressed2D** format is shown below.

```
// Create from pre-existing arrays
const int m = 3, n = 3, nnz = 5;
double values[] = { 1, 2, 3, 4, 5 };
int indices[]   = { 0, 2, 1, 1, 2 };
int row_ptr[]   = { 0, 2, 3, 5 };
compressed2D<double> twodA(m,n,nnz,values,row_ptr,indices);
row< compressed2D<double> > A(twodA);
```

```
// Create from scratch
row< compressed2D<double> > B(m, n);
B(0,0) = 1;  B(0,2) = 2;
B(1,1) = 3;  B(2,1) = 4;
B(2,2) = 5;
```

5.6 Adapter Classes

An adapter class is one which modifies the interface and/or behavior of some base class. Adapters can be used for a wide range of purposes. They can be used to modify the interface of a class to fit the interface expected by some client code [14], or to restrict the interface of a class. The STL stack adapter is a good example of this. It wraps up a container, and restricts the operations allowed to `push()`, `pop()`, and `top()`. A good example of an adapter that modifies the behavior of a class without changing its interface is the `reverse_iterator` of the STL.

An adapter class can be implemented with inheritance or aggregation (containment). The use of inheritance is nice since it can reduce the amount of code that must be written for the adapter, though there are some circumstances when it is not appropriate. One such situation is where the type to be adapted could be a built-in type (a pointer for example). This is why the `reverse_iterator` of the STL is not implemented with inheritance.

It turns out that the adapter pattern is highly applicable to numerical linear algebra. It can be used as a powerful tool to cut down on the combinatorial explosion characteristic of basic linear algebra implementations. Looking back on some of the reasons for the combinatorial explosion:

- Permutations of scaling and striding in algorithms.
- Row and column orientations for matrix types.
- Multitude of matrix types, such as banded, triangular and symmetric.

In this section we will show how each of these issues can be handled with adapters.

Scaling Adapters. For performance reasons basic linear algebra routines often need to incorporate secondary operations into the main algorithm, such as scaling a vector while adding to another vector. The `daxpy()` BLAS routine ($y \leftarrow \alpha x + y$) is a typical example, with its `alpha` parameter.

```
void daxpy(int n, double* dx, double alpha, int incx,
           double* dy, int incy);
```

The problem with the BLAS approach is that it becomes necessary to provide many different versions of the algorithm within the `daxpy()` routine, to handle special cases with respect to the value of `alpha`. If `alpha` is 1 then it is not necessary to perform the multiplication. If `alpha` is 0 then the daxpy() can return immediately. When there are two or more arguments in the interface that can be scaled, the permutations of cases result in a large amount of code to be written.

The MTL uses a family of iterator, vector, and TwoD container adapters to solve this problem. In the MTL, only one version of each algorithm is written, and it is written without regard to scaling. There are no scalar arguments in MTL routines. Instead, the vectors and matrices can optionally be modified with adapters so that they are transparently (from the point of view of the algorithm) scaled as their elements are accessed. If the scalar value is set at compile-time, we can rely on the compiler to optimize and create the appropriate specialized code. The scaling adapters should also be coded to handle the case were the specialization needs to happen at run time, though this is more complicated and is a current area of research. Note that the BLAS algorithms do all specialization at run time, which causes some unnecessary overhead for the dispatch in the situations where the specialization could have been performed at compile time.

Class scale_iterator. The first member of the scaling adapter family is the **scale_iterator**. It is an iterator adapter that multiplies each element by a scalar as the iterator dereferences. Below is an example of how the **scale_iterator** could be used in conjunction with the STL transform algorithm. The transform algorithm is similar to the daxpy() routine; the vectors x and y are being added and the result is going into vector z. The elements from x are being scaled by alpha. The operation is $z \leftarrow \alpha x + y$. Note that this example is included to demonstrate a concept, it does not show how one would typically scale a vector using MTL. The MTL algorithms take vector objects as arguments, instead of iterators, and there is a **scaled1D** adapter that will cause a vector to be scaled.

```
double alpha;
std::vector<double> x, y, z;
// set the lengths of x, y, and z
// fill x, y and set alpha
typedef std::vector<double>::iterator iter;
scale_iterator<iter> start(x.begin(), alpha);
scale_iterator<iter> finish(x.end());
// z = alpha * x + y
std::transform(start, finish, y.begin(), z.begin(), plus<int>())
```

The following example is an excerpt from the scale_iterator implementation. The base iterator is stored as a data member (instead of using inheritance) since the base iterator could possibly be just a basic pointer type, and not a class type. The scalar value is passed into the iterator's constructor and also stored as a data member. The scalar is then used in the dereference operator*() to multiply the element from the vector. The increment operator++() merely calls the underlying iterator's method.

```
template <class Iterator>
class scale_iterator {
public:
  scale_iterator(const Iterator& x, const value_type& a)
    : current(x), alpha(a) { }
  value_type operator*() const { return alpha * *current; }
  scale_iterator& operator++ () { ++current; return *this; }
  // ...
protected:
  Iter current;
  value_type alpha;
};
```

At first glance one may think that the scale_iterator introduces overhead which would have significant performance implications for inner loops. In fact modern compilers will inline the operator*(), and propagate the scalar's value to where it is used if it is a constant (known at compile time). This results in code with no extra overhead.

Class scaled1D. The scaled1D class wraps up any OneD class, and it uses the scale_iterator to wrap up the OneD iterators in its begin() and end() methods. The main job of this class is to merely pass along the scalar value (alpha) to the scale_iterator. An excerpt from the scaled1D class is given below.

```
template <class Vector>
class scaled1D {
public:
  typedef scale_iterator<Vector::const_iterator> const_iterator;
  scaled1D(const Vector& r, value_type a) : rep(r), alpha(a) { }
  const_iterator begin() const {
    return const_iterator(rep.begin(), alpha); }
  const_iterator end() const {
    return const_iterator(rep.end(), alpha); }
  const_reference operator[](int n) const {
    return *(begin() + n); }
protected:
  Vector rep;
  value_type alpha;
};
```

The helper template function scaled() is provided to make the creation of scaled vectors easier.

```
template <class Vector, class T>
scaled1D<Vector> scaled(const Vector& v, const T& a) {
  return scaled1D<Vector>(v,a);
}
```

The example below shows how one could perform the scaling step in a Gaussian elimination. In this example the second row of matrix A is scaled by 2, using a combination of the MTL `vecvec::copy` algorithm, and the `scaled1D` adapter.

```
// [[5.0,5.5,6.0],
//   [2.5,3.0,3.5],
//   [1.0,1.5,2.0]]
double scalar = A(0,0) / A(1,0);
Matrix::RowVectorRef row = A.row_vector(1);
vecvec::copy(scaled(row, scalar), row);
// [[5.0,5.5,6.0],
//  [5.0,6.0,7.0],
//  [1.0,1.5,2.0]]
```

As one might expect, there is also a scaling adapter for matrices in the MTL, the `scaled2D` class. It builds on top of the `scaled1D` in a similar fashion to the way the `scaled1D` is built on the `scale_iterator`.

Striding Adapters. A similar problem to the one of scaling is that of striding. Many times a user wishes to operate on a vector that is not contiguous in memory, but at constant strides such as a row in a column oriented matrix. Again a library writer would need to add an argument to each linear algebra routine to specify the striding factor. Also, the algorithms would need to handle striding by 1 different from other striding factors for performance reasons. Again this causes the amount of code to balloon, as is the case for the BLAS. This problem can also be handled with adapters. The MTL has a strided iterator and strided vector adapter. The implementation of the strided adapters is very similar to that of the scaled adapters.

Matrix Orientation. The job of a matrix orientation adapter is to map the *major* and *minor* aspects of a TwoD container to the corresponding *row* or *column* of a matrix. There are two orientation adapters, the `row` and `column` classes. The mapping they perform involves both member functions and typedefs. For instance, a TwoD container has a `begin_major()` function. The `row` adapter must map this to the `begin_row()` method, while the `column` adapter maps this to the `begin_column()` method. A TwoD container also has internal typedefs such as `MinorVector`. The `row` adapter maps this type to `ColumnVector` while the `column` adapter maps this to `RowVector`. The requirements for the orientation adapters is given in Table 9. Both the `row` and `column` adapters must fulfill these requirements. An excerpt from the `row` orientation class is listed below.

```
template <class TwoD>
class row : public TwoD {
public:
  typedef typename TwoD::major_iterator row_2Diterator;
```

expression	return type	note
`X::value_type`	T	
`X::reference`	lvalue of T	
`X::const_reference`	const lvalue of T	
`X::pointer`	pointer to `X::reference`	
`X::size_type`	unsigned integral	
`X::difference_type`	signed integral	
`X::TwoD`	TwoD Container	
`X::orientation`	orientation tag	row or column?
`X::shape`	shape tag	banded, etc.
`X::tranpose_type`		used by `trans()`
`X::scaled_type`		used by `scaled()`
`X::row_2Diterator`	TwoD Iterator	also const version
`X::column_2Diterator`		also const version
`X::RowVector`	OneD Container	also const and
`X::ColumnVector`		reference versions

`X u;`
`X();`
		default constructor
		`a.nrows() == 0`
		`a.ncols() == 0`

`X u(a);`		copy
`X u = a;`		assignment
`X u(size_type m,` `size_type n)`		create an $m \times n$ matrix
`a.begin_rows();`	`row_2Diterator`	
`a.end_rows();`	`row_2Diterator`	
`a.begin_columns();`	`column_2Diterator`	
`a.end_columns();`	`column_2Diterator`	
`a(i,j)`	`reference` `const_reference` for a const a	element access
`a.row_vector(i)`	`RowVectorRef`	
`a.column_vector(j)`	`ColumnVectorRef`	
`a.row()`	`size_type`	row dimension
`a.column()`	`size_type`	column dimension
`a.sub_matrix(i,j,m,n)`	`X`	create sub matrix

Table 9. Orientation adapter requirements.

```
  typedef typename TwoD::minor_iterator column_2Diterator;
  typedef typename TwoD::MajorVector RowVector;
  typedef typename TwoD::MinorVector ColumnVector;
  typedef column<TwoD> transpose_type;

  row(int nrows, int ncols) : TwoD(nrows, ncols) { }
  const_reference operator()(int row, int col) const {
    return TwoD::operator()(row,col); }
  row_2Diterator begin_rows() { return begin_major(); }
  row_2Diterator end_rows() { return end_major(); }
  column_2Diterator begin_columns() { return begin_minor(); }
  column_2Diterator end_columns() { return end_minor(); }
  RowVectorRef row_vector(int i) { return major_vector(i); }
  ColumnVectorRef column_vector(int j) {
    return minor_vector(j); }
};
```

There are often situations when one wishes to operate on the transpose of a matrix for one particular operation. Many linear algebra packages provide this capability through an extra parameter in the interface that causes a dispatch to a specialized version of the operation. The only difference in the specialized versions is the mapping of the row and column index of an element to where it is stored in memory. As discussed in the beginning of this section, the mapping of indices in an MTL matrix is controlled by the orientation adapter, not hard coded into the subroutine. Therefore, by merely switching the type of orientation adapter associated with the matrix, one can make the normal algorithm operate on the transpose of the matrix. The `trans()` helper function is provided so that it is convenient to the user to switch the orientation of a matrix. The implementation of `trans()` and an example of its use are presented below.

```
// definition of trans() helper function
template <class Orien>
Orien::transpose_type trans(Orien A) {
  typedef Orien::transpose_type trans_type;
  return trans_type(A);
}
// use of trans() function
typedef ... Matrix;
Matrix A(M,N), B(N,M);
// fill A ...
// B <- A'
matmat::copy(trans(A), B);
```

Shape Adapters. Another interesting use of adapters in the MTL is to create banded and triangle shaped matrices. The `banded` class provides the core functionality, and is the base class for the `triangle`, `symmetric`, and `hermitian` shapes.

A band of a matrix is defined by the number of diagonals above the main diagonal (the super diagonals), and the number of diagonals below the main diagonal (the sub diagonals). These two numbers are combined in a pair to form the bandwidth. For example, an upper triangular matrix of size $M \times N$ has a bandwidth of $(0, N-1)$.

The main work performed in the `banded` class (and its `banded_vector`) is that of adjusting the `begin()` and `end()` methods for each row/column to take into account the banded shape of the matrix. This is somewhat complicated by the wish to handle both `packed` and `unpacked` storage types. A banded matrix with `packed` storage format only stores the elements in the band. The `unpacked` storage format has a full matrix underneath, but restricts access to just the elements in the band. The reason for the `unpacked` storage format is that one may have a general matrix and wish to perform actions on just the upper or lower triangle (as is the case in the LU factorization described in Section 4.1).

For the packed case, the `begin()` and `end()` methods for each `banded_vector` simply call the underlying methods of the original vector. For the unpacked case, the `begin()` and `end()` methods have to offset according to the bandwidth. The bulk of MTL code for the `banded` adapter was constructed at the `TwoD` level, without respect to row and column orientation. This allows the single `banded` class to be used with both row and column oriented matrices. Figure 11 gives some of the implementation of the `banded_vector` class.

Triangle and Symmetric Shape Adapter. The `triangle` and `symmetric` adapter classes derive from the `banded` class. They merely specialize the `banded` matrix to a particular band shape (upper triangle, unit-lower triangle, etc.). In addition, the `symmetric` adapter maps all element accesses to the same half of the matrix. Most of the functionality is already provided in the `banded` class, so the implementation for each of these adapters is very concise. The implementation of the `triangle` class is given in Figure 12.

6 High Performance

We have presented many levels of abstraction, and a comprehensive set of algorithms for a variety of matrices, but this elegance matters little if high performance cannot be achieved. In this section we will discuss recent advances in languages and compilers that allow abstractions to be used with little or no performance penalty. Furthermore, we will present a comprehensive set of abstractions, the Basic Linear Algebra Instruction Set (BLAIS) and Fixed Algorithm Size Template (FAST) sub-libraries, that have the specific purpose of generating optimized code. The generation of optimized code is made possible with template meta-programming [36].

There is a common perception that the use of abstraction hurts performance. This is due to a particular set of language features that are used to

```
template <class OneD, class Packing>
class banded_vector {
public:
  class iterator : public OneD::iterator {
    ...
  };
  banded_vector(OneDRef x, size_type si, size_type fi)
    : oned(x), first_index(si), last_index(fi) { }
  iterator begin() { return begin(Packing()); }
  iterator end() { return end(Packing()); }
protected:
  // the versions of begin() and end()
  //   for unpacked storage format
  iterator begin(unpacked) {
    return iterator(oned.begin() + first_index, first_index); }
  iterator end(unpacked) {
    int offset = oned.size() - last_index;
    return iterator(oned.end() - offset, first_index); }

  // the versions of begin() and end()
  //   for packed storage format
  iterator begin(packed) {
    return iterator(oned.begin(), first_index); }
  iterator end(packed) {
    return iterator(oned.end(), first_index); }

  // the index of the first element of the vector
  size_type first_index;
  // the index of the last element of the vector
  size_type last_index;
  // the underlying OneD container
  OneDRef oned;
};
```

Fig. 11. An excerpt from the banded-vector implementation.

```
template <class Orien, class Uplo, class Packing = unpacked>
class triangle : public banded<Orien, Packing> {
public:
  triangle(int m, int n)
    : super(m, n, uplo_.bandwidth(m, n)) { }
  bool is_upper() const { return uplo_.is_upper(); }
  bool is_lower() const { return ! uplo_.is_upper(); }
  bool is_unit() const { return uplo_.is_unit(); }
protected:
  Uplo uplo_;
};

// Some of the Uplo classes:

struct upper {
  typedef lower transpose_type;
  pair<int,int> bandwidth(int m, int n) {
    return make_pair(0, n - 1);
  }
  bool is_upper() const { return true; }
  bool is_unit() const { return false; }
};

struct unit_lower {
  typedef unit_upper transpose_type;
  pair<int,int> bandwidth(int m, int n) {
    return make_pair(m, -1);
  }
  bool is_upper() const { return false; }
  bool is_unit() const { return true; }
};
```

Fig. 12. Triangle shape implementation with some of the Uplo classes.

create abstractions and how those language features are implemented. The language features that are used to create abstractions are listed below.

Procedures. The basic building block of abstraction is the procedure call. For each abstraction level, one needs a set of functions, an interface for the abstraction level. Traditionally a procedure call has incurred a significant overhead—copying parameters to the stack, etc. Many compilers are now able to inline procedure calls to remove this performance penalty.

Classes. The main tool for data encapsulation is the `class` language construct. It allows arbitrarily complex sets of data to be grouped together and hidden within an abstraction. Classes (and structures in general) interfere with the register allocation algorithms of many compilers. Optimizing compilers map the local variables of a function to machine registers. This can drastically reduce the number of loads and stores necessary since the registers are being used to cache data from memory. Many compilers do not recognize that this optimization can also be applied to objects on the stack. They specify a load or store for each access to a data item within the object, which kills performance for codes like STL and MTL which use iterators — objects that have to be accessed over and over again within the inner loops of the code.

A typical example of the problem with small objects comes up in the use of complex numbers in C++. We illustrate the problem with the code below which calculates the sum of a complex vector.

```
complex<double> a, b[N];

// take the sum of vector b
for (int i = 0; i < N; ++i)
  a += b[i];
```

We have written out the pseudo-assembly code for the loop, see Figure 13. A load or store results from each access to a complex number. Therefore each loop iteration includes 4 loads and 2 stores. We show the pseudo-assembly code for the loop after register allocation has been applied, which maps the complex number a to registers (in this case R3 and R5). This version only does 2 loads and 0 stores inside the loop. In modern processors memory access is more expensive than ALU operations, so the reduction in the number of loads and stores has a large impact on overall performance.

Polymorphism. The language feature that enables generic programming is polymorphism. It enables an algorithm to work with many data types and data structures instead of just one in particular. The first language feature in C++ that enabled polymorphism was the virtual function call, which allows a function call to dispatch to a specialized version based on the object type. With virtual functions, the object type is not known until run time, so the dispatch happens during program execution. This is called dynamic polymorphism. A disadvantage of the run time dispatch is that it adds some overhead to the cost of a normal function call. Even worse, virtual function calls interfere with inlining. Virtual function calls cannot be inlined because the object type is not know at compile time, when the inlining optimization is applied, and the compiler cannot decide which specialized version to dispatch to.

Two important advances have been introduced into C++ compilers that remove the performance penalties associated with abstraction. The first is a language feature and the second is a compiler optimization.

```
// pseudo-assembly code (unoptimized)
looptop: CMP i N
         BRZ loopend
         LOAD a.real R3
         LOAD b[i].real R4
         ADD R3 R4 R5
         STORE R5 a.real
         LOAD a.imag R3
         LOAD b[i].imag R4
         ADD R3 R4 R5
         STORE R5 a.imag
         JMP looptop
loopend:

// pseudo-assembly code (optimized)
         LOAD a.real R3
         LOAD a.imag R5
looptop: CMP i N
         BRZ loopend
         LOAD b[i].real R4
         ADD R3 R4 R3
         LOAD b[i].imag R6
         ADD R5 R6 R5
         JMP looptop
loopend: STORE R3 a.real
         STORE R5 a.imag
```

Fig. 13. Optimized and unoptimized pseudo-assembly code for calculation of the sum of a complex vector.

Static Polymorphism. The addition of templates to the C++ language creates a way for functions to be selected at compile time based on object type. With static polymorphism the object type is known at compile time, which enables the compiler to hard code the dispatch decision. As a result template functions can be inlined in the same way as regular functions. In addition, many C++ compilers have improved their ability to inline functions in general, to the point where one can know with a relatively high degree of confidence that if a function is labeled `inline` that it is really being inlined. Even extremely complicated layers of functions can be completely flattened out. We prove this with the performance achieved by the BLAIS and FAST libraries.

Lightweight Object Optimization. The performance penalty associated with the use of classes and structures can be solved with a relatively straightforward optimization (though the implementation is difficult). Each object is removed from the code, and it is replaced with its individual parts. This happens in a recursive fashion until there are only basic data types (integers, floats, etc.). Then each reference through an object to one of its parts is replaced with a direct reference to the part. Note that this is only applied to objects on the stack (local variables). The Kuck and Associates C++ compiler [17] performs this optimization, which is also known as scalar replacement of aggregates [24]. The end result is that the data items within the objects are mapped appropriately to machine registers.

With the use of template functions, and with lightweight object optimization, it is now possible to introduce abstractions with no performance penalty. This allowed us to design MTL in a generic fashion, composing containers and allowing a mix and match model with the algorithms. At this time we know that the compilers that perform both of these optimizations include Kuck and Associates Inc. C++ compiler and the SGI C++ compiler.[1] The egcs compiler (though it has great language support) is missing the lightweight object optimization as well as several other optimizations. We will be conducting a complete survey of current compilers' optimization levels in the near future.

Even after the "abstraction" barrier has been removed, there are yet more optimizations that need to be applied to achieve high performance. As alluded to in the introduction, achieving high performance on modern microprocessors is a difficult task, requiring many complex optimizations. Today's compiler can aid somewhat in this area, though to achieve "ideal" performance one must still hand optimize the code.

Compiler Optimizations: Unrolling and Instruction Scheduling. Modern compilers can do a great job of unrolling simple loops and scheduling instructions, but typically only for specific (recognizable) cases. There are many ways, especially in C and C++ to interfere with the optimization process. The MTL containers and algorithms are designed to result in code that is easy for the compiler to optimize. Furthermore, the *iterator* abstraction makes inter-compiler portability possible, since it encapsulates how looping is performed. This is discussed below in Section 6.1.

Algorithmic Blocking. To obtain high performance on a modern microprocessor, an algorithm must properly exploit the associated memory hierarchy and pipeline architecture. Todays compilers are not able to apply all of these transformations, so the programmer must apply some optimizations by hand. To make matters worse, the transformations are somewhat machine dependent. The number of registers, size of cache, and other machine characteristics affect the blocking sizes. This makes it difficult to express high performance

[1] We use the C++ compiler from Kuck and Associates, Inc. (KAI) for development. We have found that the easiest way to check if a function is inlined is to inspect the intermediate C code that KAI's compiler generates. With the proper use of compiler flags we have found that the KAI compiler does a reliable job. Hopefully, in the future, compilers will give more direct feedback on the optimizations that they are performing.

algorithms in a portable fashion. Our solution to this problem is discussed in Section 6.2.

6.1 High-Performance Iterators

Iterators control how looping is performed, and therefore their design can make a large difference to the performance of a particular code. The biggest concern here is identifying what kind of loops the underlying backend compiler will optimize (perform unrolling, instruction scheduling, etc.). The design space includes whether to increment a pointer or increment an integer offset, and it also includes whether to use the less-than or not-equal operator for the loop termination condition. One would think modern compilers should produce equally good code for all of these cases, but this is not the case. There can be a factor of 2 or more difference depending on the type of loop used. The four variations on the traversal method can be seen in the following example loop, which computes a dot product of two vectors.

```
int i;  double* x, *y, *xp, *yp;
// integer, != operator
for (i = 0; i != N; ++i)
  tmp += x[i] * y[i];
// integer, < operator
for (i = 0; i < N; ++i)
  tmp += x[i] * y[i];
// pointer, != operator
yp = y;
for (xp = x; xp != x + N; ++xp, ++yp)
  tmp += *x + *y;
// pointer, < operator
yp = y;
for (xp = x; xp < x + N; ++xp, ++yp)
  tmp += *x + *y;
```

Table 10 shows the variations in performance on a loop (dot product) for three different computer architectures/compilers. The dot product computation was chosen because there are no aliasing issues and it includes the typical add/multiply floating point operation. The native C compiler was used for each machine with maximum optimization flags turned on.

From the Table 10 we can surmise that by choosing to increment an integer offset, and using the less-than comparison operator, we can achieve top performance with all of the compilers tested. This result differs from the findings of PHiPAC [3], which suggests that one should always use the not-equal operator because it is more efficiently implemented in some architectures. Our experience has shown that the architecture implementation is not as important as whether the compiler knows how to optimize a loop that uses a not-equal comparison operator. Of course, this test did not include all

iterator type	comparison type	
	<	!=
UltraSPARC 30		
integer	180.085	44.6187
pointer	180.319	44.4693
RS6000 590		
integer	47.7829	47.78
pointer	47.7595	20.8623
R10000		
integer	102.512	106.101
pointer	81.2678	72.648

Table 10. The effect of iterator and comparison operator choice on performance (in Mflops) for dot product on Sun C, IBM XLC, and SGI C compilers.

C compilers, but it does give us a warning that we ought to write our code in such a way as to make it easy to change the operator used.

In C++ this is easy to do with an extra layer of abstraction. Instead of using a particular operator for each loop, we call the `not_at()` template function. The `not_at()` function then invokes the proper comparison operator. Now there is only one line that needs to change if the compiler or architecture has a particular operator preference. Another reason for using the `not_at()` method is that if one wants to make algorithms generic, then the less-than operator should not be used since most iterator types do not support less-than. This is why the not-equal operator is instead used for most loops in the STL. The `not_at()` function solves this problem by allowing the less-than operator to be used for `random_access_iterators` and the not-equal to be used for all others. The code below shows the implementation of the `not_at()` family of functions. The last `not_at()` function dispatches to either the first or second version of `not_at()` depending on the iterator's category type. The dispatch happens at compile time, and all the functions are easily inlined by a modern C++ compiler. This technique has been referred to as external polymorphism.

```
template <class RandomIter1, class RandomIter2>
bool not_at(const RandomIter1& a, const RandomIter2& b,
            random_access_iterator_tag) {
  return a < b;
}
template <class Iter1, class Iter2, class AnyTag>
inline bool not_at(const Iter1& a, const Iter2& b, AnyTag) {
  return a != b;
}
template <class Iter1, class Iter2>
inline bool not_at(const Iter1& a, const Iter2& b) {
  typedef typename iterator_traits<Iter1>::iterator_category Cat;
```

```
  return not_at(a, b, Cat());
}
```

Now that we have a way of picking out the proper operator, and know from the experiment that an integer offset should be incremented instead of a pointer, we are ready to implement the MTL iterators. An excerpt from the `dense_iterator` class is shown below. This is the iterator that is used in the `dense1D` container . We use the integer `pos` to keep track of the iterator position, and then use it as an offset in the `operator*` method. This implementation results in an iterator that produces loops that are easier to optimize for the compilers we have tested. The typical implementation of an iterator for a contiguous memory container such as a vector is to just use a pointer. But as the discussion above points out, this is not always the best choice. One nice thing about the iterator abstraction is that we are not forced into a particular implementation, we can change the implementation at any time without affecting the code using the iterator.

```
template <class Iter> class dense_iterator {
public:
  dense_iterator(Iter s, size_type i) : start(s), pos(i) { }
  size_type index() const { return pos; }
  reference operator*() const { return *(start + pos);  }
  self& operator++ () { ++pos; return *this; }
protected:
  Iter start;
  size_type pos;
};
template <class Iter>
bool operator < (const dense_iterator<Iter>& x,
                 const dense_iterator<Iter>& y)
  { return x.index() < y.index(); }
```

6.2 High-Performance Kernels — Metaprogramming

The bane of portable high-performance numerical linear algebra is the need to tailor key routines to specific execution environments. To obtain high performance on a modern microprocessor, an algorithm must make efficient use of the cache, registers, and the pipeline architecture (typically through careful loop blocking and structuring). Ideally, one would like to be able to express high performance algorithms in a portable fashion, but there is not enough expressiveness in languages such as C or Fortran to do so. This is because the blocking done at the lowest level, for registers and the pipeline, affects the number and type of instructions that must be in the inner loop, as is shown in the example below. The variation of the number of operations in the loop cannot be expressed directly in C or Fortran. Recent efforts (PHiPAC [3], ATLAS [39]) have resorted to going outside the language, i.e., to custom code generation systems to gain the kind of flexibility needed to generate the inner loop in a portable fashion. The BLAIS and FAST libraries use the template metaprogramming features of C++ to express flexible unrolling and blocking factors for inner loops.

```
// need to unroll by two for machine X
for (int i = 0; i < N; i += 2) {
  y[i]   += a * x[i];
  y[i+1] += a * x[i+1];
}
// need to unroll by three for machine Y
for (int i = 0; i < N; i += 3) {
  y[i]   += a * x[i];
  y[i+1] += a * x[i+1];
  y[i+2] += a * x[i+2];
}
```

In this section we describe a collection of high-performance kernels for basic linear algebra, called the Basic Linear Algebra Instruction Set (BLAIS) and the Fixed Algorithm Size Template (FAST) Library. The kernels encapsulate small fixed size computations to provide building blocks for numerical libraries in C++. The sizes are templated parameters of the kernels, so they can be easily configured to a specific architecture for portability. In this way the BLAIS delivers the power of such code generation systems as PHiPAC [3] and ATLAS [39]. BLAIS has a simple and elegant interface, so that one can write flexible-sized block algorithms without the complications of a code generation system.

The BLAIS specification contains *fixed size* algorithms with functionality equivalent to that of the Level-1, Level-2, and Level-3 BLAS [19,12,11]. The BLAIS routines themselves are implemented using the FAST library, which contains general purpose fixed-size algorithms equivalent in functionality to the generic numerical algorithms in the Standard Template Library (STL) [34]. In the following sections, we describe the implementation of the FAST algorithms and then show how the BLAIS are constructed from them. Next, we demonstrate how the BLAIS can be used as high-level instructions (kernels) to construct a dense matrix-matrix product. Finally, experimental results show that the performance obtained by our approach can equal and even exceed that of vendor-tuned libraries.

6.3 Fixed Algorithm Size Template (FAST) Library

The FAST Library includes generic algorithms such as `transform()`, `for_each()`, `inner_product()`, and `accumulate()` that are found in the STL. The interface closely follows that of the STL. All input is in the form of *iterators* (generalized pointers). The only difference is that the loop-end iterator is replaced by a *count template* object. The example below demonstrates the use of both the STL and FAST versions of `transform()` to realize an `AXPY`-like operation ($y \leftarrow x + y$).

The `first1` and `last1` parameters are iterators for the first input container (indicating the beginning and end of the container, respectively). The `first2` parameter is an iterator indicating the beginning of the second input container. The `result` parameter is an iterator indicating the start of the output container. The `binary_op` parameter is a function object that combines the elements from the first and second input containers into the result containers.

```
int x[4] = {1,1,1,1}, y[4] = {2,2,2,2};

// STL
template <class InIter1, InIter2, OutIter, BinaryOp>
OutIter transform(InIter1 first1, InIter1 last1, InIter2 first2,
                  OutIter result, BinaryOp binary_op);

std::transform(x, x + 4, y, y, plus<int>());

// FAST
template <int N, class InIter1, class InIter2,
          class OutIter, class BinOp>
OutIter transform(InIter1 first1, cnt<N>, InIter2 first2,
                  OutIter result, BinOp binary_op);

fast::transform(x, cnt<4>(), y, y, plus<int>());
```

The difference between the STL and FAST algorithms is that STL accommodates containers of arbitrary size, with the size being specified at run time. FAST also works with containers of arbitrary size, but the size is fixed at compile time. In the example below we show how the FAST `transform()` routine is implemented. We use a tail-recursive algorithm to achieve complete unrolling — there is no actual loop in the FAST `transform()`. The template-recursive calls are inlined, resulting in a sequence of `N` copies of the inner loop statement. This technique, called template metaprograms, has been used to a large degree in the Blitz++ Library and is explained in [8,36].

```
// The general case
template <int N, class InIter1, class InIter2,
          class OutIter, class BinOp>
inline OutIter
fast::transform (InIter1 first1, cnt<N>, InIter2 first2,
                 OutIter result, BinOp binary_op) {
  *result = binary_op (*first1, *first2);
  return transform(++first1, cnt<N-1>(), ++first2,
                   ++result, binary_op);
}

// The N = 0 case to stop template recursion
template <class InIter1,class InIter2,class OutIter,class BinOp>
```

```
inline OutIter
fast::transform(InIter1 first1,cnt<0>,InIter2 first2,
                OutIter result,BinOp binary_op){ return result; }
```

6.4 Basic Linear Algebra Instruction Set (BLAIS)

The BLAIS library is implemented directly on top of the FAST Library, as a thin layer that maps generic FAST algorithms into fixed-size mathematical operations. There is no added overhead in the layering because all the function calls are inlined. Using the FAST library allows the BLAIS routines to be expressed in a very simple and elegant fashion. The following discussions looks at one example from each of the levels of the BLAIS library: vector-vector, matrix-vector, and matrix-matrix.

Vector-Vector Operations. The `add()` routine is typical of the BLAIS vector-vector operations. It is implemented in terms of a FAST algorithm, in this case `transform()`. The implementation of the BLAIS `add()` is listed below. The `add()` function is implemented with a `class` and its constructor in order to provide a simple syntax for its use.

```
template <int N> struct add {
  template <class Iter1, class Iter2> inline
  add(Iter1 x, Iter2 y) {
    typedef typename iterator_traits<Iter1>::value_type T;
    fast::transform(x, cnt<N>(), y, y, plus<T>());
  }
};
```

The example below shows how the `add()` routine can be used. The comment shows the resulting code after the call to `add()` is inlined. Note that only one `add()` routine is required to provide any combination of scaling or striding. This is made possible through the use of the scaling and striding adapters, as discussed in Section 5.6. Any resulting overhead is removed by inlining and lightweight object optimizations [17]. The `scl(x, a)` call below automatically creates the proper `scale_iterator` out of `x` and `a`.

```
double x[4], y[4];
fill(x, x+4, 1); fill(y, y+4, 5);
double a = 2;
vecvec::add<4>(scl(x, a), y);

// the compiler expands add() to:
// y[0] += a * x[0];
// y[1] += a * x[1];
// y[2] += a * x[2];
// y[3] += a * x[3];
```

Matrix-Vector Operations. To illustrate the BLAIS matrix-vector operations we look at the BLAIS matrix-vector multiply. The algorithm simply carries out a vector add operation for each column of the matrix. Using the same technique as in the FAST library, we write a fixed depth recursive algorithm which becomes inlined by the compiler. The implementation is listed below.

```
// General Case
template <int M, int N>
struct mult {
  template <class AColIter, class IterX, class IterY> inline
  mult(AColIter A_2Diter, IterX x, IterY y) {
    vecvec::add<M>(scl((*A_2Diter).begin(), *x), y);
    mult<M, N-1>(++A_2Diter, ++x, y);
  }
};
// N = 0 Case
template <int M>
struct mult<M, 0> {
  template <class AColIter, class IterX, class IterY> inline
  mult(AColIter A_2Diter, IterX x, IterY y) {
    // do nothing
  }
};
```

Matrix-Matrix Operations. The most important of the BLAIS matrix-matrix operations is the matrix-matrix multiply. This algorithm builds on the BLAIS matrix-vector multiply. The code looks very similar to the matrix vector multiply, except that there are three integer template arguments (`M`, `N`, and `K`), and the inner "loop" contains a call to `matvec::mult()` instead of `vecvec::add()`. Remember that the BLAIS matrix-matrix operation is intended to be used in the inner loop (literally) of algorithms, and the BLAIS operation is completely inlined and expanded. Therefore it is perfectly fine to build the matrix-matrix multiply out of the matrix-vector operation since at this low level cache blocking issues are not a factor.

6.5 BLAIS in a General Matrix-Matrix Product

A typical use of the BLAIS kernels would be to construct linear algebra subroutines for arbitrarily sized objects. The fixed-size nature of the BLAIS routines make them well-suited to perform register-level blocking within a hierarchically blocked matrix-matrix multiplication. Blocking (or tiling) is a well known optimization that increases the reuse of data while it is in cache and in registers, thereby reducing the memory bandwidth requirements and increasing performance.

The code below shows the inner most set of blocking loops for a matrix-matrix multiply. The constants `BFM`, `BFN`, and `BFK` are blocking factors chosen so that c can fit into the registers. The blocking factors describe the size and shape of the submatrices, or blocks, that the matrix is divided into. [2]

```
for (jj = 0; jj < N; jj += BFN)
  for (ii = 0; ii < M; ii += BFM) {
    copy_block<double,BFM,BFN> c(C+ii*N+jj, BFM, BFN, N);
    for (kk = 0; kk < K; kk += BFK) {
      light2D<double> a(A+ii*K+kk, BFM, BFK);
      light2D<double> b(B + kk*N + jj, BFK, BFN);
      matmat::mult<BFM,BFN,BFK>(a, b, c);
    }
  }
```

A Configurable Recursive Matrix-Matrix Multiply. To obtain the highest performance in a matrix-matrix multiply code, algorithmic blocking must be done at each level of the memory hierarchy. A natural way to formulate this is to write the algorithm in a recursive fashion, where each level of recursion performs blocking for a level of the memory hierarchy.

We take this approach in the MTL algorithm. The size and shapes of the blocks at each level are determined by the ***blocking adapter***. Each adapter contains the information for the next level of blocking. In this way the recursive algorithm is determined by a recursive template data structure (which is set up at compile time). The setup code for the matrix-matrix multiply is shown below. This example blocks for just one level of cache, with 64 x 64 sized blocks. The small 4 x 2 blocks fit into registers. Note that these numbers would normally be constants that are set in a header file.

```
template <class MatA, class MatB, class MatC>
void matmat::mult(MatA& A, MatB& B, MatC& C) {
  MatA::RegisterBlock<4,1> A_L0;  MatA::Block<64,64> A_L1;
  MatB::RegisterBlock<1,2> B_L0;  MatB::Block<64,64> B_L1;
  MatC::CopyBlock<4,2> C_L0;      MatC::Block<64,64> C_L1;
  matmat::__mult(block(block(A, A_L0), A_L1),
                 block(block(B, B_L0), B_L1),
                 block(block(C, C_L0), C_L1));
}
```

The recursive algorithm is listed below. There is a TwoD iterator (`A_k`, `B_k`, and `C_i`) for each matrix, as well as 1-D iterator (`A_ki`, `B_kj`, and `C_ij`). The matrices have been wrapped up with `blocked` matrix adapters, so that dereferencing the OneD iterator results in a submatrix. The recursive call is then made on the submatrices `A_block`, `*B_kj`, and `*C_ij`.

---

[2] Excessive code bloat is not a problem in MTL because the complete unrolling is only done for very small sized blocks

```
template <class MatA, class MatB, class MatC>
void matmat::__mult(MatA& A, MatB& B, MatC& C) {
  A_k = A.begin_columns(); B_k = B.begin_rows();
  while (not_at(A_k, A.end_columns())) {
    C_i = C.begin_rows(); A_ki = (*A_k).begin();
    while (not_at(C_i, C.end_rows())) {
      B_kj = (*B_k).begin(); C_ij = (*C_i).begin();
      MatA::Block A_block = *A_ki;
      while (not_at(B_kj, (*B_k).end())) {
        __mult(A_block, *B_kj, *C_ij);
        ++B_kj; ++C_ij;
      } ++C_i; ++A_ki;
    } ++A_k; ++B_k;
  }
}
```

The bottom-most level of recursion is implemented with a separate function that makes the calls to the BLAIS matrix-matrix multiply, and "cleans up" the leftover edge pieces. Since the recursion depth is fixed at compile time, the whole algorithm can be inlined by the compiler. The code below shows such a kernel without the edge cleanup operations.

```
template <class MatA, class MatB, class MatC>
void matmat::__mult(MatA& A, MatB& B, MatC& C) {
  A_k = A.begin_rows(); C_k = C.begin_rows();
  while (not_at(A_k, A.end_rows())) {
    B_j = B.begin_columns(); C_kj = (*C_k).begin()
    while (not_at(B_j, B.end_columns())) {
      B_ji = (*B_j).begin(); A_ki = (*A_k).begin();
      MatC::Block C_block = *C_kj;
      while (not_at(B_ji, (*B_j).end())) {
        blais_matmat::mult(*A_ki, *B_ji, C_block);
        ++B_ji; ++A_ki;
      }
      // cleanup of K left out
      ++B_j; ++C_kj;
    }
    // cleanup of N left out
    ++A_k; ++C_k;
  }
  // cleanup of M left out
}
```

Optimizing Cache Conflict Misses. Besides blocking, there is another important optimization that can be done with matrix-matrix multiply code. Typically utilization of the level-1 cache is much lower than one might expect due to cache conflict misses. This is especially apparent for large matrices in direct mapped and low associativity caches. The way to minimize this problem is to copy the block of matrix A being accessed into a contiguous section of memory [18]. This allows the code to use blocking sizes closer to the size of the L-1 cache without inducing as many cache conflict misses.

It turns out that this optimization is straightforward to implement in our recursive matrix-matrix multiply. We already have block objects (submatrices `A_block`, `*B_j`, and `*C_j`). We modify the constructor for these objects to make a copy to a contiguous part of memory, and the destructor to copy the block back to the original matrix. This is especially nice since the optimization does not clutter the algorithm code, but instead the change is encapsulated in the `copy_block` matrix class.

7 Performance Experiments

In the following set of experiments we present some performance results comparing MTL with other available libraries (both public domain and vendor supplied). The algorithms timed were the dense matrix-matrix multiplication, the dense matrix-vector multiplication, and the sparse matrix-vector multiplication.

Dense Matrix-Matrix Multiplication. Figure 14 shows the dense matrix-matrix product performance for MTL, Fortran BLAS, the Sun Performance Library, TNT [28], and ATLAS [39], all obtained on a Sun UltraSPARC 170E. The experiment shows that the MTL can compete with vendor-tuned libraries (on an algorithm that tends to get extra attention due to benchmarking). The MTL and TNT executables were compiled using Kuck and Associates C++ (KCC) [17], in conjunction with the Solaris C compiler. ATLAS was compiled with the Solaris C compiler and the Fortran BLAS (obtained from Netlib) were compiled with the Solaris Fortran 77 compiler. All possible compiler optimization flags were used in all cases. The cache was cleared between each trial of the experiment. To demonstrate portability across different architectures and compilers, Figure 14 also compares the performance of MTL with ESSL [16] on an IBM RS/6000 590. In this case, the MTL executable was compiled with the KCC and IBM xlc compilers.

Dense and Sparse Matrix-Vector Multiplication. To demonstrate genericity across different data structures and data types, Figure 15 shows performance results obtained using the same generic matrix-vector multiplication algorithm for dense and for sparse matrices, and compares the performance to that obtained with non-generic libraries. The dense matrix-vector performance of MTL is compared to the Netlib BLAS (Fortran), the Sun Performance Library, and TNT [28]. The sparse matrix-vector performance for MTL is compared to SPARSKIT [30] (Fortran), NIST [29](C), and TNT (C++). The sparse matrices used were from the MatrixMarket [27] collection. The cache was not cleared between each matrix-vector timing trial. The focus of this experiment is on the pipeline behavior of the algorithm. If the cache is cleared, the bottle neck becomes memory bandwidth and differences in pipeline behavior can not be seen. Blocking for cache is not as important for matrix-vector multiplication because there is no data reuse of the matrix.

Performance Analysis of Matrix-Matrix Multiplication. The presence (and absence) of different optimization techniques in the various implementations of the matrix-matrix multiplication can readily be seen in Figure 14 (the UltraSPARC comparison) and manifest themselves quite strongly as a function of matrix size. In the region from $N = 10$ to $N = 256$, performance is dominated by register usage and pipeline performance. "Unroll and jam" techniques [4,40] are used to attain high levels of performance in this region. In the region from 256 to approximately 1024, performance is dominated by data locality. Loop blocking for cache is used to attain high levels of performance here. Finally, for matrix sizes larger than approximately $N = 1024$, performance can be affected by conflict misses in the cache. The results for ATLAS and Fortran BLAS fall precipitously at this point. To attain good performance in the face of conflict misses (in low associativity caches) block-copy techniques as described in [18] are used. Note that performance effects are cumulative. For instance, the Netlib BLAS do not use any of the techniques listed above for performance enhancement. As a result, performance is poor initially and continues to degrade as different effects come into play.

8 Future Work

There are two particular areas where the user interface to MTL can be improved. First, we would like to use operator overloading to provide a more natural mathematical notation. Expression templates [37] are a promising technology in this regard and their use within the MTL framework warrants further investigation. Second, we intend to simplify matrix definition and construction through the use of of a second layer of templates that hide the underlying nested matrix template structure (as described in [7]). Along these lines, we are also interested in integrating MTL with Blitz++ [38], a high-performance C++ library for multi-dimensional arrays and tensor operations.

We would like to broaden the number of matrix types supported in the MTL, including diagonally oriented matrices, Hermitian matrices, hierarchical matrices, and various types of blocked (super-node) sparse matrices. In addition, we will continue to add numerical types that can be used with MTL, such as intervals and variable precision (an extended precision class is currently distributed with MTL).

Some of the algorithms in MTL need further optimization, especially in the area of sparse matrix operations. Although, current performance is on par with Fortran sparse matrix libraries such as SPARSKIT [30], there are

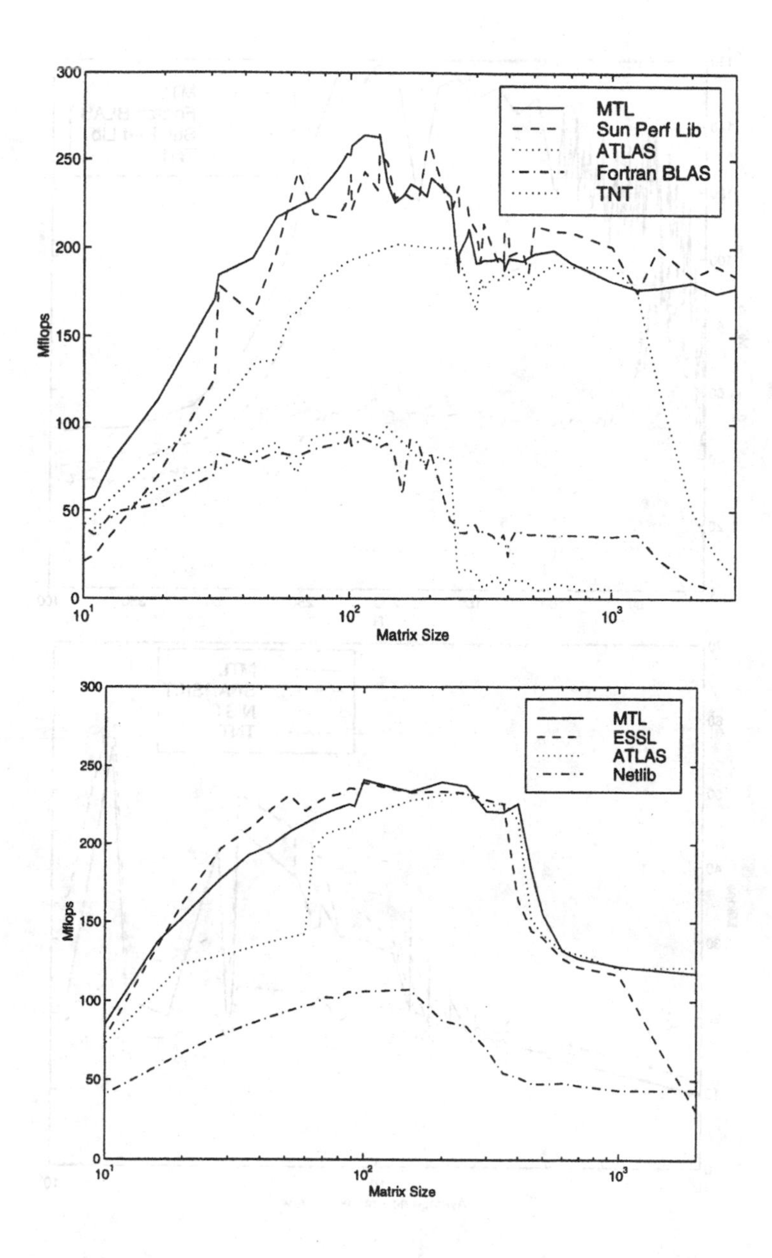

Fig. 14. Performance comparison of generic dense matrix-matrix product with other libraries on Sun UltraSPARC (upper) and IBM RS6000 (lower).

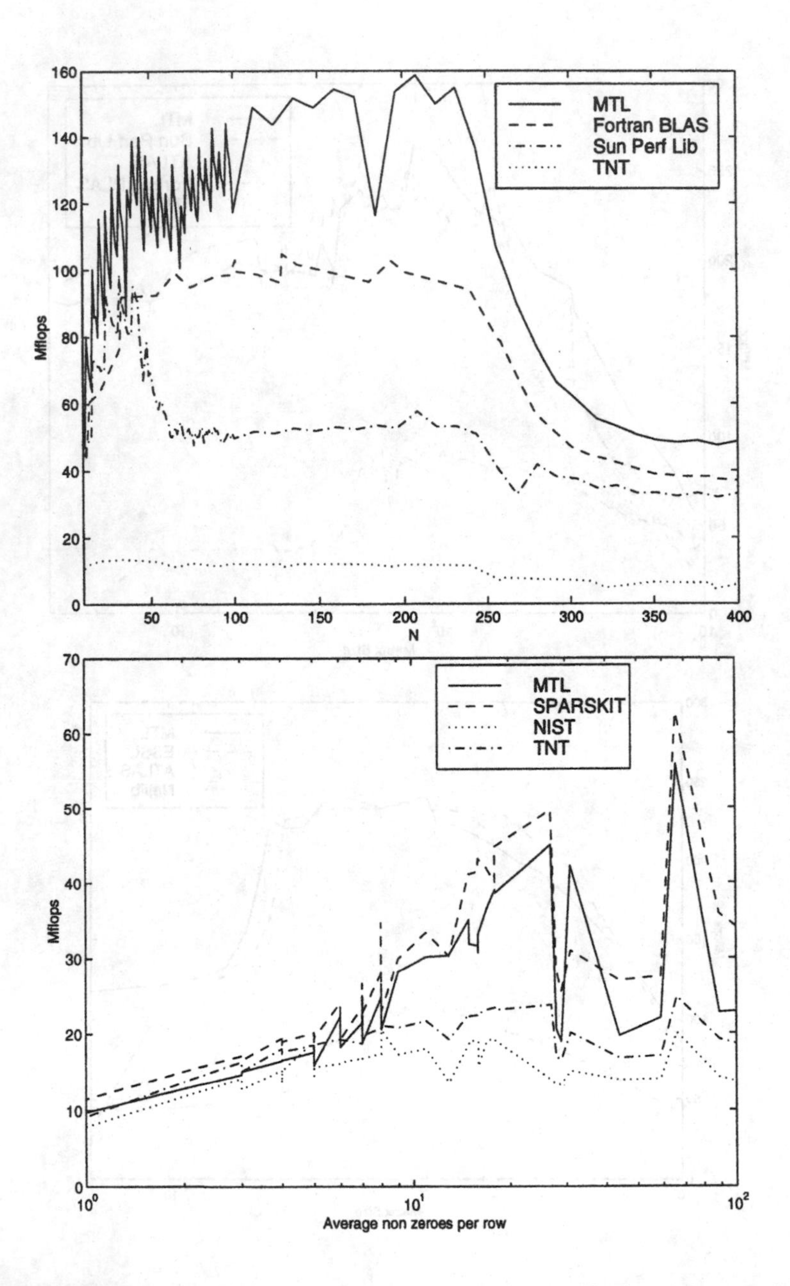

Fig. 15. Performance of generic matrix-vector product applied to column-oriented dense (upper) and row-oriented sparse (lower) data structures compared with other libraries on Sun UltraSPARC.

more advanced techniques that can be used to improve performance (e.g., by ordering the matrix properly or by collecting and forming super-nodes). We also intend to incorporate sparse direct solvers (with the associated graph algorithms) into MTL.

Currently, the approach to tuning blocked algorithms for a particular architecture consists of exhaustively searching the parameter space to find the blocking factors with best performance. By using optimization results from the compiler community ([4,40], e.g.) to generate optimal loop configurations, the tuning process can be made significantly more efficient.

The Matrix Template Library provides a good basis for creating higher level numerical linear algebra libraries. We have constructed an iterative methods library, and will continue to construct other solvers such as Newton's method. With the performance benefits of the MTL, it would be good to gradually replace LAPACK functionality with versions that use MTL (though we now have the intermediate solution of an MTL to LAPACK interface).

Another area that will be explored is better language support for high performance generic programming and metaprogramming. C++ is a good first step and it provides many features necessary to support these programming paradigms, but it was not designed with generic programming or metaprogramming in mind. As a result, implementation is at times cumbersome and restrictive.

Finally, in the continuing search for ever higher performance, there is an obvious need to integrate MTL into parallel execution environments, such as distributed memory (e.g., using MPI) and shared memory parallel machines (e.g., using the multi-threaded programming model).

Acknowledgments

This work was supported by NSF grants ASC94-22380 and CCR95-02710. The authors would like to express their appreciation to Tony Skjellum and Puri Bangalore for numerous helpful discussions. Jeff Squyres and Kinis Meyer also deserve thanks for their help on parts of the MTL. Thanks goes to Brian McCandless for paving the way with the first version of MTL.

References

1. E. Anderson, Z. Bai, C. Bischoff, J. Demmel, J. Dongarra, J. DuCroz, A. Greenbaum, S. Hammarling, A. McKenney, and D. Sorensen. LAPACK: A portable linear algebra package for high-performance computers. In *Proceedings of Supercomputing '90*, pages 1–10. IEEE Press, 1990.
2. M. H. Austern. *Generic Programming and the STL*. Professional computing series. Addison-Wesley, 1999.
3. J. Bilmes, K. Asanovic, J. Demmel, D. Lam, and C.-W. Chin. Optimizing matrix multiply using PHiPAC: A portable, high-performance, ANSI C coding methodology. Technical Report CS-96-326, University of Tennessee, May 1996. Also available as LAPACK working note 111.

4. S. Carr and Y. Guan. Unroll-and-jam using uniformly generated sets. In *Proceedings of the 30th Annual IEEE/ACM International Symposium on Microarchitecture (MICRO-97)*, pages 349–357, Los Alamitos, Dec.1–3 1997. IEEE Computer Society.
5. S. Carr and K. Kennedy. Blocking linear algebra codes for memory hierarchies. In D. C. Dongarra, Jack, Messina, Paul, Sorensen and R. G. Voigt, editors, *Proceedings of the 4th Conference on Parallel Processing for Scientific Computing*, pages 400–405, Philadelphia, PA, USA, Dec. 1989. SIAM Publishers.
6. L. Carter, J. Ferrante, and S. F. Hummel. Hierarchical tiling for improved superscalar performance. In *Proceedings of the 9th International Symposium on Parallel Processing (IPPS'95*, pages 239–245, Los Alamitos, CA, USA, Apr. 1995. IEEE Computer Society Press.
7. K. Czarnecki. *Generative Programming: Principles and Techniques of Software Engineering Based on Automated Configuration and Fragment-Based Component Models*. PhD thesis, Technische Universitat, Ilmenau, Germany, 1998.
8. K. Czarnecki and U. Eisenecker. *Generative Programming: Methods, Techniques and Applications*. Addison-Wesley, 1999. to appear.
9. K. Czarnecki, U. Eisenecker, R. Glück, D. Vandevoorde, and T. L. Veldhuizen. Generative programming and active libraries. In *TBA of Lecture Notes in Computer Science*, Dagstuhl-Seminar on Generic Programming, 1998.
10. J. Dongarra, J. Bunch, C. Moler, and G. Stewart. *LINPACK Users Guide*. Philadelphia, PA, 1978.
11. J. Dongarra, J. D. Croz, I. Duff, and S. Hammarling. A set of level 3 basic linear algebra subprograms. *ACM Transactions on Mathematical Software*, 16(1):1–17, 1990.
12. J. Dongarra, J. D. Croz, S. Hammarling, and R. Hanson. Algorithm 656: An extended set of basic linear algebra subprograms: Model implementations and test programs. *ACM Transactions on Mathematical Software*, 14(1):18–32, 1988.
13. C. Forum. Working paper for draft proposed international standard for information systems – programming language C++. Technical report, American National Standards Institute, 1995.
14. E. Gamma, R. Helm, R. Johnson, and J. Vlissides. *Design Patterns: Elements of Reusable Object-Oriented Software*. Addison-Wesley Professional Computing Series. Addison Wesley, 1995.
15. G. H. Golub and C. F. V. Loan. *Matrix Computations*. Johns Hopkins, 3rd edition, 1996.
16. IBM. *Engineering and Scientific Subroutine Library, Guide and Reference*, 2nd edition, 1992.
17. Kuck and Associates. *Kuck and Associates C++ User's Guide*.
18. M. S. Lam, E. E. Rothberg, and M. E. Wolf. The cache performance and optimizations of blocked algorithms. In *ASPLOS IV*, April 1991.
19. C. Lawson, R. Hanson, D. Kincaid, and F. Krogh. Basic linear algebra subprograms for fortran usage. *ACM Transactions on Mathematical Software*, 5(3):308–323, 1979.
20. B. C. McCandless. The role of abstraction in high performance linear algebra. Master's thesis, University of Notre Dame, 1998.
21. B. C. McCandless and A. Lumsdaine. The role of abstraction in high-performance computing. In *Scientific Computing in Object-Oriented Parallel Environments*. ISCOPE, December 1997.

22. K. Mens, C. Lopes, B. Tekinerdogan, and G. Kiczales. Aspect-oriented programming. *Lecture Notes in Computer Science*, 1357:483–490, 1998.
23. MTL home page: `http://www.lsc.nd.edu/research/mtl`
24. S. Muchnick. *Advanced Compiler Design and Implementation.* Morgan Kaufmann, 1997.
25. D. R. Musser and A. Saini. *STL tutorial and Reference Guide.* Addison-Wesley, Reading, 1996.
26. Netlib BLAS http://www.netlib.org/blas/index.html.
27. NIST. MatrixMarket. http://gams.nist.gov/MatrixMarket/.
28. R. Pozo. *Template Numerical Toolkit (TNT) for Linear Algebra.* National Insitute of Standards and Technology.
29. K. A. Remington and R. Pozo. *NIST Sparse BLAS User's Guide.* National Institute of Standards and Technology.
30. Y. Saad. SPARSKIT: a basic tool kit for sparse matrix computations. Technical report, NASA Ames Research Center, 1990.
31. J. Siek, A. Lumsdaine, and L.-Q. Lee. Generic programming for high performance numerical linear algebra. In *Proceedings of the SIAM Workshop on Object Oriented Methods for Inter-operable Scientific and Engineering Computing (OO'98).* SIAM Press, 1999.
32. B. Smith, J. Boyle, Y. Ikebe, V. Klema, and C. Moler. *Matrix Eigensystem Routines: EISPACK Guide.* New York, NY, second edition, 1970.
33. A. A. Stepanov. Generic programming. *Lecture Notes in Computer Science*, 1181, 1996.
34. A. A. Stepanov and M. Lee. The Standard Template Library. Technical Report X3J16/94-0095, WG21/N0482, ISO Programming Language C++ Project, May 1994.
35. B. Stroustrup. *The C++ Programming Language.* Addison Wesley, 3rd edition, 1997.
36. T. Veldhuizen. Using C++ template metaprograms. *C++ Report*, May 1995.
37. T. L. Veldhuizen. Expression templates. *C++ Report*, 7(5):26–31, June 1995. Reprinted in C++ Gems, ed. Stanley Lippman.
38. T. L. Veldhuizen. Arrays in Blitz++. In *Proceedings of the 2nd International Scientific Computing in Object-Oriented Parallel Environments (ISCOPE'98)*, Lecture Notes in Computer Science. Springer-Verlag, 1998.
39. R. C. Whaley and J. Dongarra. Automatically tuned linear algebra software (ATLAS). Technical report, University of Tennessee and Oak Ridge National Laboratory, 1997.
40. M. E. Wolf and M. S. Lam. A data locality optimising algorithm. In B. Hailpern, editor, *Proceedings of the ACM SIGPLAN '91 Conference on Programming Language Design and Implementation*, pages 30–44, Toronto, ON, Canada, June 1991. ACM Press.

Blitz++: The Library that Thinks it is a Compiler

Todd L. Veldhuizen

Indiana University Computer Science Department, Bloomington, USA
Email: tveldhui@acm.org

Abstract. Blitz++ provides dense numeric arrays for C++ with performance on par with Fortran. It does so by using "template techniques" (expression templates and template metaprograms). In addition to fast performance, Blitz++ arrays provide many nice notations and features not available in Fortran 90.

1 Introduction

The goal of Blitz++ is to provide a good "base environment" for scientific computing in C++. This chapter describes arrays in Blitz++, which are the most important part of the library. Blitz++ arrays have been influenced by Fortran 90, High-Performance Fortran, APL, the Math.h++ class library [4], A++/P++ [5], and POOMA [3]. Blitz++ provides some novel features, such as tensor notation, multicomponent arrays, stencil objects, and irregular subarrays.

The important features of arrays are described from a user's perspective in Sections 2 to 8. Some implementation issues (for example, expression templates and loop transformations) are explored in Section 9. In Section 10, benchmark results are presented for various platforms and compilers. Finally, lessons learned about using C++ are summarized in Section 11.

Blitz++ is freely available under an open source license. Blitz++ may be found on the web at:

http://seurat.uwaterloo.ca/blitz/

Included with the Blitz++ distribution are example programs, a testsuite, benchmarks, and a detailed user's guide.

Blitz++ is written in ISO/ANSI C++, which causes portability problems since few compilers (as of 1999) fully implement the C++ standard. Blitz++ is supported for KAI C++, DECcxx, egcs, and has been successfully compiled with Intel C++ and C++ Builder 4. More compilers will be supported as they achieve compliance with the standard.

2 Arrays

Multidimensional arrays in Blitz++ are provided by a single array class, called `Array<T_numtype,N_rank>`. This class provides a dynamically allocated N-dimensional array, with reference counting, arbitrary storage ordering, subarrays and slicing, flexible expression handling, and many other useful features.

2.1 Template Parameters

The Array class takes two template parameters:

- T_numtype is the numeric type to be stored in the array. T_numtype can be an integral type (bool, char, int, etc.), floating point type (float, double, long double), complex type (e.g. complex<float>) or any user-defined type with a default constructor.
- N_rank is the **rank** (or dimensionality) of the array. This must be a positive integer.

When no constructor arguments are provided, the array is empty, and no memory is allocated. To create an array which contains some data, users may provide the size of the array as constructor arguments:

```
Array<double,2> y(4,4);   // A 4x4 array of double
```

The contents of a newly-created array are garbage. To initialize the array, one can write:

```
y = 0;
```

and all the elements of the array will be set to zero. Arrays may also be initialized using a comma-delimited list of values.[1] This code initializes y to contain a 4x4 identity matrix:

```
y = 1, 0, 0, 0
    0, 1, 0, 0
    0, 0, 1, 0
    0, 0, 0, 1;
```

2.2 Array Types

The Array<T,N> class supports a variety of array types:

- Arrays of scalar types, such as Array<int,1> and Array<float,3>
- Complex arrays, such as Array<complex<float>,2>.
- Arrays of user-defined types. If you have a class called Polynomial, then Array<Polynomial,2> is an array of Polynomial objects. Any operators which are defined on Polynomial are available on arrays of Polynomial.

[1] Blitz++ overloads the comma operator to accomplish this.

– Nested homogeneous arrays using `TinyVector` and `TinyMatrix`, in which each element is a fixed-size vector or array. (`TinyVector` is the Blitz++ lightweight vector class, and `TinyMatrix` is the lightweight matrix class. They improve performance by avoiding dynamic memory allocation for small vectors and matrices.) For example,
`Array< TinyVector<float,3>,3>` is a three-dimensional vector field.
– Nested heterogeneous arrays, such as `Array<Array<int,1>,1>`, in which each element is a variable-length array. However, note that overloaded operators and math functions do not yet work for such arrays.

2.3 Storage Orders

Blitz++ is very flexible about the way arrays are stored in memory. The default storage format is row-major, C-style arrays whose indices start at zero. Fortran-style arrays can also be created: such arrays are stored in column-major order, and have indices which start at one. To create a Fortran-style array, this syntax may be used:

```
Array<int,2> A(3, 3, fortranArray);
```

The `fortranArray` parameter tells the `Array` constructor to use a 2-dimensional Fortran-style array format. This last parameter may also be an instance of class `GeneralArrayStorage<N>`, which encapsulates information about how an array is laid out in memory. By altering the contents of a `GeneralArrayStorage<N>` object, users can customize array layout: the dimensions can be ordered arbitrarily and stored in ascending or descending order, and the starting indices can be arbitrary.

3 Creating Array Objects

Blitz++ provides dozens of constructors for array objects. Here are the most commonly used versions:

```
Array<float,2> A;
Array<float,2> B(3,4);
Array<float,2> C(3,4,fortranArray);
Array<float,2> D(Range(2,5),Range(6,8));
```

The array `A` is empty; it may be allocated using the `resize` method, or be used as an alias for another array using the `reference` method. `B` is a 3x4 array, stored in C array style (row-major, 0-based indices). `C` is a 3x4 Fortran array (column-major, 1-based indices). `D` is a row-major array defined over the indices `2..5` and `6..8`.

Versions of the constructors used for B, C, and D are available for arrays of up to 11 dimensions. For example,

```
Array<float,2> F(3,5,7,9,fortranArray);
```

creates a 4-dimensional array in Fortran storage format, with shape 3x5x7x9. Arrays may be constructed from pre-existing data, for example:

```
// Make a 2x2 array
double data[] = 1, 2, 3, 4 ;
Array<double,2> A(data, shape(2,2), neverDeleteData);
```

This feature makes it easier for Blitz++ to interoperate with other libraries and languages. Blitz++ lets you provide arbitrary starting indices and strides for pre-existing data, making it possible to turn any dense array into a Blitz++ array object. The enum parameter `neverDeleteData` indicates that Blitz++ should not delete the array data when it is no longer needed. Alternative parameters are `deleteDataWhenDone` (the memory *is* deleted by Blitz++ when no longer needed) and `duplicateData` (a copy of the data is made).

3.1 Reference Counting

Blitz++ arrays use reference counting. When new arrays are created, a memory block is allocated. The `Array` object acts like a handle for this memory block. A memory block may be shared among multiple `Array` objects – for example, when subarrays and slices are taken. The memory block tracks how many `Array` objects are referring to it. When a memory block is orphaned – when no `Array` objects are referring to it – it automatically deletes itself and frees the allocated memory. For example, this code results in A, B, and C all sharing the same data (Figure 1):

```
Array<float,2> A(4,4);
Array<float,2> B = A(Range::all(), Range(1,3));
Array<float,1> C = A(2, Range::all());
```

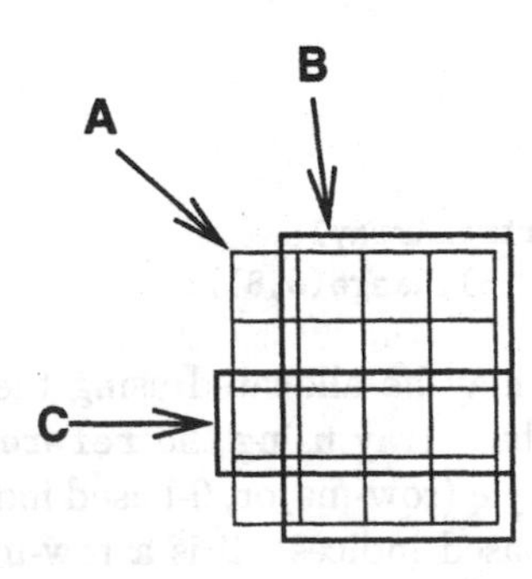

Fig. 1. Three array objects sharing the same block of data.

3.2 Subarrays

Subarrays may either be the same rank as the original array, or a lesser rank (a *slice*). As a running example, we will consider the three dimensional array pictured in Figure 2, which has index ranges (0..7, 0..7, 0..7). Shaded portions of the array show regions which have been obtained by indexing, creating a full-rank subarray, and slicing.

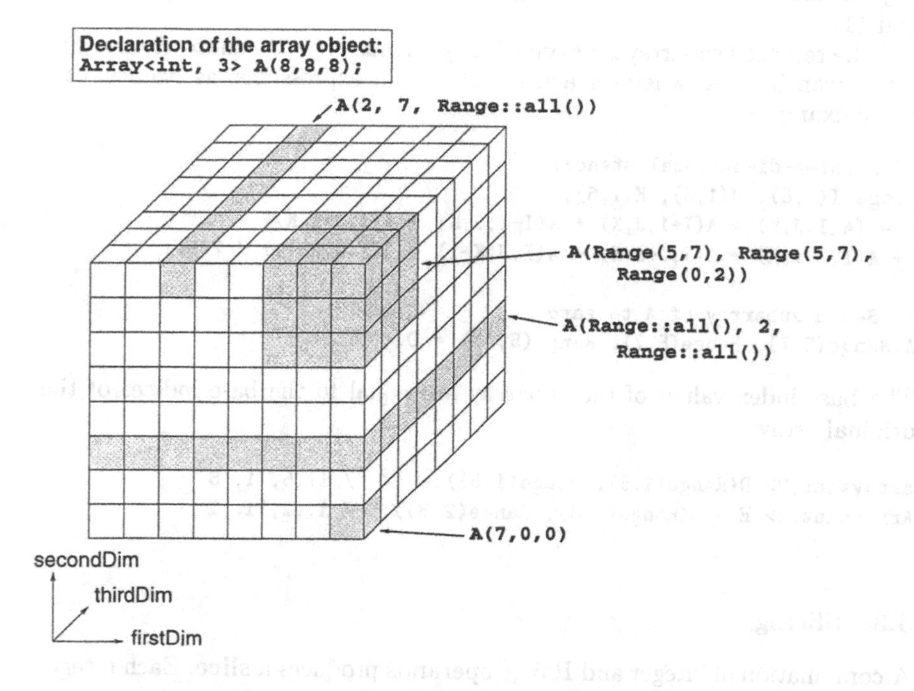

Fig. 2. Examples of array indexing, subarrays, and slicing.

Indexing. There are two ways to get a single element from an array. The simplest is to provide a list of integer indices to `operator()`:

```
A(7,0,0) = 5;
cout << "A(7,0,0) = " << A(7,0,0) << endl;
```

This version of indexing is available for arrays of rank one through eleven. Alternately, one can use operands of type `TinyVector<int,N_rank>` where `N_rank` is the rank of the array object:

```
TinyVector<int,3> index;
index = 7, 0, 0;
A(index) = 5;
cout << "A(7,0,0) = " << A(index) << endl;
```

Subarrays. Full-rank subarrays are obtained by providing Range operands to an array's operator(). A Range object represents a set of regularly spaced index values. For example:

```
Array<int,3> B = A(Range(5,7), Range(5,7), Range(0,2));
```

The object B now refers to elements (5..7,5..7,0..2) of the array A. Range objects can also have a stride: Range(5,11,2) represents the indices 5, 7, 9, and 11.

The returned subarray is of type Array<T_numtype,N_rank>. Hence subarrays can be used wherever arrays can be: in expressions, as lvalues, etc. Some examples:

```
// A three-dimensional stencil
Range I(1,6), J(1,6), K(1,6);
B = (A(I,J,K) + A(I+1,J,K) + A(I-1,J,K) + A(I,J+1,K)
  + A(I,J-1,K) + A(I,J+1,K) + A(I,J,K+1) + A(I,J,K-1)) / 7.0;

// Set a subarray of A to zero
A(Range(5,7), Range(5,7), Range(5,7)) = 0.;
```

The base index values of the subarray are equal to the base indices of the original array:

```
Array<int,2> D(Range(1,5), Range(1,5));      // 1..5, 1..5
Array<int,2> E = D(Range(2,3), Range(2,3)); // 1..2, 1..2
```

3.3 Slicing

A combination of integer and Range operands produces a **slice**. Each integer operand reduces the rank of the array by one. For example:

```
Array<int,2> F = A(Range::all(), 2, Range::all());
Array<int,1> G = A(2,            7, Range::all());
```

Range and integer operands can be used in any combination, for arrays up to rank 11.

3.4 Debug Mode

The Blitz++ library has a debugging mode which is enabled by defining the preprocessor symbol BZ_DEBUG. In debugging mode, programs run quite slowly since Blitz++ does many precondition and bounds checks. When a problem is detected, Blitz++ reports it to the standard error stream. For example, this piece of code attempts to access an element of a 4x4 array which doesn't exist:

```
Array<complex<float>, 2> Z(4,4);

// Since this is a C-style array, the valid index
// ranges are 0..3 and 0..3
Z(4,4) = complex<float>(1.0, 0.0);
```

When the code is run with debug mode enabled, the out of bounds index is detected and an error message results:

```
[Blitz++] Precondition failure: Module array.h line 1070
Array index out of range: (4, 4)
Lower bounds: [          0          0 ]
Upper bounds: [          3          3 ]

Assertion failed: __EX, file array.h, line 1070
```

Precondition failures send their error messages to the standard error stream (`cerr`). After displaying the error message, `assert(0)` is invoked.

4 Expressions

Array expressions in Blitz++ are implemented using the *expression templates* technique [37]. Prior to expression templates, use of overloaded operators meant generating temporary arrays, which caused huge performance losses. In Blitz++, temporary arrays are never created. Section 9 describes how Blitz++ uses expression templates to avoid temporary arrays.

4.1 Expression Operands

An expression can contain a mix of array objects, scalars, stencil operators and index placeholders (described later). Arrays with different storage formats (for example, C-style and Fortran-style) can be mixed in the same expression. Blitz++ will handle the different storage formats automatically. Arrays of different numeric types may also be mixed in an expression: type promotion follows the standard C rules, with some modifications to handle complex numbers and user-defined types.

4.2 Operators

These binary operators are supported:

`+ - * / % > < >= <= == != && || ^ & |`

These unary operators are supported:

`- ~ !`

The operators `> < >= <= == != && || !` result in a bool-valued expression. All operators are applied *elementwise*. For example, if `A` and `B` are two-dimensional arrays, `A*B` means `A(i,j)*B(i,j)`, not matrix multiplication.

4.3 Assignment Operators

These assignment operators are supported:

= += -= *= /= %= ^= &= |= >>= <<=

An array object should appear on the left side of the operator. The right side can be:

- A scalar value of type T_numtype
- An array of appropriate rank, possibly of a different numeric type
- An array expression, with appropriate rank and shape

4.4 Index Placeholders

Blitz++ provides objects called *index placeholders* which represent array indices. There is a distinct index placeholder type associated with each dimension of an array. The types are called `firstIndex`, `secondIndex`, `thirdIndex`, ..., `tenthIndex`, `eleventhIndex`. Index placeholders can be used to initialize arrays using a function of the array indices. Here is an example which fills an array with a sampled sine wave:

```
Array<float,1> A(16);
firstIndex i;
A = sin(2 * M_PI * i / 16.);
```

For convenience, the namespace `blitz::tensor` contains pre-defined index placeholders `i`, `j`, `k`, and so on. Here is a two-dimensional example:

```
Array<float,2> B(4,4);
using namespace blitz::tensor;
B = 1.0 / (1 + i + j);
```

Instead of the `using namespace` directive, one can alternately preface index placeholders with `tensor::`, for example:

```
Array<float,2> B(4,4);
B = 1.0 / (1 + tensor::i + tensor::j);
```

4.5 Single-Argument Math Functions

All math functions are applied *element-wise*. For example, this code–

```
Array<float,2> A, B;    //
A = sin(B);
```

results in `A(i,j) = sin(B(i,j))` for all (i,j). These math functions are available on all platforms:

```
abs  acos  asin   atan ceil  cexp
cos  cosh  csqrt  exp  fabs  floor
log  log10 pow[n] sin  sinh  sqr
sqrt tan   tanh

atan2 pow
```

The functions pow2, pow3, ..., pow8 are power functions implemented by optimal multiplication trees. On platforms which provide the IEEE and System V math libraries, these math functions may be available:

```
acosh asinh atanh  _class   cbrt  expm1
erf   erfc  finite ilogb    isnan itrunc
j0    j1    lgamma logb     log1p nearest
rint  rsqrt trunc  uitrunc  y0    y1
copysign    drem   fmod     hypot nextafter
remainder   scalb  unordered
```

4.6 Tensor Notation

Blitz++ arrays support a tensor-like notation. Here is an example of real-world tensor notation:

$$A^{ijk} = B^{ij}C^{k}$$

A is a rank 3 tensor (a three dimensional array), B is a rank 2 tensor, and C is a rank 1 tensor. The above expression sets $A(i,j,k) = B(i,j) * C(k)$.

To implement this product using Blitz++, one uses index placeholders (Section 4.4) as tensor indices:

```
Array<float,3> A(4,4,4);
Array<float,2> B(4,4);
Array<float,1> C(4);

using namespace blitz::tensor;  // Access to i, j, k
A = B(i,j) * C(k);
```

The index placeholder arguments tell an array how to map its dimensions onto the dimensions of the destination array. For example, the tensor expression

$$C^{ijk} = A^{ij}x^{k} - A^{jk}y^{i}$$

would be coded using Blitz++ as:

```
using namespace blitz::tensor;
C = A(i,j) * x(k) - A(j,k) * y(i);
```

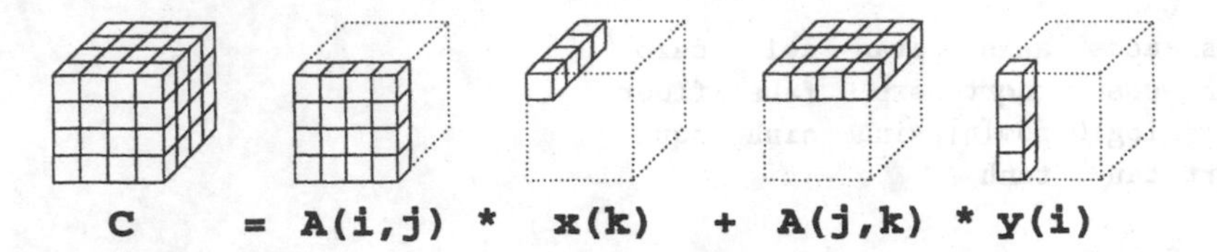

Fig. 3. Tensor notation.

This tensor expression can be visualized as shown in Figure 3.

Index placeholders may not be used on the left-hand side of an expression. If the indices must be reordered, it must be done on the right-hand side. In real-world tensor notation, repeated indices imply a contraction (or summation). For example, this tensor expression computes a matrix-matrix product:

$$C^{ij} = A^{ik} B^{kj}$$

The repeated k index is interpreted as meaning

$$c_{ij} = \sum_k a_{ik} * b_{kj} \tag{1}$$

In Blitz++, repeated indices do *not* imply contraction. If you want to contract (sum along) an index, you must use the sum() function:

```
Array<float,2> A, B, C;   // ...
using namespace blitz::tensor;
C = sum(A(i,k) * B(k,j), k);
```

The sum() function is an example of an *array reduction*, described in the next section.

4.7 Array Reductions

Blitz++ arrays support two forms of reduction:

- Reductions which transform an array into a scalar (for example, summing the elements). These are referred to as **complete reductions**.
- Reducing an N dimensional array (or array expression) to an N-1 dimensional array expression. These are called **partial reductions**.

4.8 Complete Reductions

Complete reductions transform an array (or array expression) into a scalar. Here are some examples:

```
Array<float,2> A(3,3);
A = 0, 1, 2,
    3, 4, 5,
    6, 7, 8;
cout << sum(A) << endl              // 36
     << min(A) << endl              // 0
     << count(A >= 4) << endl;      // 5
```

Here are the available complete reductions:

sum() Summation (may be promoted to a higher-precision type)
product() Product
mean() Arithmetic mean (promoted to floating-point type if necessary)
min() Minimum value
max() Maximum value
minIndex() Index of the minimum value (`TinyVector<int,N_rank>`)
maxIndex() Index of the maximum value (`TinyVector<int,N_rank>`)
count() Counts the number of times the expression is logical true (int)
any() True if the expression is true anywhere (bool)
all() True if the expression is true everywhere (bool)
first() First location at which the expression is logical true.

The reductions `any()`, `all()`, and `first()` have short-circuit semantics: the reduction will halt as soon as the answer is known. For example, if you use `any()`, scanning of the expression will stop as soon as the first true value is encountered.

4.9 Partial Reductions

Partial reductions transform an array (or array expression) to an N-1 dimensional array expression. Here is an example which computes the sum of each row of a two-dimensional array:

```
Array<float,2> A;     // ...
Array<float,1> rs;    // ...
using namespace blitz::tensor;

rs = sum(A, j);
```

The reduction `sum()` takes two arguments:

- The first argument is an array or array expression.

– The second argument is an index placeholder (Section 4.4) indicating the dimension over which the reduction is to occur.

Reductions have an **important restriction**: It is currently only possible to reduce over the *last* dimension of an array or array expression. Reducing a dimension other than the last would require Blitz++ to reorder the dimensions to fill the hole left behind. For example, in order for this reduction to work:

```
Array<float,3> A;   // ...
Array<float,2> B;   // ...
secondIndex j;

// Reduce over dimension 2 of a 3-D array?
B = sum(A, j);
```

Blitz++ would have to remap the dimensions so that the third dimension became the second. It is not currently smart enough to do this.

However, there is a simple workaround which solves some of the problems created by this limitation: you can do the reordering manually, prior to the reduction:

```
B = sum(A(i,k,j), k);
```

Writing `A(i,k,j)` interchanges the second and third dimensions, permitting you to reduce over the second dimension.

All of the reductions described in the previous section (sum, product, mean, min, max, minIndex, maxIndex, count, any, all, first) may be used as partial reductions. The result of a reduction is an array expression, so reductions can be used in array expressions:

```
Array<int,3> A;
Array<int,2> B;
Array<int,1> C;   // ...

using namespace blitz::tensor;

B = mean(pow2(A), k);        // **

// Do two reductions in a row
C = sum(sum(A, k), j);           // ***
```

The expression `**` calculates the mean sum-of-squares in each column of the three-dimensional array `A`. The result is a two-dimensional array expression, which is stored in `B`. The expression `***` calculates the sum of each `j-k` plane in the three-dimensional array `A`, and stores the result in the one-dimensional array `C`.

4.10 Where Statements

Blitz++ provides the `where` function as an array expression version of the `?:` operator. The syntax is:

```
where(array-expr1, array-expr2, array-expr3)
```

Wherever `array-expr1` is true, `array-expr2` is returned. Where `array-expr1` is false, `array-expr3` is returned. For example, suppose one wanted to sum the squares of only the positive elements of an array. This can be implemented using a where function:

```
double posSquareSum = sum(where(A > 0, pow2(A), 0));
```

5 Stencils

The stencil notation used in Blitz++ was developed in collaboration with the POOMA team [3] to provide a nicer notation for finite differencing.

5.1 Motivation: A Nicer Notation for Stencils

Suppose we wanted to implement the 3-D acoustic wave equation using finite differencing. Here is how a single iteration would look using subarray syntax:

```
Range I(1,N-2), J(1,N-2), K(1,N-2);

P3(I,J,K) = (2-6*c(I,J,K)) * P2(I,J,K)
   + c(I,J,K)*(P2(I-1,J,K) + P2(I+1,J,K) + P2(I,J-1,K)
   + P2(I,J+1,K) + P2(I,J,K-1) + P2(I,J,K+1)) - P1(I,J,K);
```

This is a bit awkward. With stencil objects, the implementation becomes:

```
// Create arrays ci, P1i, P2i, P3i which are views of
// the "interior" of c, P1, P2, P3
Range I(1,N-2), J(1,N-2), K(1,N-2);
Array<float,3> ci = c(I,J,K), P1i = P1(I,J,K),
       P2i = P2(I,J,K), P3i = P3(I,J,K);

// Stencil equation
P3i = 2 * P2i + ci * Laplacian3D(P2i) - P1i;
```

This is clearly a big improvement in readability. And it comes at no performance cost, since the resulting code is very similar to the version without stencil objects.[2]

[2] In fact, performance can be much better for complicated stencils, since only one pointer is needed per stencil operator. This is in contrast to the subarray style, which requires one pointer for each array operand. Fewer pointers means less chance of registers being spilled.

5.2 Stencil Operators

Stencil operators assume an array represents evenly spaced data points separated by a distance of h. A 2nd-order accurate operator has error term $O(h^2)$; a 4th-order accurate operator has error term $O(h^4)$.

All stencil operators are *expression-templatized*, meaning they can be used in array expressions.

Stencils follow a naming convention. In the stencil name `central12`, the word `central` means the stencil is a central difference; the numeral 1 means a first derivative; the numeral 2 means 2nd-order accuracy. The stencil `central34` is a central difference, 3rd derivative with 4th-order accuracy.

Here are the simple differences:

```
central12    central22    central32    central42
central14    central24    central34    central44
forward11    forward21    forward31    forward41
forward12    forward22    forward32    forward42
backward11   backward21   backward31   backward41
backward12   backward22   backward32   backward42
```

Some of these stencils have associated factors by which they must be multiplied to convert them into derivatives; for example, the stencil `central24` must be multiplied by $12h^2$. All stencils are provided in normalized versions `central12n`, `central22n`, ... which have factors of h, h^2, h^3 or h^4 as appropriate (i.e. no integer factors).

These stencil operators must be given both an array and a dimension: for example, `central12(A,secondDim)` represents a first-derivative of the array `A` in the second dimension direction.

Laplacian Operators.

Laplacian2D(A) : 2nd order accurate, 2-dimensional Laplacian.
Laplacian3D(A) : 2nd order accurate, 3-dimensional Laplacian.
Laplacian2D4(A) : 4th order accurate, 2-dimensional Laplacian.
Laplacian3D4(A) : 4th order accurate, 3-dimensional Laplacian.

Gradient Operators. These return `TinyVectors` of the appropriate numeric type and length:

grad2D(A) : 2nd order, 2-dimensional gradient (vector of first derivatives), generated using the central12 operator.
grad3D(A) : 2nd order, 3-dimensional gradient, using central12 operator.
grad2D4(A) : 4th order, 2-dimensional gradient, using central14 operator.
grad3D4(A) : 4th order, 3-dimensional gradient, using central14 operator.

Curl Operators. The curl operators return three-dimensional `TinyVectors` of the appropriate numeric type:

curl(Vx,Vy,Vz) : 2nd order curl operator using the central12 operator.
curl4(Vx,Vy,Vz) : 4th order curl operator using the central14 operator.

There are also curl operators which operate over vector fields.

Divergence Operators. The divergence operators return a scalar value.

div(Vx,Vy,Vz) : 2nd order div operator using the central12 operator.
div4(Vx,Vy,Vz) : 4th order div operator using the central14 operator.

There are also divergence operators which operate over vector fields.

Mixed Partial Derivatives.

mixed22(A,dim1,dim2) : 2nd order accurate, 2nd mixed partial derivative.
mixed24(A,dim1,dim2) : 4th order accurate, 2nd mixed partial derivative.

In addition to the stencil operators provided by Blitz++, users may define their own. The syntax for this is straightforward and is detailed in the user's guide.

5.3 Stencil Objects

Blitz++ also provides *stencil objects*, which are intended for computationally intense stencils involving multiple arrays.

Stencil objects are declared using the macros `BZ_DECLARE_STENCIL1`, `BZ_DECLARE_STENCIL2`, etc. The number suffix is how many arrays are involved in the stencil (in the above example, 4 arrays– P1, P2, P3, c – are used, so the macro `BZ_DECLARE_STENCIL4` is invoked). These macros expand into class definitions.

The first argument is a name for the stencil object. Subsequent arguments are names for the arrays on which the stencil operates. After the stencil declaration, the macro `BZ_END_STENCIL` must appear. In between the two macros, there may be multiple assignment statements, if/else/elseif constructs, function calls, loops, etc. Here are some simple examples:

```
BZ_DECLARE_STENCIL2(smooth2D,A,B)
  A = (B(0,0) + B(0,1) + B(0,-1) + B(1,0) + B(-1,0)) / 5.0;
BZ_END_STENCIL

BZ_DECLARE_STENCIL8(prop2D,E1,E2,E3,M1,M2,M3,cE,cM)
```

```
  E3 = 2 * E2 + cE * Laplacian2D(E2) - E1;
  M3 = 2 * M2 + cM * Laplacian2D(M2) - M1;
BZ_END_STENCIL

BZ_DECLARE_STENCIL3(smooth2Db,A,B,c)
  if ((c > 0.0) && (c < 1.0))
    A = c * (B(0,0) + B(0,1) + B(0,-1) + B(1,0)
      + B(-1,0)) / 5.0 + (1-c)*B;
  else
    A = 0;
BZ_END_STENCIL
```

Currently, a stencil can take up to 11 array parameters. The notation `A(i,j,k)` reads the element at an offset `(i,j,k)` from the current element in array A. A term of the form `A` just reads the current element of A.

Stencil objects may invoke *stencil operators*, described in the previous section.

5.4 Applying a Stencil

The syntax for applying a stencil is:

```
applyStencil(stencilname(),A,B,C...,F);
```

where `stencilname` is the name of the stencil, and `A,B,C,...,F` are the arrays on which the stencil operates.

Blitz++ interrogates the stencil object to find out how large its footprint is. It only applies the stencil over the region of the arrays where it will not overrun the boundaries.

6 Multicomponent and Complex Arrays

Multicomponent arrays have elements which are vectors. Examples of such arrays are vector fields, colour images (which contain, say, RGB tuples), and multi-spectral images. Complex-valued arrays can also be regarded as multicomponent arrays, since each element is a 2-tuple of real values.

Here are some examples of multicomponent arrays:

```
// A 3-dimensional array; each element is a
// length 3 vector of float
Array<TinyVector<float,3>,3> A;

// A complex 2-dimensional array
Array<complex<double>,2> B;

// A 2-dimensional image containing RGB tuples
```

```
struct RGB24
  unsigned char r, g, b;
;

Array<RGB24,2> C;
```

6.1 Extracting Components

Blitz++ provides special support for such arrays. The most important is the ability to extract a single component. For example:

```
Array<TinyVector<float,3>,2> A(128,128);
Array<float,2> B = A.extractComponent(float(), 1, 3);
B = 0;
```

The call to `extractComponent` returns an array of floats; this array is a view of the second component of each element of A. The arguments of `extractComponent` are: (1) the type of the component (in this example, float); (2) the component number to extract (numbered 0, 1, ... N-1); and (3) the number of components in the array.

In the above example, the line `B = 0` sets the second element of each element of `A` to zero. This is because `B` is a view of `A`'s data, rather than a new array.

This is a tad messy, so Blitz++ provides a handy shortcut using `[]`:

```
Array<TinyVector<float,3>,2> A(128,128);
A[1] = 0;
```

The number inside the square brackets is the component number. However, for this operation to work, Blitz++ has to already know how many components there are, and what type they are. It knows this already for `TinyVector` and `complex<T>`. If you use your own type, though, you must tell Blitz++ this information using the macro `BZ_DECLARE_MULTICOMPONENT_TYPE()`. This macro has three arguments:

```
BZ_DECLARE_MULTICOMPONENT_TYPE(T_element, T_componentType,
    numComponents)
```

`T_element` is the element type of the array. `T_componentType` is the type of the components of that element. `numComponents` is the number of components in each element.

An example will clarify this. Suppose we wanted to make a colour image, stored in 24-bit HSV (hue-saturation-value) format. We can make a class `HSV24` which represents a single pixel:

```
#include <blitz/array.h>
```

```
using namespace blitz;

class HSV24
public:
    // These constants will makes the code below cleaner;  we
    // can refer to components by name, rather than number.

    static const int hue=0, saturation=1, value=2;

    HSV24()
    HSV24(int hue, int saturation, int value)
      : h_(hue), s_(saturation), v_(value)


    // Some other stuff here, obviously


private:
    unsigned char h_, s_, v_;
;
```

Right after the class declaration, we will invoke the macro BZ_DECLARE_-MULTICOMPONENT_TYPE to tell Blitz++ about HSV24:

```
// HSV24 has 3 components of type unsigned char
BZ_DECLARE_MULTICOMPONENT_TYPE(HSV24, unsigned char, 3);
```

Now we can create HSV images and modify the individual components:

```
int main()

    Array<HSV24,2> A(128,128);   // A 128x128 HSV image
    ...

    // Extract a greyscale version of the image
    Array<unsigned char,2> A_greyscale = A[HSV24::value];

    // Bump up the saturation component to get a
    // pastel effect
    A[HSV24::saturation] *= 1.3;

    // Brighten up the middle of the image
    Range middle(32,96);
    A[HSV24::value](middle,middle) *= 1.2;
```

6.2 Special Support for Complex Arrays

Since complex arrays are used frequently, Blitz++ provides two special methods for getting the real and imaginary components:

```
Array<complex<float>,2> A(32,32);

real(A) = 1.0;
imag(A) = 0.0;
```

The function `real(A)` returns an array view of the real component; `imag(A)` returns a view of the imaginary component. Special math functions are provided for complex arrays: `cexp`, `polar`, and `conj`.

6.3 Zipping Together Expressions

Blitz++ provides a function `zip()` which lets you combine two or more expressions into a single multicomponent expression. For example, you can combine two real expressions into a complex expression:

```
int N = 16;
Array<complex<float>,1> A(N);
Array<float,1> theta(N);

  ...

A = zip(cos(theta), sin(theta), complex<float>());
```

The above line is equivalent to:

```
for (int i=0; i < N; ++i)
   A[i] = complex<float>(cos(theta[i]), sin(theta[i]));
```

7 Indirection

Indirection is the ability to modify or access an array at a set of selected index values. Blitz++ provides several forms of indirection:

- **Using a list of array positions**: this approach is useful if you need to modify an array at a set of scattered points.
- **Cartesian-product indirection**: as an example, for a two-dimensional array you might have a list I of rows and a list J of columns, and you want to modify the array at all (i,j) positions where $i \in I$ and $j \in J$. This is a **Cartesian product** of the index sets I and J.
- **Over a set of strips**: for efficiency, you can represent an arbitrarily-shaped subset of an array as a list of one-dimensional strips. This is a useful way of handling **Regions Of Interest** (ROIs).

In all cases, Blitz++ expects a Standard Template Library container. STL containers are used because they are widely available and provide easier manipulation of "sets" than Blitz++ arrays. For example, one can easily expand

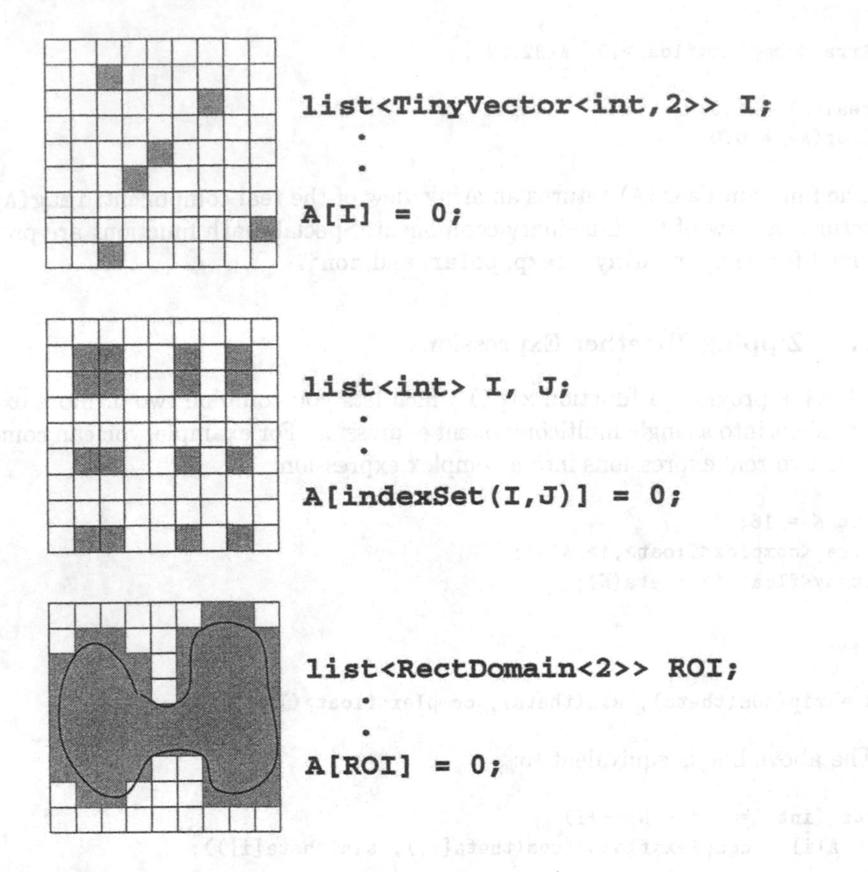

Fig. 4. Three types of indirection (from top to bottom): using a list of array positions; Cartesian-product indirection; lists of strips.

and merge sets which are stored in STL containers; doing this is not so easy with Blitz++ arrays, which are designed for numerical work.

The `[]` operator is overloaded on arrays so that the syntax `A[C]` (where `A` is an array and `C` is an STL container) provides an indirect view of the array. So far, this indirect view may only be used as an lvalue (i.e. on the left-hand side of an assignment statement).

7.1 Indirection Using Lists of Array Positions

The simplest kind of indirection uses a list of points. For one-dimensional arrays, you can just use an STL container of integers. Example:

```
Array<int,1> A(5), B(5);
A = 0;
B = 1, 2, 3, 4, 5;
```

```
vector<int> I;
I.push_back(2);
I.push_back(4);
I.push_back(1);

A[I] = B;
```

After this code, the array A contains `[ 0 2 3 0 5 ]`.

Note that arrays on the right-hand-side of the assignment must have the same shape as the array on the left-hand-side (before indirection). In the statement `A[I]=B`, A and B must have the same shape, not I and B.

For multidimensional arrays, one can use an STL container of `TinyVector<int,N_rank>` objects. Example:

```
Array<int,2> A(4,4), B(4,4);
A = 0;
B = 10*tensor::i + tensor::j;

typedef TinyVector<int,2> coord;

list<coord> I;
I.push_back(coord(1,1));
I.push_back(coord(2,2));

A[I] = B;
```

After this code, the array A contains:

```
0  0  0  0
0 11  0  0
0  0 22  0
0  0  0  0
```

The `tensor::i` notation is explained in Section 4.4.

7.2 Cartesian-Product Indirection

The Cartesian product of the sets I, J and K is the set of (i,j,k) tuples for which $i \in I$, $j \in J$, and $k \in K$.

Blitz++ implements Cartesian-product indirection using an **adaptor** which takes a set of STL containers and iterates through their Cartesian product. Note that the Cartesian product is never explicitly created. You create the Cartesian-product adaptor by calling the function:

```
template<class T_container>
indexSet(T_container& c1, T_container& c2, ...)
```

The returned adaptor can then be used in the [] operator of an array object. Here is a two-dimensional example:

```
Array<int,2> A(6,6), B(6,6);
A = 0;
B = 10*tensor::i + tensor::j;

vector<int> I, J;
I.push_back(1);
I.push_back(2);
I.push_back(4);

J.push_back(0);
J.push_back(2);
J.push_back(5);

A[indexSet(I,J)] = B;
```

After this code, the A array contains:

```
 0  0  0  0  0  0
10  0 12  0  0 15
20  0 22  0  0 25
 0  0  0  0  0  0
40  0 42  0  0 45
 0  0  0  0  0  0
```

All the containers used in a Cartesian product must be the same type (e.g. all `vector<int>` or all `set<int>`), but they may be different sizes. Singleton containers (containers containing a single value) are fine.

7.3 Indirection with Lists of Strips

You can also do indirection with a container of one-dimensional **strips**. This is useful when you want to manipulate some arbitrarily-shaped, well-connected subdomain of an array. By representing the subdomain as a list of strips, you allow Blitz++ to operate on vectors, rather than scattered points; this is much more efficient.

Strips are represented by objects of type `RectDomain<N>`, where `N` is the dimensionality of the array. The `RectDomain<N>` class can be used to represent any rectangular subdomain, but for indirection it is only used to represent strips.

You create a strip by using this function:

```
RectDomain<N> strip(TinyVector<int,N> start,
    int stripDimension, int ubound);
```

The `start` parameter is where the strip starts; `stripDimension` is the dimension in which the strip runs; ubound is the last index value for the strip. For example, to create a 2-dimensional strip from (2,5) to (2,9), one would write:

```
TinyVector<int,2> start(2,5);
RectDomain<2> myStrip = strip(start,secondDim,9);
```

Here is a more substantial example which creates a list of strips representing a circle subset of an array:

```
const int N = 7;
Array<int,2> A(N,N), B(N,N);
typedef TinyVector<int,2> coord;

A = 0;
B = 1;

double centre_i = (N-1)/2.0;
double centre_j = (N-1)/2.0;
double radius = 0.8 * N/2.0;

// circle will contain a list of strips which represent a
// circular subdomain.

list<RectDomain<2> > circle;
for (int i=0; i < N; ++i)

  double jdist2 = pow2(radius) - pow2(i-centre_i);
  if (jdist2 < 0.0)
    continue;

  int jdist = int(sqrt(jdist2));
  coord startPos(i, int(centre_j - jdist));
  circle.push_back(strip(startPos, secondDim,
    int(centre_j + jdist)));


// Set only those points in the circle subdomain to 1
A[circle] = B;
```

After this code, the A array contains:

```
0  0  0  0  0  0  0
0  0  1  1  1  0  0
0  1  1  1  1  1  0
0  1  1  1  1  1  0
0  1  1  1  1  1  0
0  0  1  1  1  0  0
0  0  0  0  0  0  0
```

8 Creating Arrays of a User Type

You can use the Array class with types you have created yourself, or types from another library. If you want to do arithmetic on the array, whatever operators you use on the arrays have to be defined on the underlying type.

For example, here is a simple class which provides fixed-point numbers in the interval [0,1]. Fixed-point numbers have a fixed exponent, but the mantissa can take on any value. They are sometimes used in financial calculations where an exact representation is needed.

```
#include <blitz/numinquire.h>     // for huge()

class FixedPoint {
public:
    // The type to use for the mantissa
    typedef unsigned int T_mantissa;

    FixedPoint() { }

    explicit FixedPoint(T_mantissa mantissa)
    { mantissa_ = mantissa; }

    FixedPoint(double value)
    {
        // Create a FixedPoint number from a double
        // value, by multiplying by the largest value
        // of "unsigned int" (typically 2^32 or 2^64):
        assert((value >= 0.0) && (value <= 1.0));
        mantissa_ = value * huge(T_mantissa());
    }

    FixedPoint operator+(FixedPoint x)
    { return FixedPoint(mantissa_ + x.mantissa_); }

    double value() const
    { return mantissa_ / double(huge(T_mantissa())); }

private:
    T_mantissa mantissa_;
};

ostream& operator<<(ostream& os, const FixedPoint& a)
{
    os << a.value();
    return os;
}
```

The function huge(T) returns the largest representable value for type T; in the example above, it is equal to UINT_MAX.

The `FixedPoint` class declares three useful operations: conversion from `double`, addition, and outputting to an `ostream`. We can use all of these operations on an `Array<FixedPoint>` object:

```
#include <blitz/array.h>

using namespace blitz;

int main()
{
    // Create an array using the FixedPoint class:

    Array<FixedPoint, 2> A(4,4), B(4,4);

    A = 0.5, 0.3, 0.8, 0.2,
        0.1, 0.3, 0.2, 0.9,
        0.0, 0.2, 0.7, 0.4,
        0.2, 0.3, 0.8, 0.4;

    B = A + 0.05;

    cout << "B = " << B << endl;

    return 0;
}
```

Note that the array `A` is initialized using a comma-delimited list of `double`; this makes use of the constructor `FixedPoint(double)`. The assignment `B = A + 0.05` uses `FixedPoint::operator+(FixedPoint)`, with an implicit conversion from `double` to `FixedPoint`. Formatting the array `B` onto the standard output stream is done using the output operator defined for `FixedPoint`.

Here is the program output:

```
B = 4 x 4
         0.55         0.35         0.85         0.25
         0.15         0.35         0.25         0.95
         0.05         0.25         0.75         0.45
         0.25         0.35         0.85         0.45
```

9 Implementation Overview of Blitz++

Of the many features of C++, it is operator overloading which makes it most attractive for scientific computing.

Unfortunately, straightforward use of overloaded operators for arrays leads to very poor performance. Why is this? In C++, operators always produce intermediate results. If the intermediate results are arrays, then evaluating an expression such as

```
Vector<double> a(N), b(N), c(N), d(N);
// ...

a = b + c + d;
```

results in code similar to:

```
double* t1 = new double[N];
for (int i=0; i < N; ++i)
    t1[i] = b[i] + c[i];
double* t2 = new double[N];
for (int i=0; i < N; ++i)
    t2[i] = t1[i] + d[i];
for (int i=0; i < N; ++i)
    a[i] = t2[i];
delete [] t2;
delete [] t1;
```

For very small arrays, the overhead of `new` and `delete` results in poor performance – often 1/10th that of C. For medium-size arrays which fit entirely in cache, the overhead of extra loops and extra cache memory accesses reduces performance by roughly 30-50% for small expressions. For large arrays, it is the temporary arrays which cause poor performance – all that extra data must be shipped between main memory and cache. For stencil expressions, it is possible to get 1/7 or 1/27 the performance of C or Fortran, depending on the size of the stencil.

Blitz++ avoids these performance problems by using the expression templates technique. For a full discussion of expression templates, see [6,2]. The basic idea of expression templates is to avoid the cost of intermediate arrays by having operators return parse trees. For example, an expression such as `b+c+d` does not return an array; instead, it returns a parse tree which represents the expression `b+c+d`. Only when this parse tree is assigned to an array does it get evaluated.

The tricky part is that these parse trees are constructed at compile time, by exploiting the ability of C++ templates to perform compile-time computations. Here is a simple example which gives the general flavour of the technique: we can create a template class Expr which represents a node in an expression parse tree:

```
template<typename Op, typename Operand1, typename Operand2>
class Expr { ... };

class plus { ... };
class minus { ... };
```

An expression such as `(b+c)+d` can then be represented by the type

```
Expr<plus, Expr<plus, Array, Array>, Array>
```

This expression encodes a parse tree of `(b+c)+d` as a type. To automatically build such types, one can overload operators, for example:

```
template<class T>
Expr<plus, T, Array> operator+(T, Array)
{
    return Expr<plus, T, Array>();
}
```

Consider how this parses `(b+c)+d`:

```
a = (b + c) + d;
  = Expr<plus, Array, Array>() + d;
  = Expr<plus, Expr<plus, Array, Array>, Array>();
```

In a real implementation one would store data inside the parse tree, such as pointers to the array data. The next step is to use the parse tree to generate an efficient evaluation routine: one which uses only one loop, and no temporary arrays. This is too complicated to describe here, so the curious reader is referred to [6,2] for details.

9.1 Optimizing Array Expressions

As described in the previous section, Blitz++ uses expression templates to parse array expressions and generate implementation code at compile time. Blitz++ is able to perform a number of optimizations in the generated code. As a running example, consider this piece of code, which is a naive implementation of the expression `A = B + C`:

```
for (int i=0; i < N; ++i)
  for (int j=0; j < N2; ++j)
    for (int k=0; k < N3; ++k)
      A(i,j,k) = B(i,j,k) + C(i,j,k);
```

Blitz++ does not actually generate nested loops such as these. Instead, it uses a data structure to represent the outer loops; only the innermost loop is written as a conventional loop. This allows the same code to be reused for arrays of one, two, three, or many dimensions. The cost of manipulating this data structure is usually small compared to the time spent in the innermost loop. Since the loops are represented by a data structure, loop transformations (see e.g. [8]) can be performed easily:

Loop interchange and reversal: In the above example, the memory layout of the arrays `A`, `B` and `C` is unknown at compile time. If they are column-major, this code will be very inefficient because of poor data locality (the cache lines would be aligned along the `i` direction, but the arrays are traversed in the k direction). To avoid this problem, Blitz++ selects a traversal order at run-time such that the arrays are traversed in memory-storage order. This requires rearranging the order of the loops, which is simple because of the data-structure representation.

Hoisting stride calculations: The inner loop of the above code fragment would expand to contain many stride calculations. Although some compilers will hoist some of these calculations out of the inner loop, others will not. Blitz++ generates code which has only the essential stride calculations in the inner loop.

Collapsing inner loops: Suppose that in the above code fragment, `N3` is quite small. Loop overhead and pipeline effects will conspire to cause poor performance. The solution is to convert the three nested loops into a single loop. At runtime, Blitz++ collapses the inner loops whenever possible. It does this with a small test at the beginning of expression evaluation, which rearranges the loop data structure.

Partial unrolling: Many compilers partially unroll inner loops to expose low-level parallelism. For compilers which will not, Blitz++ does this unrolling itself. A preprocessor flag results in the innermost loop being manually unrolled at compile time.

Common stride optimizations: Blitz++ tests at run-time to see if all the arrays in an expression have a unit or common stride. If so, a faster version of the innermost loop is used, which eliminates some stride arithmetic.

Tiling: At compile time, Blitz++ tests whether the array expression contains a stencil. If so, it generates code which uses tiling to ensure good cache use.

10 Performance Results

Figure 5 shows performance of the Blitz++ classes `Array` and `Vector` for a DAXPY operation on the Cray T3E-900 using KAI C++. The Blitz++ classes achieve the same performance as Fortran 90 (the Fortran 77 compiler is no longer supported on the T3E, and is actually slower). The flags used for the fortran compiler were `-O3,aggress,unroll2,pipeline3`.

Table 1 shows median performance of Blitz++ arrays on 21 loop kernels. Performance is reported as a fraction of Fortran performance: > 100 is faster, and < 100 is slower. The fastest native Fortran compiler was used, with typical optimization switches (`-O3,-Ofast`). The loop kernels and makefiles are available as part of the Blitz++ distribution.

11 Lessons Learned

One of the contributions of Blitz++ has been showing that domain-specific abstractions (such as dense arrays) do not have to be built into compilers to achieve high performance. With sufficiently powerful language features

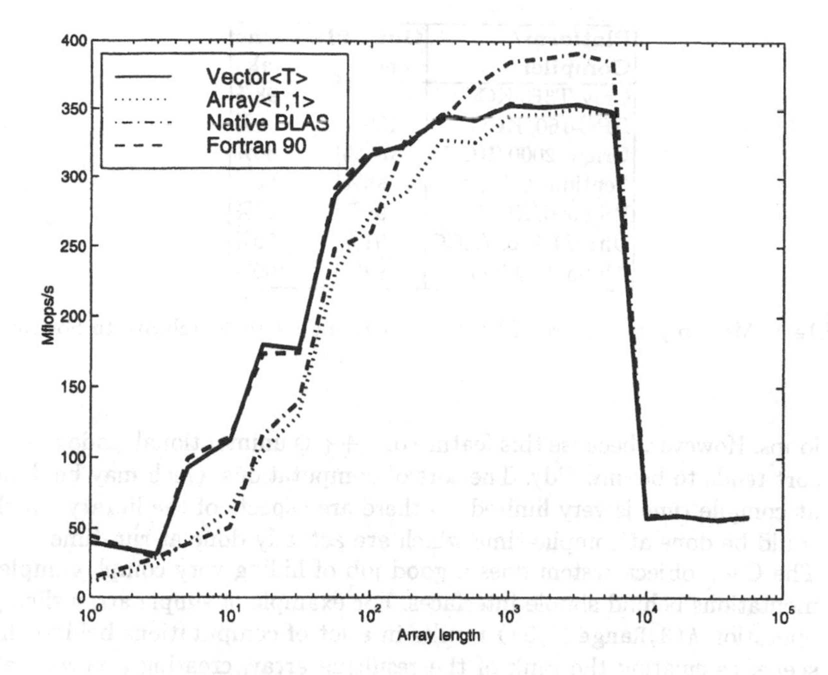

Fig. 5. DAXPY benchmark on the Cray T3E (single PE).

(such as templates in C++), it is possible to build libraries which both define abstractions and control how they are optimized. This is the idea of an *Active Library* [7,1].

The most important idea of an active library is that responsibility for high-level optimizations is shifted from the compiler to the library. For example, Blitz++ relies on the compiler to do low-level optimizations such as instruction scheduling and register allocations. However, Blitz++ itself performs optimizations specific to dense multidimensional arrays, such as loop transformations.

Blitz++ is able to provide high-performance, high-abstraction arrays because of these C++ features:

- C++ has operator overloading, which provides a limited form of *syntactic abstraction.* By exploiting operator overloading, Blitz++ is able to make scientific codes resemble the high-level physics or math they implement. There are several aspects of the library which would benefit from a more powerful macro system (for example, the syntax for defining stencil objects is quite clumsy).
- Templates let Blitz++ provide a single array class which works for any numeric type, and any dimension. Templates were enormously useful in providing a general-purpose library.
- Templates also provide the ability to perform computations at compile time. Blitz++ uses this ability to parse array expressions and unroll small

Platform/ Compiler	Out of cache	In-cache (peak)
Cray T3E/KCC	96%	98%
HPC-160/KCC	100%	95%
Origin 2000/KCC	88%	80%
Pentium II/egcs	98%	80%
RS 6000/KCC	94%	97%
UltraSPARC/KCC	91%	79%
Alpha/DECcxx	98%	69%

Table 1. Median performance of Blitz++ on 21 loop kernels, relative to Fortran.

loops. However, because this feature of C++ is unintentional, code of this sort tends to be unwieldy. The sort of computations which may be done at compile time is very limited, so there are aspects of the library which could be done at compile-time which are actually done at run-time.

– The C++ object system does a good job of hiding very complex implementations behind simple interfaces. For example, a simple array slicing operation `A(3,Range(2,5))` results in a lot of computations behind the scene: calculating the rank of the resulting array, creating a new array object, updating reference counts, checking array bounds, etc. All of this is hidden from the user.

Blitz++ did not achieve great performance right from the start: it took careful tuning for each compiler and platform. One of the frustrating aspects of this tuning process was that compiled C++ code bears little resemblance to the source code: after preprocessing, template instantiation, and turning objects into low-level code, the code has been transformed so much that it is hard to understand. Another difficulty in tuning Blitz++ was the inability of many compilers to optimize away small temporary objects. The KAI C++ compiler was an exception.

Acknowledgments

The design of Blitz++ has benefitted from discussions with the POOMA team at Los Alamos National Laboratory [3]. Numerous suggestions and patches for Blitz++ have been contributed by its users. The expression templates technique was invented independently by David Vandevoorde.

This work was supported in part by the Director, Office of Computational and Technology Research, Division of Mathematical, Information, and Computational Sciences of the U.S. Department of Energy under contract number DE-AC03-76SF00098. This research used resources of NERSC and the Advanced Computing Laboratory (LANL) which are supported by the Office of Energy Research of the U.S. Department of Energy, and of ZAM (Research Centre Jülich, Germany).

References

1. Krzysztof Czarnecki, Ulrich Eisenecker, Robert Glück, David Vandevoorde, and Todd L. Veldhuizen. Generative Programming and Active Libraries. In *Proceedings of the 1998 Dagstuhl-Seminar on Generic Programming*, volume TBA of *Lecture Notes in Computer Science*, 1998. (in review).
2. Scott W. Haney. Beating the abstraction penalty in C++ using expression templates. *Computers in Physics*, 10(6):552–557, Nov/Dec 1996.
3. Steve Karmesin, James Crotinger, Julian Cummings, Scott Haney, William Humphrey, John Reynders, Stephen Smith, and Timothy Williams. Array design and expression evaluation in POOMA II. In *ISCOPE'98*, volume 1505. Springer-Verlag, 1998. Lecture Notes in Computer Science.
4. Thomas Keffer and Allan Vermeulen. *Math.h++ Introduction and Reference Manual.* Rogue Wave Software, Corvallis, Oregon, 1989.
5. Rebecca Parsons and Daniel Quinlan. A++/P++ array classes for architecture independent finite difference computations. In *Proceedings of the Second Annual Object-Oriented Numerics Conference (OON-SKI'94)*, pages 408–418, April 24–27, 1994.
6. Todd L. Veldhuizen. Expression templates. *C++ Report*, 7(5):26–31, June 1995. Reprinted in C++ Gems, ed. Stanley Lippman.
7. Todd L. Veldhuizen and Dennis Gannon. Active libraries: Rethinking the roles of compilers and libraries. In *Proceedings of the SIAM Workshop on Object Oriented Methods for Inter-operable Scientific and Engineering Computing (OO'98)*. SIAM Press, 1998.
8. Michael Wolfe. *High Performance Compilers for Parallel Computing.* Addison-Wesley, 1995.

The Design of Sparse Direct Solvers using Object-Oriented Techniques

Florin Dobrian[1], Gary Kumfert[1], and Alex Pothen[2]

[1] Department of Computer Science, Old Dominion University, USA
Email: {dobrian,kumfert,pothen}@cs.odu.edu
[2] ICASE, NASA Langley Research Center, USA
Email: pothen@icase.edu

Abstract. We describe our experience in designing object-oriented software for sparse direct solvers. We discuss Spindle, a library of sparse matrix ordering codes and OBLIO, a package that implements the factorization and triangular solution steps of a direct solver. We discuss the goals of our design: managing complexity, simplicity of interface, flexibility, extensibility, safety, and efficiency. High performance is obtained by carefully implementing the computationally intensive kernels and by making several tradeoffs to balance the conflicting demands of efficiency and good software design. Some of the missteps that we made in the course of this work are also described.

1 Introduction

We design and implement object-oriented software for solving large, sparse systems of linear equations by direct methods. Sparse direct methods solve systems of linear equations by factoring the coefficient matrix, employing graph models to control the storage and work required. Sophisticated algorithms and data structures are needed to obtain efficient direct solvers. This is an active area of research, and new algorithms are being developed continually. Our goals are to create a laboratory for quickly prototyping new algorithmic innovations, and to provide efficient software on serial and parallel platforms.

Object-oriented techniques have been applied to iterative solvers such as those found in Diffpack [22,10] and PETSc [5,27]. However, the application of object-oriented design to direct methods has not received the attention it deserves. Ashcraft and Liu have designed an object-oriented code called SMOOTH to compute fill-reducing orderings [3]. Ashcraft et. al. have created an object-oriented package of direct and iterative solvers called SPOOLES, see [4,30],which we discuss in Section 9. Both SMOOTH and SPOOLES are written in C. George and Liu have implemented object-oriented user interfaces in Fortran 90 and C++ for the SPARSPAK library, and they have discussed their design goals in [17].

We are interested in object-oriented design of direct solvers since we need to perform experiments to quickly prototype new algorithms that we design.

We are also interested in extending our work to parallel and out-of-core computations, and to a variety of computer architectures. While several direct solvers are currently available, most of them are designed as "black boxes"—difficult to understand and adapt to new situations. We needed a robust and extensible platform to serve as a baseline, and a framework to support inclusion of new components as they are developed.

This project is the result of a balanced approach between sound software design and the "need for speed." We achieve the second goal by carefully implementing highly efficient algorithms. Our success here can be quantifiably measured by comparison to other implementations, both in terms of quality of the results and the amount of resources needed to compute it. Determining how well we achieve the first goal of good software design is much more subjective. In this arena, we argue on the grounds of *usability*, *flexibility*, and *extensibility*.

The research presented here tells the story of two separate but inter-related projects, parts of two PhD theses. The first project, Spindle, is a library of sparse matrix ordering algorithms including a variety of fill-reducing orderings. The second project, OBLIO, handles the remaining steps in solving a linear system of equations: matrix factorization and triangular solves. During development, ideas developed in one project were incorporated into the other when they were found to be suitable. As the projects matured, several design methodologies and programming techniques proved themselves useful twice. Currently the two projects have converged on most major design issues, though they are still maintained separately.

We present the work as a whole and concentrate on the commonalities of object-oriented design that have proved themselves in both projects. As this is ongoing research, we also present some equally good ideas that may not apply to both the ordering and solver codes, or that have not been incorporated into both yet. We find this convergence of design strategies somewhat surprising and most beneficial.

We explain the algorithms we chose and discuss the key features of the implementations. We also show the important tradeoffs that were made; often sacrificing more elegant applications of object-oriented techniques for very real improvements in efficiency. We include some missteps that were made in the course of our research and the lessons learned from the endeavor. Our code is written in C++.

Background information about direct methods is presented in Section 2. We identify the primary goals for our implementations in Section 3. The actual design of the software is presented in Section 4, with special emphasis on the ordering code in Section 5 and the factorization code in Section 6. We present the results of our software in Section 7. Finally, we discuss the lessons learned in course of our research in Section 9.

2 Background

The direct solution of a sparse symmetric linear system of equations $Ax = b$ can be described in three lines, corresponding to the three main computational steps: `order`, `factor` and `solve`,

$$P_1 = \text{order}(A), \tag{1}$$

$$(L, D, P_2) = \text{factor}(A, P_1), \tag{2}$$

$$x = \text{solve}(L, D, P_2, b). \tag{3}$$

Here A is a sparse, symmetric, positive definite or indefinite matrix; L is its lower triangular factor, and D is its (block) diagonal factor, satisfying $A = LDL^T$; P_1 and P_2 are permutation matrices; b is the known right-hand-side vector; and x is an unknown solution vector. We illustrate this three-step "black box" scheme in Figure 1.

Of the three steps, computing the factorization is usually the most time consuming step. For the rest of this section, we will discuss the significance of these three computational steps and their interplay.

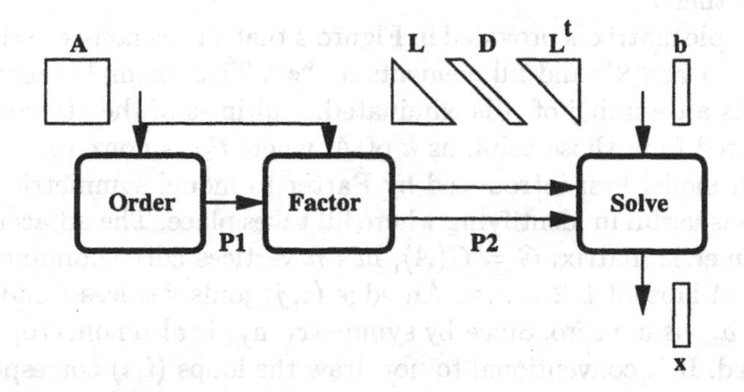

Fig. 1. A "black box" formulation of a sparse direct solver.

2.1 Fill and the Elimination Forest

The first step of the factorization of a symmetric positive definite matrix A can be described by the equation

$$A \equiv A_0 \equiv \begin{pmatrix} a_{11} & a_1^T \\ a_1 & A_1 \end{pmatrix}$$

$$= \begin{pmatrix} 1 & 0^T \\ l_1 & I_{n-1} \end{pmatrix} \begin{pmatrix} a_{11} & 0^T \\ 0 & A_1 \end{pmatrix} \begin{pmatrix} 1 & l_1^T \\ 0 & I_{n-1} \end{pmatrix}.$$

By multiplying the factors, we see that $l_1 = (1/a_{11})a_1$, and $A_1 = \overline{A_1} - a_{11}l_1l_1^T$. Hence the first column of L (the subdiagonal elements) is obtained by dividing the corresponding elements of A by the first diagonal element; and the matrix that remains to be factored, A_1, is obtained by subtracting the term $a_{11}l_1l_1^T$ from the corresponding submatrix of A. At the end of the first step, we have computed the first column (and row) of the factors L and D. The remainder of the factorization can be computed recursively by applying this elimination process to the first column of the submatrix that remains to be factored at each step.

Computing the factors of a symmetric indefinite matrix is a little more involved, since we have to consider the elimination of two columns (and rows) at a step if the first diagonal element of the submatrix to be factored is zero or small. We do not discuss the details of factoring an indefinite matrix here.

We observe from the equations above that the factor L has nonzero elements corresponding to all subdiagonal nonzero positions in A; in addition, for $j > i$, if the ith and jth positions of l_1 are nonzero, then the (i, j) element of A_1 is nonzero even when A has a zero element in that position. These additional nonzeros introduced by the factorization are called *fill* elements. Fill increases both the storage needed for the factors and the work required to compute them.

An example matrix is provided in Figure 2 that shows non-zeros in original positions of A as "×" and fill elements as "•". This example incurs two fill elements. As a column j of A is eliminated, multiples of the jth column of L are subtracted from those columns k of A, where l_{kj} is nonzero.

A graph model first introduced by Parter to model symmetric Gaussian elimination is useful in identifying where fill takes place. The adjacency graph of the symmetric matrix, $G = G(A)$, has n vertices corresponding to the n columns of A labeled $1, 2, \ldots, n$. An edge (i, j) joins vertices i and j in G if and only if $a_{i,j}$ is nonzero. Since by symmetry, $a_{j,i}$ is also nonzero, the graph is undirected. It is conventional to not draw the loops (i, i) corresponding to the n nonzero diagonal elements a_{ii}.

The example in Figure 2 illustrates the graph model of elimination. A sequence of elimination graphs, G_k, represents the fill created in each step of the factorization. The initial elimination graph is the adjacency graph of the matrix, $G_0 = G(A)$. At each step k, let v_k be the vertex corresponding to the kth column of L_k to be eliminated. The elimination graph at the next step, G_{k+1}, is obtained by adding edges needed to make all vertices adjacent to v_k pairwise adjacent to each other, and then removing v_k and all edges incident on v_k. The inserted edges are *fill edges* in the elimination graph. This process is repeated until all the vertices are eliminated. We denote by F the filled matrix $L + L^T$. The filled graph $G(F)$ (the adjacency graph of F) contains all the edges of $G(A)$ together with the filled edges.

When a column j updates another column k during the elimination, dependencies are created between the columns j and k. An important data structure that captures these dependencies is an *elimination tree*, or more

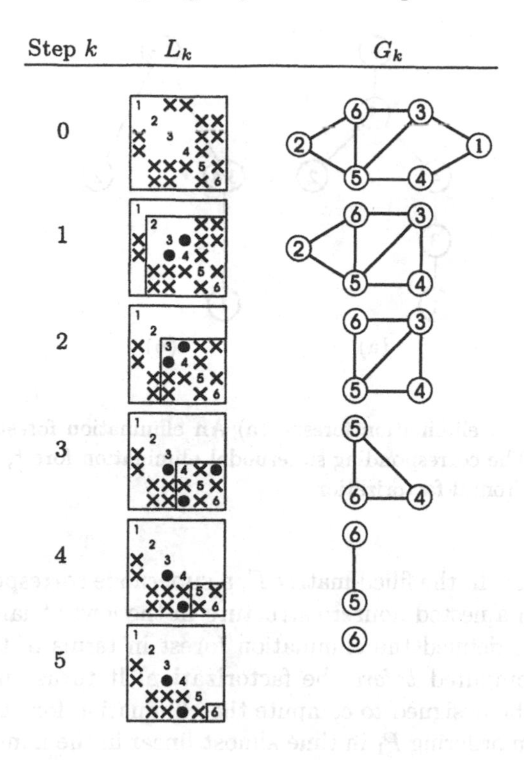

Fig. 2. Examples of factorization and fill. For each step k in the factorization, the nonzero structure of the factor L_k and the associated elimination graph G_k are shown.

generally an *elimination forest.* The reader unfamiliar with elimination trees will find a comprehensive survey in Liu [24]. The critical detail for our purposes is that elimination forests minimally represent the dependencies in the factorization. Each column in the matrix is represented as a node in the forest. The *parent* of a column j is the smallest row index of a subdiagonal nonzero element (the first subdiagonal nonzero) in column j. Note that the definition of the elimination forest employs the factor and not the original matrix A. In our example, the parent of column three is four, and $l_{4,3}$ is a fill element.

An important concept that permits us to perform block operations in the factorization is that of a *supernode.* Supernodes are groups of adjacent vertices that have identical higher numbered neighbors in the filled graph $G(F)$. We group vertices into supernodes using the elimination forest; a supernode consists of a set of vertices that (1) forms a path in elimination forest, and (2) have the same higher numbered neighbors. In our example, the reader can verify that vertices 3 and 4, and vertices 5 and 6 form supernodes. This grouping of vertices into supernodes can then be used to define a supernodal

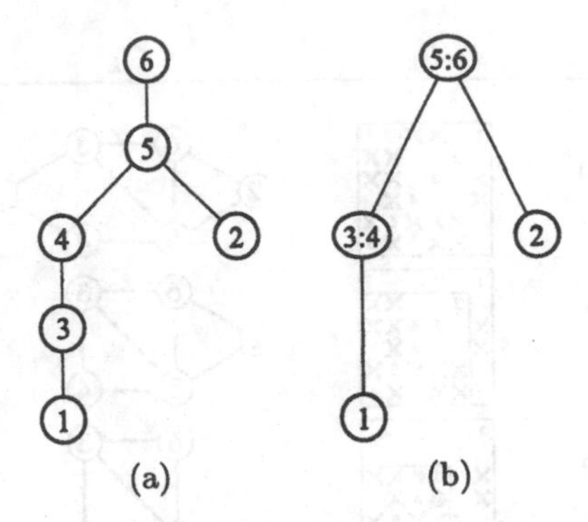

Fig. 3. Examples of elimination forests. (a) An elimination forest of the example in Figure 2. (b) The corresponding supernodal elimination forest, which is used in supernodal multifrontal factorization.

elimination forest. In the filled matrix F, a supernode corresponds to a group of columns with a nested nonzero structure in the lower triangle.

Although we defined the elimination forest in terms of the factor L, in practice it is computed *before* the factorization. It turns out that efficient algorithms can be designed to compute the elimination forest from the given matrix A and an ordering P_1 in time almost linear in the number of nonzeros in A. The elimination forest is then used in symbolic factorization algorithms to predict where fill occurs in the factor. Fill entries can only occur between a node in the elimination forest and its ancestors.

The elimination forest corresponding to the example in Figure 2 is a single tree shown in Figure 3. It can be shown that any postordering of the elimination forest leaves the fill unchanged.

2.2 Factorization

The example in Figure 2 illustrates a *right-looking* factorization, where updates from the current column being eliminated are propagated immediately to the submatrix on the right. In a *left-looking* factorization, the updates are not applied right away; instead, before a column k is eliminated, all updates from previous columns of L are applied together to column k of A. Another approach is a *multifrontal* factorization, in which the updates are propagated from a descendant column j to an ancestor column k via all intermediate vertices on the elimination tree path from j to k.

OBLIO currently includes a multifrontal factorization code. In the multifrontal method, we view the sparse matrix as a collection of submatrices, each of which is dense. Each submatrix is partially factored; the factored columns are stored in the factor L, while the updates are propagated to the parent of the submatrix in the elimination forest. The multifrontal method and supernodal left-looking algorithms are suited to modern RISC architectures since the regular computations in the dense submatrices make better use of cache and implicit indexing.

We provide a high-level description of the multifrontal algorithm here, and the implementation details will be presented in Section 6. The multifrontal factorization is computed in post-order on the elimination forest. The algorithm creates a dense submatrix called a *frontal matrix* for each supernode. A frontal matrix is a symmetric dense matrix with columns and rows corresponding to the groups of columns to be eliminated and all the rows in which these columns have nonzeros. The frontal matrix is partially factored corresponding to the columns to be eliminated. The remainder of the frontal matrix forms a dense update matrix that will be stacked. The factored columns are copied into the sparse factor L. When all the children of a parent vertex in the elimination forest have been eliminated, the parent retrieves the update matrices of its children from the stack, and adds them into its own frontal matrix. The parent then has been "fully assembled"; it can then partially factor its frontal matrix, and propagate an update matrix to its parent in the elimination forest.

We illustrate the multifrontal factorization process in Figure 4. At each step, k, we assemble the frontal matrix, F_k, by adding the entries in the kth rows and columns of A and any update matrices, U_k. These contributions to F_k vary in size, so special attention must be paid to indices; hence we call this operation *scatter-add*, and denote it by "$\oplus$". The frontal matrices then factor all columns that are "fully assembled." This looks identical to how L_k was generated in Figure 2, except that here, the frontal and update matrices are dense, not sparse. After the appropriate columns are factored from F_k, it is split into an update matrix U_k and a factored column C_k, which is scatter-added to the sparse factor L_k.

2.3 Ordering

The order in which the factorization takes place can change the structure of the elimination forest, and therefore greatly influences the amount of fill produced in the factorization. To control fill, the coefficient matrix is symmetrically permuted by rows and columns in attempts to reduce the storage(memory) and work(flops) required for factorization. Given the example in Figure 2, the elimination order $\{2, 6, 1, 3, 4, 5\}$ produces only one fill element, instead of the two shown. This is the minimum amount of fill for this example.

If A is indefinite, the initial permutation may have to be modified during factorization for numerical stability. This detail is captured in equation 2 and in Figure 1 by the distinction between the two permutations, P_1 and P_2. The first ordering, P_1, reduces fill and the second ordering, P_2, is a modification of P_1 to preserve numerical stability. In general, any modification of P_1 increases fill and, hence, the work and storage required for factorization.

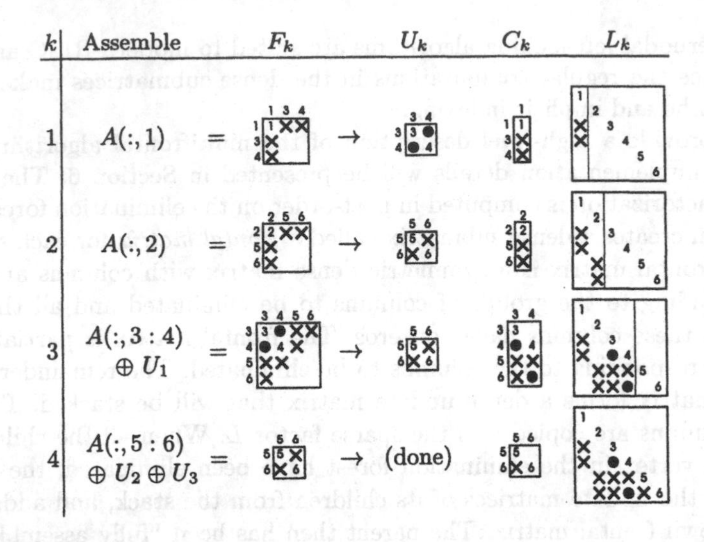

Fig. 4. **Example of multifrontal factorization. For each step k in the factorization, the frontal matrix F_k is assembled from the appropriate column(s) of A and any update matrices from the children of supernode k in the supernodal elimination forest. The fully assembled columns are factored; the factor columns C_k are scatter-added to the sparse triangular factor L. Corresponding to the remaining columns in the factor matrix, a new update matrix U_k is computed.**

If A is positive definite, Cholesky factorization is numerically stable for any symmetric permutation of A.

Fill reduction is handled before the factorization step by ordering algorithms. Finding the elimination order that produces the minimum amount of fill is a well-known NP-complete problem. The heuristics used to compute fill-reducing orderings fall mainly into two classes: divide-and-conquer algorithms that recursively partition the graph and prevent fill edges joining two vertices in different partitions; and greedy algorithms that attempt to control fill in a local manner.

Spindle is a library of symmetric sparse matrix ordering algorithms — including most of the current "bottom-up" fill-reducing orderings. Though Spindle has native support for many matrix file formats, it does not implement a full matrix class and relies entirely on the graph model. Therefore while we talked about eliminating columns for the factorization, it is more natural to discuss eliminating vertices from the graph in this context. These greedy approaches choose which vertex to eliminate next based on some approximation of the fill its elimination will induce. An upper bound on the fill created by eliminating a vertex of degree d is $d(d-1)/2$; hence in the minimum degree algorithm a vertex whose degree (or some related quantity) is minimum is eliminated next.

Abbreviation.	Algorithm Name
MMD	Multiple Minimum Degree [23]
AMD	Approximate Minimum Degree [1]
AMF	Approximate Minimum Fill [29]
MMDF	Modified Minimum Deficiency [26]
MMMD	Modified Multiple Minimum Degree [26]
AMMF	Approximate Minimum Mean Local Fill [28]
AMIND	Approximate Minimum Increase in Neighbor Degree [28]

Table 1. Some of the algorithms in the Minimum Priority family.

Multiple Minimum Degree (MMD) is an enhancement of the minimum degree algorithm that eliminates maximal independent sets of vertices at once to keep the cost of updating the graph low. Many more enhancements are necessary to obtain a practically efficient implementation. A survey article by George and Liu [16] provides all of the details for MMD. There have been several contributions to the field since this survey. A list of algorithms that we implement in our library and references are in Table 1.

2.4 Further Information

This "breathless introduction" to sparse direct solvers is intended to help the reader to understand the basic computational problems in this area. For readers who desire further information, we recommend the following papers: the MMD ordering algorithm, [16]; the AMD ordering algorithm, [1]; the multifrontal method, [25]; elimination trees, [24]; and symmetric indefinite solvers, [2]. The books [12,15] were written earlier than these articles, nevertheless they would be helpful too.

3 Goals

Our design and implementation were guided by several principal goals we wanted to achieve. We list them briefly here, and then discuss each one in detail.

- Managing Complexity
- Simplicity of Interface
- Flexibility
- Extensibility
- Safety
- Efficiency

3.1 Managing Complexity

Writing code for sparse direct solvers is a challenging task because the design requires sophisticated data structures and algorithms. To manage such complexity, proper abstractions must be used. Some of the earlier solvers were written in Fortran 77, which has very limited support for abstraction. As a consequence, the computation is expressed in terms of the implementation. For example, a sparse matrix is commonly represented as several arrays and the algorithms are a collection of subroutines which read from and write to these arrays. Since there is no support for abstract data types in the language, these subroutines tend to have long argument lists. Additionally, since Fortran 77 lacks dynamic memory allocation, all arrays are global, and are reused many times for different purposes. All these characteristics make the code extremely difficult to understand. Some of the more recent codes are written in C, but they tend to be influenced by the earlier codes in Fortran 77. For instance, the collection of subroutines is often the same.

In contrast to earlier codes, we wish to express the computation in terms of the mathematical formulation. This means programming in terms of vectors, trees, matrices, and permutations. This is possible in C++ because the language supports abstraction through classes, inheritance, and polymorphism. By programming at a higher level of abstraction, we are able to implement code faster and to relegate minute details to lower layers of the abstraction.

3.2 Simplicity of Interface

Simplicity is also achieved by good abstraction, but in a different sense. Whereas with complexity management we want to build layers of abstraction, here we focus on intuitive components with a minimal interface. Recall that the computation is formulated in terms of sparse matrices, vectors and permutations and in terms of algorithms like orderings, factorizations and triangular solves. Accordingly, we provide abstractions for such entities. The code for a solver that uses our libraries becomes then a simple driver that reads the inputs to the algorithms, runs the algorithms, and writes their outputs.

3.3 Flexibility

The key to achieving flexibility in our codes is decoupling. We distinguish between data structures and algorithms, and design separate abstractions for structural entities such as sparse matrices, permutations and vectors on one hand, and for ordering and factorization algorithms on the other hand. One can and should expect to swap in different ordering or factorization objects in the solver in much the same way that components in a stereo system can be swapped in and out. In addition, one has the possibility of choosing the number of times different components are used. Ordering and factorization are performed only once in a series of systems with the same coefficient matrix with different right hand sides for example, while the triangular solves must be repeated for every system in the series. A similar situation occurs with iterative refinement, where triangular solves must be repeated for a single run of the ordering and factorization algorithms. It should also be possible to solve several systems of equations simultaneously, and to have them at various stages of their solution process.

Swapping components happens not only with data structures and algorithms; more generally, we want to swap smaller components within a larger one. For instance, factorization is composed of a couple of distinct phases (symbolic and numerical). Solvers for positive definite and indefinite problems differ only in the numerical part. To switch from a positive definite solver to an indefinite solver, we only have to swap the components that perform the numerical factorization. By splitting a factorization algorithm in this way we also provide the option of performing only the symbolic work for those who wish to experiment with the symbolic factorization algorithms.

The design challenge here is that flexibility and simplicity are at odds with one another. The simplest interface to a solver would be just a black box; one throws a coefficient matrix and a right hand side vector in one end and produces the solution out the other end of the black box. While very easy to use, it would not be flexible at all. On the other hand, a very flexible implementation can expose too many details, making the learning curve steeper for users. In general, we provide multiple entry points to our code and let the users decide which one is appropriate for their needs.

3.4 Extensibility

Whereas flexibility allows us to push different buttons and interchange components, extensibility allows us to create new components and alter the effects of certain buttons. This software is part of ongoing research and gets regularly tested, evaluated, and extended.

The best techniques we found for ensuring extensibility in our codes were to enforce decoupling and to provide robust interfaces. These practices also encourage code reuse. In the following sections, we will show an example of how two types of dense matrices in **OBLIO**, *frontal* and *update*, are used in numerical factorization and inherit characteristics from a dense matrix. We also present an example of how polymorphism is used in the ordering algorithms of **Spindle**.

Extensibility is not an automatic feature of a program written in an object-oriented language. Rather, it is a disciplined choice early in the design. In our implementations, we have very explicit points where the code was designed to be extended. Our intent is to keep the package open to other ordering and factorization algorithms. We are interested in extending these codes to add functionality to solve new problems (unsymmetric factorizations, incomplete factorizations, etc.) and to enable the codes to run efficiently in serial, parallel (with widely differing communication latencies), and out-of-core environments.

3.5 Safety

We are concerned with two major issues concerning safety: protecting the user from making programming mistakes with components from our codes (compile-time errors), and providing meaningful error handling when errors are detected at run-time. Earlier we argued that the simplicity of the interface enhances the usability of the software, we add here that usability is further enhanced by safety.

Compile-time safety is heavily dependent on features of the programming language. Any strongly typed language can adequately prevent users from putting square pegs in round holes. That is, we can prevent users from passing a vector to a matrix argument.

Run-time errors are more difficult to handle. Structural entities such as sparse matrices and permutations are inputs for factorization algorithms. When such an algorithm is run, it should first verify that the inputs are valid, and second that the inputs correspond to each other.

The biggest difficulty about error handling is that while we, the library writers, know very well *how* to detect the error conditions when they occur, we must leave it to the user of the library to determine *what* action should be taken. Error handling is an important and all too often overlooked part of writing applications of any significant size. Because we are writing a multi-language based application and provide multiple interfaces (Fortran77, C, Matlab, C++) we immediately rejected using C++ exception handling.

The way in which errors are handled and reported in OBLIO and Spindle differ slightly in the details, but the strategies are similar. In both codes, the classes are self-aware, and are responsible for performing diagnostics to confirm that they are in a valid state. Instances that are in invalid states are responsible for being able to report, upon request, what errors were detected.

3.6 Efficiency

An efficient piece of software is one that makes judicious use of resources. To achieve efficiency, data structures and algorithms must be implemented carefully. Previous direct solver packages tend to be highly efficient, using compact data structures and algorithms that are intimately intertwined in the code. Decoupling the data structures from the algorithms and requiring them to interact through high-level interfaces can add significant computational overhead. Choosing robust and efficient interfaces can be quite involved. The compromise between flexibility and efficiency is determined by these interfaces.

Consider the means by which an algorithmic object accesses the data stored in a structural object: e.g., a factorization operating on a sparse matrix.

A rigorous object-oriented design requires full encapsulation of the matrix, meaning that the factorization algorithm must not be aware of its internal representation.

In practice, sparse matrix algorithms must take advantage of the storage format as an essential optimization. This is often done in object-oriented libraries like Diffpack and PETSc by "data-structure neutral" programming (see [6] and [5] respectively). This is accomplished by providing an abstract base class for a matrix or vector, and deriving concrete implementations for each data-layout: compressed sparse row, compressed column major, AIJ (row index, column index, value) triples, blocked compressed sparse, etc. Then each concrete derived class must implement its own basic linear algebra functions. Given t types of matrices and n basic linear algebra subroutines for each, these libraries provide $t \times n$ virtual functions.

Our goal is somewhat narrower: provide a solver that is as fast as any other solver written in any other language, but is more usable, flexible, and extensible, because we use object-oriented design and advanced programming paradigms. Our algorithms are more complicated than a matrix-vector multiply, and providing a set of algorithms for each possible representation of the matrix is unreasonable. Even if we designed general algorithms that operate on matrices regardless of their layout, we could not make any guarantees about their performance. We are more concerned about adding new algorithms, not adding more matrix formats. Our codes apply to general, sparse, symmetric matrices that must be laid out in a specific way for efficient computations, and we provide tools to convert to this format. Specific formats are needed because we are forced to make a tradeoff: weaken the encapsulation to increase the performance.

In terms of running time, it is also important to realize which components of the code are the most expensive and to focus on efficiently implementing them. Usually, a sparse direct solver spends most of the time in the factorization step. A triply nested loop performs the bulk of the computation. Although most of the time is spent here, the code for the factorization represents just a tiny piece of the whole program. The rest of the code does not account for much of the whole time but it is certainly more sophisticated than this kernel. We use a layered approach, in which the focus is on software design in the higher layers and on efficiency in the lower ones. To get the best speed, we use a C subset or Fortran 77 code for kernels such as the triply nested loop inside the factorization.

Finally, we restricted our use of some specific object-oriented features. For example, we avoid operator overloading for matrices and vectors because this leads to the creation of several temporary objects and unnecessary data movement[1].

---

[1] The reader is invited to see Chapter 2, by Todd Veldhuizen, in the present book, which explains some solutions to this problem in detail.

4 Design

This section tackles the major issues in designing our solver. We begin by explaining the base classes in the inheritance hierarchy in Section 4.1. Then in Section 4.2, we discuss the specialization and composition of our data structure and algorithm classes. We also discuss how we use iterators to get the algorithms and data structures to interact in Section 4.3. The important problem of I/O and handling multiple (often disparate) sparse matrix file formats is discussed in Section 4.4. Then we focus on implementation details specific to each package. The discussion of Spindle details in Section 5 highlights extensibility in using polymorphic fill reducing ordering algorithms. The corresponding section for OBLIO is Section 6.

4.1 Base Classes

We have organized most of our heavy-weight classes into a single inheritance tree. The two main branches are `DataStructure` and `Algorithm`. Each matrix, permutation, and graph class inherits from `DataStructure` a common set of state information, interfaces for validating its state, and (optionally) resources to support persistent objects. Similarly classes that perform computational services such as ordering and factorization are derived from the `Algorithm` class, which provides support for algorithmic state information and the interface for running the algorithm. Both `DataStructure` and `Algorithm` are derived from a common ancestor, the `Object` class shown in Figure 5. Although this class has no physical analogy, it does provide a suite of useful services to all its descendants such as error handling and instance counting. Through inheritance from `Object`, every structural or algorithmic object remembers when the most recent error occurred, if any. This error can be retrieved on demand and communicated to other classes.

```
class Object
{
  protected:
    Error error;

  public:
    Object() {error = None;}
    Error getError(void) const {return error;}
    virtual void print(const char *description) const = 0;
};
```

Fig. 5. The Object Class.

The DataStructure class shown in Figure 6 implements the validity checking mechanism specific to all structural objects. A structural object is usually initialized as invalid. After it is fully initialized, it must be validated. Later it can be completely reset. Again, the methods for validation and resetting are pure virtual, every structural class being required to implement its own behavior.

```
class DataStructure: public Object
{
   protected:
      bool valid;

   public:
      DataStructure() {valid = False;}
      bool isValid(void) const {return valid;}
      virtual void validate(void) = 0;
      virtual void reset(void) = 0;
};
```

Fig. 6. The DataStructure class.

The Algorithm class shown in Figure 7 handles the execution of algorithmic objects. Each algorithmic object requires a certain amount of time for its execution. This information can be retrieved for later use. The execution of any algorithmic object is triggered by the same interface, only the implementation is specific. This requires the method for running algorithmic objects to be pure virtual.

There are some distinctions between Spindle and OBLIO in their implementations. Both have forms of object-persistence through the DataStructure class, though the exact mechanisms differ. Spindle defines a wider variety of states for DataStructure classes to be in, and carries a related idea of state for the Algorithm classes. In both packages, however, the non-trivial concrete classes are derived from this fundamental hierarchy.

4.2 Algorithms and Data Structures

We have seen several structural classes such as Matrix, Vector, Graph and Permutation. Instances of these classes are the inputs and outputs of algorithmic objects that order, factor, and perform the triangular solves.

The ordering algorithms, for example, require only the Graph of a Matrix as input, although there are additional options that advanced users could

```
class Algorithm: public Object
{
  protected:
    float runTime;

  public:
    Algorithm() {runTime = 0;}
    float getRunTime(void) {return runTime;}
    virtual void run(void) = 0;
};
```

Fig. 7. The Algorithm class.

set. After the algorithm is run, the results could be queried. In Spindle, all the fill-reducing ordering algorithms can produce either a `Permutation` or an `EliminationForest` (or both) upon successful execution of the algorithm.

Ordering algorithms can enter an invalid state for several reasons. The input data could have been invalid, the user could have attempted to run the algorithm without sufficient input information, a run-time error might have been detected, or there might be a problem servicing the user's output request. In all these cases, the instance of the ordering algorithm can be reset and reused with different input data.

Algorithms can also be composed to obtain more sophisticated algorithmic objects. The factorization itself is made of several components, each derived from the `Algorithm` class; we provide details in Section 6. Having these subcomputations implemented separately is useful for many reasons. We subject each component to unit testing and later combine them during integration testing. It also helps in code reuse, since the factorization algorithms for positive definite matrices and indefinite matrices share several steps.

The derivation and composition of data structures is just as rich. Multifrontal factorization requires several dense matrix types that are derived from a common class `DenseSymmetricMatrix`: these are `FrontalMatrix` and `UpdateMatrix`.

4.3 Iterators

Separating data structures from algorithms makes it necessary to create interfaces between the modules. Inspired by the Standard Template Library (STL), which also makes a distinction between data structures and algorithms, we explored the paradigm of *iterators* for this purpose. We first define the iterator construct, and then discuss how we used it. In later sections, we show examples where the iterator paradigm was applied with much success (Section 6) and where they were misapplied and reduced performance significantly (Section 9.2).

Definition of an Iterator. An *iterator* is a class that is closely associated with a particular container (data structure) class. The iterator is usually a "friend" class of the container granting it privileged access. Its purpose is to abstract away the implementational details of how individual items within the container are stored.

Assume, for example, that the list of all edges in a Graph is in the array `adjList[]` and that a second array `adjHead[]` stores the beginning index into `adjList[]` for each vertex in the graph. Then to check if vertex `i` is adjacent to vertex `j` we simply run through the arrays as in the piece of code in Figure 8.

```
bool isAdjacent( const Graph& g, int i, int j ) const
{
  for( int k = g.adjHead[i]; k < g.adjHead[i+1]; ++k ) {
    if ( g.adjList[k] == j ) {
      return true;
    }
  }
  return false;
}
```

Fig. 8. A C-like function that directly accesses the data in Graph.

This design has a flaw, in that the function assumes the layout of data in the `Graph` class and accesses it directly. Consider now a different approach where the `Graph` class creates an iterator. Conventionally, an iterator class mimics the functionality of a pointer accessing an array. The dereference operator (`operator*()`) is overloaded to access the current item in the container, and the increment operator (`operator++()`) advances the iterator to the next item in the container. Rewriting our function above, as shown in Figure 9, we define `Graph::adj_begin(int)` to create an iterator pointing to the beginning of the adjacency list of a specified vertex and `Graph::adj_end(int)` to return an iterator pointing to the end of the list[2].

The benefit of this second approach is that the function `isAdjacent` no longer assumes any information about how the data inside `Graph` is laid out. If it is indeed sequential as it was in the previous example, then the iterator

[2] Actually, it points to one past the end – a standard C++ convention.

```
bool isAdjacent( const Graph& g, int i, int j ) const
{
   for( Graph::adj_iterator it = g.adj_begin(i);
        it != g.adj_end(i); ++it ) {
     if ( *it == j ) {
       return true;
     }
   }
   return false;
}
```

Fig. 9. A C++ global function that uses an iterator to search Graph.

could be simply a pointer to an integer data type. However, if the adjacency lists are stored in a different format, e.g., a red-black tree (for faster insertions and deletions), the iterator approach still applies for accessing the adjacency lists, since it assumes a suitable iterator class has been provided.

Application of Iterators. Iterators provide a kind of "compile-time" polymorphism. They allow a level of abstraction between the data structure and the algorithm, but the concrete implementation is determined at compile time. This allows the compiler to inline function calls (often through several levels) and get very good performance[3].

There were a few difficulties in applying this technique to our problems. The most complicated aspect was that all STL containers are one-dimensional constructs. Many of our data structures, such as matrices and graphs, are two-dimensional. This was not a serious problem since we, as programmers, tend to linearize things anyway. In the example of iterators before, for instance, we used the iterator to iterate over the set of vertices adjacent to a particular vertex.

4.4 Input/Output

An important problem that we often run into is sharing problems with other researchers. Whenever we agree to generate some solutions for a client (either academia or industry) we often find that we must adapt our code to a new file format. Attempts to standardize sparse matrix file formats do exist, most notably the Harwell-Boeing Format [9], and the Matrix Market format [7]. However, we found it unreasonable to expect clients to restrict themselves to a small choice of formats. We found, in fact, that understanding the nature

[3] C++ cannot inline virtual functions.

of this problem and applying object-oriented techniques is a good exercise in preparation for the harder problems ahead.

Perhaps the easiest way to handle sparse matrix I/O is to have a member function of the matrix class to write a particular format and another to read. This is a simple solution, but it has a scalability problem. First, as the number of formats increase, the number of member functions grows and the matrix class becomes more and more cumbersome. Second, if separate matrix classes are needed then all of the I/O functions must be replicated.

The Chicken and Egg Problem. One could reasonably create separate `Matrix` and `MatrixFile` classes. The immediate problem with this design is determining which begets which. One would expect to read a matrix from a file, but it also makes sense to generate a matrix and want to save it in a particular format as shown in Figure 10.

```
// Matrix.h
#include "MatrixFile.h"

class Matrix
{
  public:
    Matrix( MatrixFile& );
    // ...
};
```

```
#include "Matrix.h"

class MatrixFile
{
  public:
    MatrixFile( Matrix& );
    // ...
};
```

Fig. 10. A first design of the Matrix and Matrixfile classes.

Such a design induces a cyclic dependency between the two objects, which is bad because such dependencies can dramatically affect the cost of maintaining code, especially if there are concrete classes inheriting from the cyclic dependency [21, pg. 224]. This is exactly the case here, since the intention is to abstract away the differences between different file formats.

A solution is to escalate the commonality between the two classes. This has the advantage that the dependencies are made acyclic, the downside is that an additional class that has no physical counterpart is introduced for purely "software design" reasons. We will call this class `MatrixBase`, which is the direct ancestor of both the `Matrix` and `MatrixFile` classes. This second design in shown in Figure 11.

Now we can derive various matrix file formats from the `MatrixFile` class, independent of the internal computer representation of the `Matrix` class. We will show later that the benefits compound when considering matrix to graph and graph to matrix conversions.

```
#include "MatrixBase.h"

class Matrix
  : public MatrixBase
{
  public:
    Matrix( MatrixBase& );
    // ...
};
```

```
#include "MatrixBase.h"

class MatrixFile
  : public MatrixBase
{
  public:
    MatrixFile( MatrixBase& );
    // ...
};
```

Fig. 11. The Matrix and Matrixfile classes derived from a MatrixBase class.

Navigating Layers of Abstraction. It is important to understand that abstractions involved around the construct we call a "matrix" come from different levels and have different purposes. To define a class `Matrix` and possibly many subclasses, care must be taken to capture the abstraction correctly. It is hard to give a formula for designing a set of classes to implement an abstract concept. However, when the abstraction is captured just right, using it in the code is natural and intuitive. Often we have found that good designs open new possibilities that we had not considered.

For the matrix object, we have identified at least two dimensions of abstraction that are essentially independent, one from the mathematical point of view, one from computer science. Along the first dimension, the mathematical one, a matrix can be sparse, dense, banded, triangular, symmetric, rectangular, real or complex, rank-deficient or full-rank, etc. From a mathematical point of view, all of these words describe properties of the matrix.

From a computer science point of view, there are different ways that these two-dimensional constructs are mapped out into computer memory, which is essentially one-dimensional. Primarily, matrix elements must be listed in either row-major or column-major order, though diagonal storage schemes have been used. For sparse matrices, indices can start counting from zero or one. Layout is further complicated by blocking, graph compression, etc.

The critical question is: in all the specifications of matrix listed above, which are specializations of a matrix and which are attributes? The answer to this question determines which concepts are implemented by subclassing and which are implemented as fields inside the class.

The answer also depends on how the class(es) will be used. Rarely will a programmer find a need to implement separate classes for full-rank and rank-deficient matrices, but it is not obvious that a programmer must implement sparse and dense matrices as separate classes either. Matlab uses the same structure for sparse and dense matrices and allows conversion between the two. On the other hand, PETSc has both sparse and dense matrices subclassed from their abstract `Mat` base class.

A third dimension of complexity comes from matrix file formats, which can be either a text or binary file, and more generally, a pipe, socket connection, or other forms of I/O streams.

Final Layout. Our major concern was to have a flexible extensible system for getting matrices in many different formats into our program at runtime. We discuss in this section how we finally organized our solution. The inheritance hierarchy of the subsystem is shown in Figure 12.

First we made non-abstract base classes `GraphBase` and `MatrixBase` which define a general layout for the data. From these, we derive `Graph` and `Matrix` classes that provide the public accessor/mutator functions; each class provides constructors from both `GraphBase` and `MatrixBase`. Furthermore, `Graph` and `Matrix` classes also inherit from the `DataStructure` class, which gives them generic data structure state, error reporting functionality, etc. In this way both can construct from each other without any cyclic dependency.

The final important piece before fitting together the entire puzzle is a `DataStream` class. This abstract base class has no ancestors. It does all of its I/O using the C style `FILE` pointers. We chose this C-style of I/O because, although it lacks the type-safety of C++ style `iostream`, it does allow us to do I/O through files, pipes, and sockets. This functionality is (unfortunately) not part of the standard C++ library.

If we try to open a file with a "`.gz`" suffix, the file object inherits from the `DataStream` class the functionality to open a `FILE` pointer that is in fact a pipe to the output of gunzip[4]. The `DataStream` class is therefore responsible for opening and closing the file, uncompressing if necessary, opening or closing the pipe or the socket, etc.; but it is an abstract class because it does not know what to do with the `FILE` once it is initialized. This class also provides the error handling services that are typical of file I/O.

To understand how all these partial classes come together to do I/O for a sparse matrix format, consider adding a new format to the library, a Matrix-Market file. To be able to read this format, we create a class `MatrixMarketFile` that inherits from `MatrixBase` and `DataStream`. This new class needs to implement two constructors based on `MatrixBase` or `GraphBase` and two virtual functions: `read( FILE * )` and `write( FILE * )` (in practice, it also implements many more accessor/modifier methods specific to the Matrix-Market format). Now we can read a Matrix-Market file, and create instances of either `Graph` or `Matrix` (or any other class that uses a `MatrixBase` in its constructor). Furthermore, from any class that inherits from `MatrixBase` or `GraphBase` we can write the object in Matrix-Market format. A graph-based file format, for instance Chaco [18], can be created using a similar inheritance hierarchy based on `GraphBase`.

---

[4] The "`.gz`" suffix indicates a file that is compressed with the GNU zip utility (`gzip`) and can be uncompressed by its complement, `gunzip`.

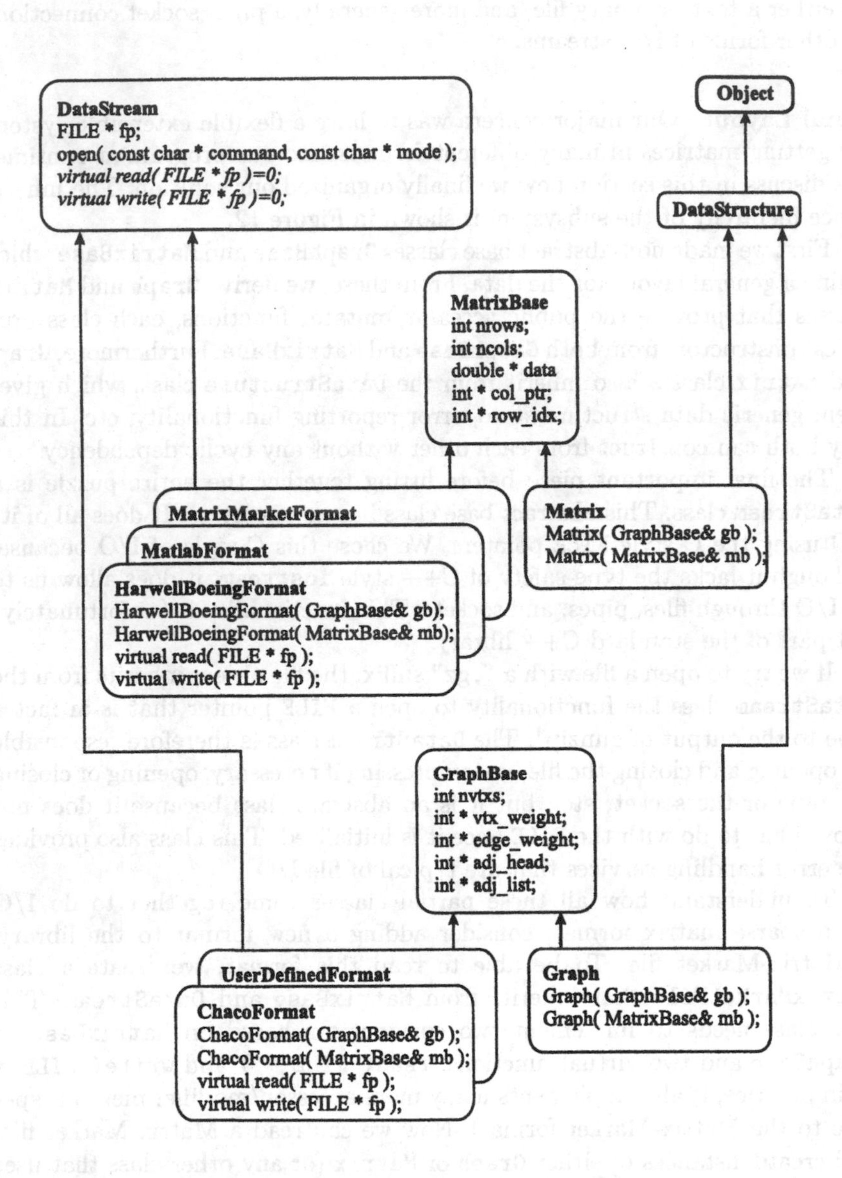

Fig. 12. A fragment of the inheritance hierarchy highlighting how multiple file formats are implemented and extended. To add an additional format, create a class that inherits from `DataStream` and one of `GraphBase` or `MatrixBase`. Then implement the two pure virtual methods inherited from `DataStream`.

5 Ordering Specifics

Spindle is a C++ library of sparse matrix ordering algorithms for reducing bandwidth, envelope, wavefront, or fill. We had earlier developed new envelope and wavefront reducing algorithms [19], but since they were implemented in C, we faced problems with complexity, scaling, and extensibility of the code. This motivated us to create Spindle. Although Spindle has several stand-alone drivers, it is intended to be used as a library, and is distributed with C, C++, Matlab, and PETSc calling interfaces[5].

In this section, we provide some examples of major concrete classes in Spindle, one from the `DataStructure` and one from the `Algorithm` branches of the inheritance hierarchy. In Section 5.1 we introduce the `QuotientGraph` class and describe its use. It will be revisited again in Section 9.2 when we discuss lessons learned. We also discuss the framework for implementing a family of algorithms in Section 5.2. We believe that this section provides a compelling example for how flexibility and extensibility can be achieved by proper design.

5.1 The Quotient Graph

A *quotient graph* [14] is an implicit representation of the sequence of elimination graphs obtained during the factorization process. The quotient graph represents (implicitly) the elimination graph at any stage in the same amount of space as the original graph, even though the ordering algorithm creates new fill edges in the elimination graph. Thus the quotient graph is immune to the effects of fill; the price to be paid for this property is greater searching costs to determine from the quotient graph the adjacency set of a node in the elimination graph. The space efficiency helps avoid dynamic storage allocation for the ordering step, since the total fill is known only at the end of this step.

The quotient graph is an augmented graph with two distinct types of vertices: nodes and enodes. Nodes represent uneliminated nodes from the original graph, suitably grouped together; enodes represent groups of eliminated nodes that are adjacent to each other. The edges in a quotient graph join either one node to another, or a node to an enode. There are no edges connecting two enodes[6].

The set of nodes adjacent to a node v in an elimination graph can be computed from the quotient graph by computing the nodes that can be reached from v in the latter graph. This reachable set of v includes two groups of nodes: First, the set of all nodes adjacent to v in the quotient graph; second, the union of the nodes that are adjacent to the enodes adjacent to v. The

[5] Fortran interfaces are also possible, but have not been implemented yet.

[6] Conceptually, an edge could connect two enodes in some intermediate state, but it would immediately be removed and the enodes would be merged into one.

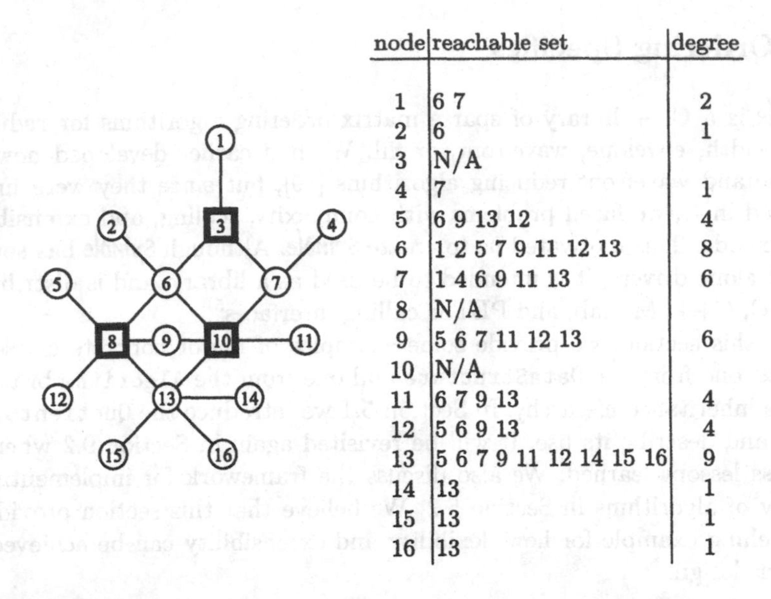

node	reachable set	degree
1	6 7	2
2	6	1
3	N/A	
4	7	1
5	6 9 13 12	4
6	1 2 5 7 9 11 12 13	8
7	1 4 6 9 11 13	6
8	N/A	
9	5 6 7 11 12 13	6
10	N/A	
11	6 7 9 13	4
12	5 6 9 13	4
13	5 6 7 9 11 12 14 15 16	9
14	13	1
15	13	1
16	13	1

Fig. 13. An example of a quotient graph. The nodes are represented as circles, and enodes as boxes. The *reachable set* of a node v is the union of the set of nodes adjacent to v, and the set of all nodes adjacent to enodes adjacent to v. The degree of a node in a quotient graph is the size of its reachable set, and not the cardinality of its adjacency set, as is usual in graphs.

union of these two sets is the reachable set of v in the quotient graph, and hence the adjacency set of v in the elimination graph. Figure 13 shows a sample quotient graph and the reachable sets and degrees for each node.

5.2 Polymorphic Fill Reducing Orderings

One example where object-oriented implementation had substantial payoffs in terms of extensibility was in our ability to construct polymorphic fill reducing orderings. Recall from Table 1 that there are several different variants of these algorithms, many of which are quite recent. Currently there is no known library containing all of these algorithms, besides Spindle. While some of these heuristics are related, others—particularly MMD and AMD—are radically different in the ways priorities are computed, the underlying graph is updated, and in the allowed and disallowed optimizations. A fundamental distinction is that MMD allows multiple elimination, while AMD is restricted to single elimination. This means that MMD allows many vertices to be eliminated between each graph update (but the eliminated nodes have to belong to non-adjacent supernodes). AMD does not require as much work per graph update, but the graph needs to be updated after every node is eliminated; this is a consequence of the way in which priorities are approximated in AMD.

Polymorphism was achieved using the Strategy Pattern [13, pg. 315]. We created a complete framework for the entire family of minimum-degree like algorithms but deferred the ability to compute the (exact or approximate) degree of a node to a separate class in its own mini hierarchy. See Figure 14 for the basic layout.

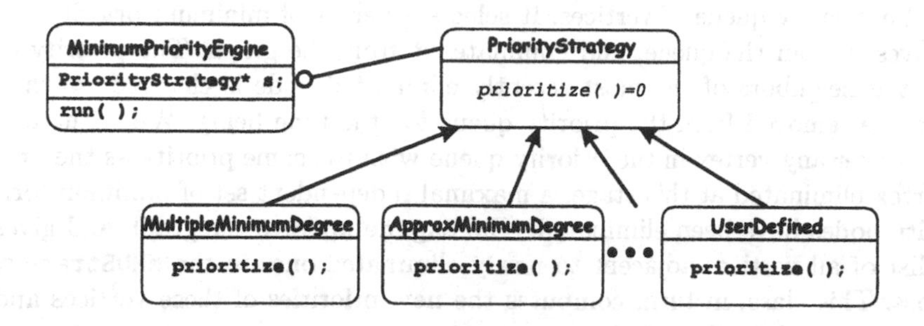

Fig. 14. The Strategy Pattern.

In this arrangement, the class **MinimumPriorityEngine** (which we will call **Engine** for short) is an algorithm that repeatedly selects a node of minimum priority from the given graph, eliminates it from the graph, and then updates the graph by adding appropriate fill edges when necessary. The catch is that it has no idea how to determine the priority of the vertices. It must rely on a `PriorityStrategy` class (`Strategy` for short), or more specifically, a specialized descendant of the `Strategy`.

This important design pattern offers several benefits. It provides a more efficient and more extensible alternative to long chains of conditional statements for selecting desired behavior. It allows new strategies to be implemented without changing the implementation of the engine. It is also more attractive than overriding a member function of the engine class directly because of the engine's overall complexity. This design also forces a separation between the work that is done and how the quality of work is measured.

There are potential drawbacks for using this pattern in general. The first drawback is that there is an increased number of classes in the library, one for each ordering algorithm. This is not a major concern, though users should be insulated from this by reasonable defaults being provided. A second drawback is the communication overhead. The calling interface must be identical for all the strategies, though individual types may not need all the information provided. A third drawback is the potential algorithmic overhead in decoupling engine and strategy. In our case, the engine could query the strategy once for each vertex that needs to be evaluated, though the virtual-function call overhead would become high. Alternatively, the engine might request all uneliminated nodes in the current quotient graph to be re-prioritized after each node is eliminated. This may result in too much work being done inside the strategy. Fortunately, in all these algorithms the only nodes whose priorities change are the ones adjacent to the most recently eliminated node. The `QuotientGraph` keeps track of this information.

For the `Engine` to work, a class must be derived from the `Strategy` abstract base class, and this class must override the pure virtual member function `computePriority`. The `Engine` is responsible for maintaining the graph and a priority queue of vertices. It selects a vertex of minimum priority, removes it from the queue, and eliminates it from the graph. The priority of all the neighbors of the most recently eliminated node is changed, so they too are removed from the priority queue for the time being. When there is no longer any vertex in the priority queue with the same priority as the first vertex eliminated at this stage, a maximal independent set of minimum priority nodes have been eliminated. The `Engine` updates the graph and gives a list of all vertices adjacent to newly eliminated ones to the `MMDStrategy` class. This class, in turn, computes the new priorities of these vertices and inserts them into the priority queue.

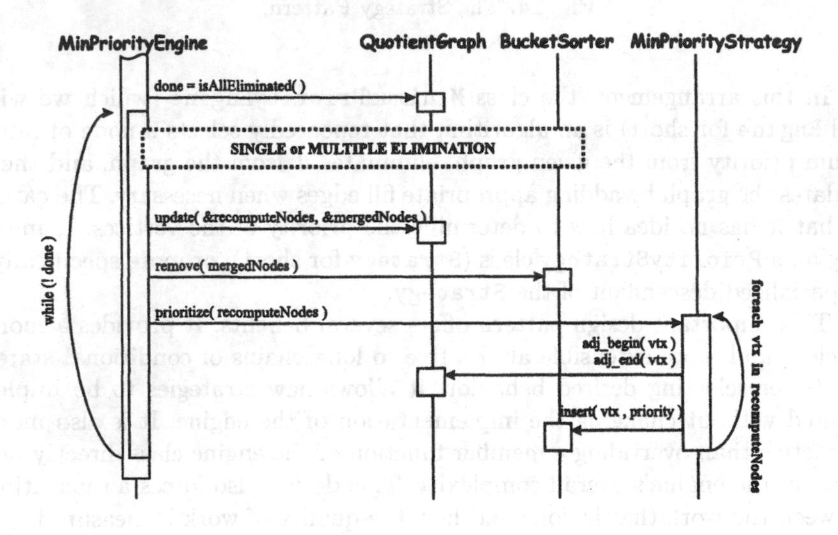

Fig. 15. Interaction of the classes in the ordering step.

To make this setup efficient, we use a `BucketSorter` class to implement the priority queue and a `QuotientGraph` class to implement the series of graphs during elimination. The interaction of these four major objects is shown in Figure 15. We hide the details of how single elimination and multiple elimination are handled. This too is determined by a simple query to the `Strategy` class. When the `QuotientGraph` is updated, it performs various types of compressions which may remove additional vertices or modify the list of vertices that need their priority recomputed. When it calls the `Strategy` to compute the priorities, it provides a `const` reference to the `QuotientGraph` for it to explore the datastructure without fear of side-effects, and the `BucketSorter` to insert the vertices in.

To implement all of the ordering algorithms in Table 1 efficiently, we had to augment the Strategy Pattern. The `QuotientGraph` is required to behave in slightly different ways when updating for single elimination algorithms (e.g., AMD) and multiple elimination algorithms (e.g., MMD). Thus the `Engine` must query the `Strategy` what type is required and set the `QuotientGraph` to behave accordingly. This is handled in the first phase of the `run()` function that is overridden from its parent `Algorithm` class.

6 Factorization Specifics

OBLIO currently focuses on the factorization and triangular solve steps, and imports orderings from other libraries such as Spindle. Here we restrict our discussion to the factorization algorithms, which are the most complex. All abstractions associated with the factorization are built incrementally, beginning from the base classes described in Section 4.

6.1 The First Structural and Algorithmic Classes

The initial abstractions result naturally from the high level formulation of the factorization, illustrated in Figure 1. We need to describe coefficient matrix objects, permutation objects and factor objects, and for these OBLIO provides the following three classes: `SparseSymmMatrix`, `Permutation`, and `SparseLwTrMatrix`.

A coefficient matrix is described by the `SparseSymmMatrix` class. Internally, the matrix data is stored using the compressed column format: for each column there is a set of numerical values and a set of row indices that correspond to these values. In addition, an index array contains the location of the first element in each column in the value and row index arrays. The `Permutation` class describes permutations by storing two maps: a map from old to new indices and another from new to old indices. The `SparseLwTrMatrix` class describes both the triangular and diagonal factors. While this abstraction may look awkward, it is done with a specific intent. Note that the diagonal elements of L do not have to be stored explicitly, which allows the elements of D to replace them. Since the factors are usually manipulated together, this unified implementation is more efficient. The storage of the data is similar to the one used in the `SparseSymmMatrix` class, with the difference that row indices are compressed according to the supernodal structure of the factors.

OBLIO provides a separate abstraction for each factorization algorithm. Currently, it supports only multifrontal factorizations, for both symmetric positive definite and indefinite problems. In the indefinite case we make use of the Duff-Reid [11] pivoting strategy now, but we plan to extend the library with other strategies too, for example Aschcraft-Grimes-Lewis [2]. Pivoting strategies are discussed in detail in the latter paper. Also, the library can be easily extended with other factorization algorithms, such as the left-looking algorithm. The current factorization classes are **PosDefMultFrtFactor**, and **IndefMultFrtFactor**.

6.2 Connecting Data Structures and Algorithms

Data structures are inputs to and outputs for algorithms. We associate data structures with algorithms by using pointers to structural objects inside algorithmic classes. The inputs to a factorization algorithm are the coefficient matrix and the sparsity preserving permutation. The outputs are the factors and the permutation that provides stability. We make the choice of composing the two permutations, so the output permutation replaces the input permutation after the factorization algorithm is run. Figure 16 describes the connection between a factorization algorithm and its input and output data structures.

```
class PosDefMultFrtFactor: public Algorithm
{
  private:
    const class SparseSymmMatrix *a;
    class Permutation *p;
    class SparseLwTrMatrix *l;
    // ...
  public:
    PosDefMultFrtFactor(const class SparseSymmMatrix,
                              class Permutation,
                              class SparseLwTrMatrix);
    //...
};
```

Fig. 16. An algorithm connects to its inputs and outputs by means of pointers to the corresponding data structures.

Inputs and outputs can be associated with an algorithm through the constructor of the algorithm class, or at a later stage (not shown in Figure 16). This provides the flexibility to change them at any time.

6.3 The Factorization in More Depth

More abstractions are needed deeper inside the factorization algorithms. First, there is the elimination forest, which guides both the symbolic and the numerical phases of the factorization. The forest stores parent, child and sibling pointers for every node. This way it can be traversed both bottom-up and top-down. The symbolic and numerical factorization proceed bottom-up (postorder) but require child and sibling information. The parent, child and sibling pointers are stored for both nodal and supernodal versions of the forest. There are also three major algorithms inside the factorization: algorithms to compute the elimination forest, the symbolic factorization, and the numerical factorization. Positive definite and indefinite solvers differ only in the numerical part, so separate numerical factorization abstractions are needed. Thus we have the classes: **ElimForest**, **CompElimForest**, **PosDefMultFrtNumFactor**, **IndefMultFrtNumFactor** and **SymFactor**.

A factorization class is based on these new classes. To define it we use composition, as shown in Figure 17.

```
class PosDefMultFrtFactor: public Algorithm
{
  private:
    // ...
    ElimForest f;
    CompElimForest elm;
    SymFactor sym;
    PosDefMultFrtNumFactor num;
  public:
    // ...
};
```

Fig. 17. Composition within a multifrontal factorization algorithm.

By splitting the factorization algorithm into its three algorithmic components, we give the user the possibility of running these components individually. This flexibility is needed in many situations. When some types of nonlinear equations are solved by successive linearization, the zero-nonzero structures remain constant while the numerical values change from one iteration to the next. In this case, only the numerical factorization algorithm needs to be run at each iteration. This splitting also aids code reuse since the algorithms for computing the elimination forest and the symbolic factorization are implemented only once but are used by both positive definite and indefinite factorizations.

Three more abstractions are required within the numerical factorization. They correspond to the frontal and update matrices and to the update stack. Since both frontal and update matrices are dense we first provide a general abstraction for dense matrices. In addition to the numerical values, frontal and update matrices must also store two types of maps: the map from local to global indices and vice versa. The corresponding classes are DenseSymmMatrix, FrontalMatrix, UpdateMatrix, and UpdateStack.

Composition is used again within the numerical factorization algorithms, as shown in Figure 18. Also, the update stack is composed of update matrices.

```
class PosDefMultFrtNumFactor: public Algorithm
{
   private:
      // ...
      FrontalMatrix fm;
      UpdateMatrix um;
      UpdateStack u;
   public:
      // ...
};
```

Fig. 18. Composition within a multifrontal numerical factorization algorithm.

Figures 19 and 20 summarize our discussion above. They provide a high level description of the OBLIO classes associated with the factorization from two perspectives, showing inheritance and composition relationships.

6.4 The Interaction between Data Structures and Algorithms

Because of the conflict between high level abstractions and efficiency, the interaction between data structures and algorithms is a major concern in OBLIO. We have made several tradeoffs here rather than adopt a uniform solution in all such situations. There are cases when efficient interfaces make algorithmic classes aware of the internal representation of structural classes. In other situations we were able to achieve more abstraction. The code in Figure 21, which is the kernel of a numerical factorization algorithm, is representative from this perspective. The loop traverses the supernodal elimination forest in postorder using iterators. For each supernode the frontal matrix assembles original numerical values from the coefficient matrix; for a non-leaf supernode, update matrices from the update stack are also assembled into the frontal matrix. Partial factorization is then performed on the frontal matrix, eliminated columns are saved in the factors, and the update matrix is pushed onto the stack.

Note that most of the execution time is spent inside the partial factorization of the frontal matrices. Accordingly, we had to make sure that this operation is performed efficiently. The factorization is a triply nested loop which needs to be carefully written. To make sure that we get the best performance, we have written it in Fortran 77 in addition to C++, since Fortran 77 compilers tend to generate more efficient code than C++ compilers.

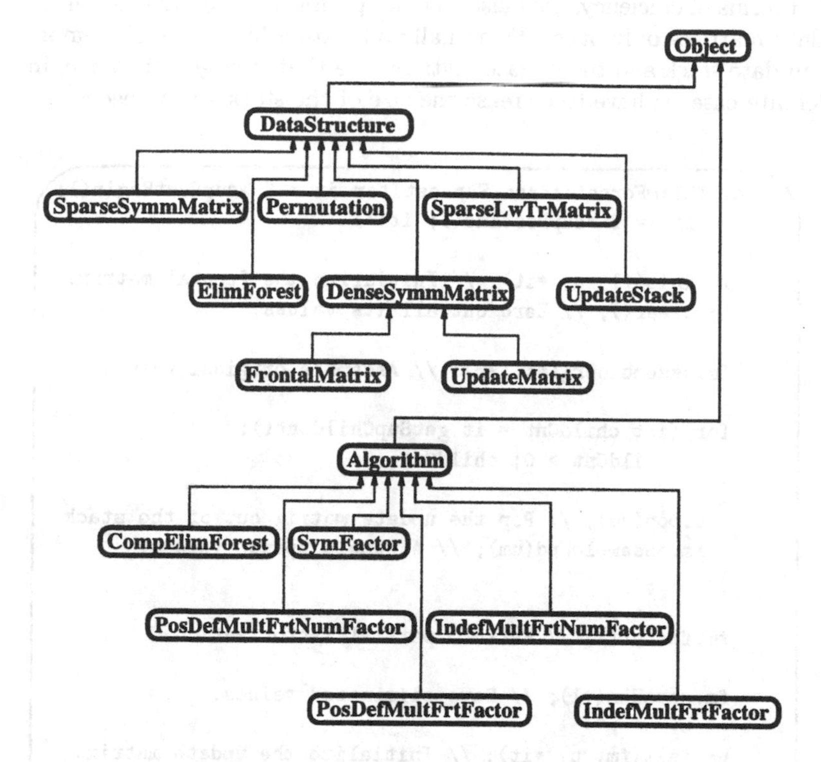

Fig. 19. Inheritance relationships.

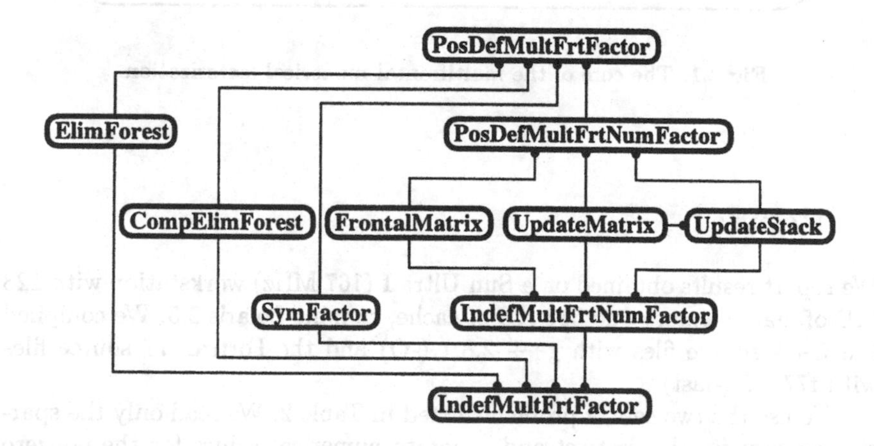

Fig. 20. Composition relationships.

In terms of efficiency, note also that we perform storage allocation for the update matrices only once. We actually allocate a big chunk of memory for the update stack and the update matrices use that storage. Of course, in the indefinite case we have to increase the size of the stack when needed.

```
for (ElimForest::constSupPostIter it = f->supPostBegin();
     it != f->supPostEnd(); it++)
{
  fm.init(*l, u, *it); // Initialize the frontal matrix.
  fm.clear(); // Zero out all its values.

  fm.assembleOrg(*a, *p); // Assemble original values.

  for (int childCnt = it.getSupChildCnt();
       childCnt > 0; childCnt--)
  {
    u.pop(um); // Pop the update matrix out of the stack.
    fm.assembleUpd(um); // Assemble update values.
  }

  fm.factor(); // Perform partial factorization.

  fm.saveElm(*l); // Save eliminated values.

  um.init(fm, u, *it); // Initialize the update matrix.
  fm.saveUpd(um); // Save update values.

  u.push(um); // Push the update matrix into the stack.
}
```

Fig. 21. The core of the multifrontal numerical factorization.

7 Results

We report results obtained on a Sun Ultra I (167 MHz) workstation with 128 MB of main memory and 512 KB of cache, running Solaris 2.5. We compiled the C++ source files with g++ 2.8.1 (-O) and the Fortran 77 source files with f77 4.2 (-fast).

We use the two sets of problems listed in Table 2. We read only the sparsity pattern for the first set and generate numerical values for the nonzero elements (both real and complex) that make the coefficient matrices positive definite. The problems in the second set are indefinite and we read the original numerical values also. To compare ordering algorithms we use the first set of problems. Table 3 shows ordering and numerical factorization times for several ordering algorithms available in Spindle, for real-valued coefficient matrices.

problem	order	nonzeros in A (subdiag.)	description
commanche	7,920	11,880	helicopter mesh
barth4	6,019	17,473	computational fluid dynamics
ford1	18,728	41,424	structural analysis
barth5	15,606	45,878	computational fluid dynamics
shuttle_eddy	10,429	46,585	model of space shuttle
ford2	100,196	222,246	structural analysis
tandem_vtx	18,454	117,448	helicopter mesh (tetrahedal)
pds10	16,558	66,550	multicommodity flow
copter1	17,222	96,921	helicopter mesh
ken13	28,632	66,586	multicommodity flow
tandem_dual	94,069	183,212	dual of helicopter mesh
onera_dual	85,567	166,817	dual of ONERA-M6 wing mesh
copter2	55,476	352,238	helicopter mesh
shell	4,815	100,631	model of venous stent
rev7a	9,176	143,586	,,
nok	13,098	806,829	,,
e20r0000	4,241	64,185	Driven cavity (Stokes flow)
e30r0000	9,661	149,416	,,
e40r0000	17,281	270,367	,,
helmholtz0	4,224	19,840	Helmholtz problem (acoustics)
helmholtz1	16,640	74,752	,,
helmholtz2	66,048	283,648	,,

Table 2. Test problems.

We focus next on one particular ordering, MMD, and look at several other properties. Table 4 lists, for the first set of problems, the number of off-diagonal nonzero entries in the factors, the numerical factorization performance and the backward error for both real and complex-valued problems. Note that performance results must be interpreted with care for sparse matrix computations, since an ordering that increases arithmetic work tends to increase performance as well. Thus one can achieve a higher performance by not preordering the coefficient matrix by a fill-reducing ordering. Performance reports are useful when the same ordering is used on different architectures, or as in this case, when looking at both real and complex valued problems. Note that while the arithmetic work of complex-valued problems is roughly four times the work in real-valued problems, performance improves only by

problem		MMD	MMMD	AMD	AMIND	AMMF
commanche	o	0.28	0.33	0.55	0.55	0.58
	f	0.24	0.22	0.25	0.23	0.26
barth4	o	0.23	0.25	0.33	0.33	0.35
	f	0.27	0.28	0.28	0.28	0.27
ford1	o	0.89	1.07	1.31	1.38	1.44
	f	0.89	0.84	0.91	0.95	0.95
ken13	o	2.26	2.72	6.01	7.44	7.32
	f	1.61	1.79	1.61	1.86	1.85
barth5	o	0.93	1.06	1.17	1.21	1.27
	f	0.93	0.83	0.94	0.88	0.88
shuttle_eddy	o	0.44	0.51	0.62	0.66	0.65
	f	0.98	0.87	0.98	0.87	0.87
ford2	o	6.67	7.86	8.05	8.78	8.81
	f	12.19	9.01	11.41	9.15	8.57
tandem_vtx	o	2.25	2.45	1.82	2.11	2.17
	f	31.64	20.95	30.02	19.05	16.34
pds10	o	89.52	103.74	3.88	4.46	4.43
	f	41.60	41.50	40.86	39.66	42.51
copter1	o	2.50	2.62	2.01	2.38	2.80
	f	42.15	34.98	39.50	31.81	31.18
tandem_dual	o	11.34	12.32	9.84	11.26	11.15
	f	256.46	205.72	224.28	146.66	121.28
onera_dual	o	10.78	11.70	8.98	10.35	10.55
	f	344.38	214.66	261.60	191.71	136.83
copter2	o	11.74	12.27	7.85	9.32	9.45
	f	413.02	259.38	364.55	234.47	218.37

Table 3. Time (seconds) needed to compute some fill-reducing orderings from the MMD family (o), and to factor the reordered matrices (f), for positive definite real-valued problems.

a factor of two. The relative residual ($\|(b - Ax)\| / (\|A\| \, \|x\|)$) is computed to report backward error.

We switch now to the second set of problems. The first two groups contain real-valued structural analysis and computational fluid dynamics (Stokes) problems. The third group contains Helmholtz problems, which are complex-valued. We order the coefficient matrices with MMD. Table 5 reports the number of off-diagonal nonzero entries in the factors, numerical factorization time, performance during numerical factorization, and the relative residual. Performance is still meaningful because we are interested in how it changes for problems within the same category. It is expected to increase with increasing number of nonzeros in the factor, and that is what we see for structures and the Stokes problem. Yet, performance decreases with increasing size for

problem	nonzeros in L (subdiagonal)	real-valued perf. (Mfl/s)	real-valued error	complex-valued perf. (Mfl/s)	complex-valued error
commanche	70,308	12.1	6e-17	37.0	1e-16
barth4	110,054	21.6	4e-17	58.3	8e-17
ford1	313,855	28.6	2e-17	72.6	3e-17
barth5	364,751	28.8	5e-17	75.2	1e-16
shuttle_eddy	390,020	37.2	4e-17	89.6	7e-17
ford2	2,463,669	32.2	2e-17	75.5	2e-17
tandem_vtx	2,708,953	36.2	9e-17	69.3	1e-16
pds10	1,634,148	27.2	2e-17	53.0	3e-17
copter1	2,470,144	32.0	2e-16	69.8	2e-16
ken13	334,470	16.3	1e-17	42.0	6e-18
tandem_dual	11,796,630	34.5	1e-16	69.9	2e-16
onera_dual	12,028,652	34.7	1e-16	69.5	2e-16
copter2	14,977,651	33.8	7e-17	67.1	1e-16

Table 4. Factorization results obtained with the positive definite code.

the Helmholtz problems. This is likely caused by increased paging due to the larger storage requirements of these problems relative to the available memory. We have observed a performance increase with size for these problems on machines with more memory (such as a Sun Ultra 60 workstation with 512 MB of memory and an IBM RS6000 workstation with 256 MB of memory). These results show that for large problems an out-of-core solver is needed to obtain high performance.

There is a detail that needs to be explained here. The indefinite numerical factorization is based on a sparsity/stability threshold which can take values between 0 and 1. A large threshold enhances stability but also generates more swapping among the columns within each supernode, and more delaying of columns from child to parent supernodes, which leads to reduced sparsity and performance. Sparsity can be preserved and performance increased by lowering the threshold, but that decreases stability. The optimal choice depends on the problem and on what tradeoffs are made. Stability can usually be reinstated from an unstable factorization by few steps of iterative refinement. The results in Table 5 are obtained with a threshold value of 0.1, and no iterative refinement.

8 Comparisons with Existing Work

While we were completing this article, we learned about another object-oriented library called SPOOLES that contains direct solvers. SPOOLES (SParse Object-oriented Linear Equations Solver) is a recent package devel-

problem	nonzeros in L (subdiagonal)	time (s)	perf. (Mfl/s)	error
shell	139,643	0.33	19.7	1e-17
rev7a	580,599	2.95	26.7	7e-17
nok	2,346,692	17.74	34.5	1e-19
e20r0000	369,843	2.47	22.1	2e-18
e30r0000	1,133,759	10.30	24.0	1e-18
e40r0000	2,451,480	27.11	27.3	1e-18
helmholtz0	129,525	0.56	66.3	3e-17
helmholtz1	619,292	4.64	58.5	4e-17
helmholtz2	3,106,177	39.45	66.9	5e-17

Table 5. Factorization results obtained with the indefinite code.

oped by Cleve Ashcraft and colleagues at the Boeing Company [30]. The code is publicly available without any licensing restrictions.

SPOOLES contains solvers for symmetric and unsymmetric systems of equations, and for least squares problems. It supports pivoting for numerical stability; it has three ordering algorithms (minimum degree, nested dissection, and multisection). Implementations on serial, shared parallel (threads), and distributed memory parallel (message-passing using MPI) environments are included. It is a robust, well documented package that is written in C using object-oriented techniques. We have learned much from its design and from our continuing interactions with Cleve Ashcraft.

SPOOLES distinguishes between data classes, which can have some trivial functionality, and algorithm classes, which handle the more sophisticated functionality. However, it does not enforce this paradigm strictly, as it has ordering classes but only factorization and solve methods.

SPOOLES employs a left-looking factorization algorithm instead of the multifrontal algorithm that we have implemented. However, with the functionality provided in SPOOLES, a multifrontal method can be quickly implemented. It employs a one-dimensional mapping of the matrix onto processors during parallel factorization, and a two-dimensional (2-D) mapping during the parallel triangular solves. It would be harder to support pivoting with a mapping of the matrix. The factorization algorithms are not as scalable as parallel algorithms that employ the 2-D mapping.

SPOOLES covers more ground than we have done, as it has solvers for symmetric and unsymmetric systems of equations, as well as over-determined systems. It does not have all of the minimum priority orderings in Spindle.

SPOOLES is implemented in C, which is generally more portable than C++. The C programming language is a procedural language and not designed to support object-oriented semantics of inheritance, automatic construction and destruction of objects, polymorphism, template containers, etc. These can be emulated in C, but it falls upon the library developer and the user to adhere to certain programming styles with no help or enforcement from the compiler.

9 Lessons Learned

9.1 General Comments

It is difficult to create general solutions to software design problems. Often a solution that seems good in a particular context becomes unsatisfactory when the context is extended. For example, the initial design of OBLIO did not decouple algorithms from data structures. It was initially implemented in C instead of C++. When OBLIO was ported to C++, the structs were converted to classes and algorithms were made methods of these classes. This introduced several difficulties. First, the association was not natural. Is an ordering algorithm supposed to be associated with a matrix or with a permutation? Second, this solution was neither flexible nor extensible. These problems led us later to the natural and general solution in which we introduced algorithmic classes.

One area where we continue to struggle for the "right" solution is with shared objects. This happens, for example, with an ordering algorithm and a coefficient matrix since the matrix exists outside the algorithm but it also has a copy inside the algorithm, as the algorithm's input. Since the matrix is built before the ordering algorithm is run, the input may just be a pointer to the matrix, an external object. What happens with the output of the ordering algorithm (the `Permutation` object) is trickier to manage.

Currently, we have a limited approach for the ownership of dynamically allocated data inside structural objects. It is not desirable to allocate separate copies for different copies of the same entity. We have used two approaches until now. In the first, the output object also exists before the algorithm is run (it can be empty) and the algorithm packs the empty class with information. The second approach lets algorithm classes construct the output classes internally and spawn ownership (and cleanup responsibilities) of only the outputs that the user requests.

The preallocation scheme is simple and makes responsibility of instance reclamation obvious. The downside of this is that an algorithm's output has to be allocated and assigned to it before the algorithm is run. It is less intuitive than running the algorithm and querying the results. Spindle's fill-reducing orderings construct the parent pointers for the elimination forest in the normal course of computing the ordering. One user may indeed want a permutation, but others may want the elimination forest. It is inelegant to force users to pre-initialize output classes that they are not even interested in.

The approach of having shared objects and explicitly transferring ownership can be complicated to use and is more prone to programmer error. This is, however, much easier to implement than another method for sharing large data structures: reference counting.

Full reference counting requires an interface class for each class that can be shared. An additional layer is needed to provide a counter that counts the number of interfaces that access the current instance. This gives more flexibility in assigning inputs and outputs (outputs can actually be built before or later). Changes are then seen by all interfaces and memory deallocation occurs correctly when the last interface is destroyed. The problem of unwanted changes by another interface is eliminated by performing "deep-copy" on write.

9.2 Judicious Application of Iterators

One disappointing endeavor was to provide an iterator class to traverse the *reachable set* of the `QuotientGraph` class. The idea was to provide an adjacency iterator for an `EliminationGraph` class, a reachable set iterator for the `QuotientGraph` class, and a collection of minimum priority algorithms that were totally unaware of whether it was operating on an elimination graph or a quotient graph. Although we were successful in implementing the `ReachableSetIter` class, its performance was so poor, that its general use was abandoned. This required the minimum priority algorithms to be specific to the `QuotientGraph` class and iterate directly over the enode and supernode adjacency lists. This increased the complexity of the interface and the coupling between data-structure and algorithm, but also significantly improved the performance. Here we explain why this idea looked good on paper, why it did not work well in practice, and why this problem is unavoidable.

Ideally, one would like to provide a class that iterates over the reachable set so that the priority computation can be implemented cleanly. In Figure 22 this class is `typedef`'ed inside the `QuotientGraph` class as `reach_iterator`. We add the additional detail that a weight might be associated with each supernode, so the degree computation sums the weights of the nodes in the reachable set.

Internally, we expected the reachable set iterator to have much in common with the C++ standard `deque::iterator`. A `deque` is a doubly ended queue that is commonly implemented as a vector of pointers to pages of items. The `deque` can easily add or remove items from each end by checking the first or last pages, and adding or deleting pages as necessary. The iterator of a `deque` need only advance to the end of a page, then jump to the next page. In the context of a quotient graph, an enode is much like a page with each having a list of items; namely supernodes.

There are, unfortunately, some critical differences. In a `deque`, the size of each page is known *a priori*, which is not true for the supernodes adjacent to an enode. Furthermore, the same supernode may be reachable through two different enodes. The reachable set need not be traversed in sorted order (as presented in Figure 13), but it cannot allow the same supernode to be counted twice through two different enodes. Furthermore, the reachable set does not include the node itself.

```
void MinimumDegree::prioritize( const QGraph& g, List& l,
                                PriorityQueue& pq )
{
  for( List::iterator it = l.begin(); it != l.end(); ++it ) {
    int i = *it;
    int degree = 0;
    for( QGraph::reach_iterator j = g.reach_begin(i);
         j != g.reach_end(i); ++j ) {
      degree += g.getNodeWeight( *j )
    }
    pq.insert( degree, i );
  }
}
```

Fig. 22. Computing the degree of all the nodes in `List` the elegant (but inefficient) way using reachable set iterators. The innocuous looking `++it` hides a cascade of `if-then-else` tests that must be performed at each call.

We were able to implement such an iterator, but there is a hidden overhead that causes the code fragment in Figure 22 to be too expensive. A reachable set is the union of several non-disjoint sets; and therefore the iterator must test at each iteration if there are any more items in the current set, if there are any more sets, and use some internal mechanism to prevent double visiting the same node in different sets.

Most of the details for the `ReachableSetIter` are not difficult, but the increment operator is excessively tedious. The problem is that the increment operator must redetermine its state at each call: Is it already at the end of the adjacency list? Are there more nodes or enodes in the current list? Once a next node has been located, has it been marked? The alternative is to evaluate the reachable set by iterating over sets of *adjacent* nodes and enodes of the quotient graph manually. This is shown in Figure 23 which is functionally equivalent to the code in Figure 22. In this case, the state information is implicit in the loop and not confined to the increment operator, which allows for the entire process to execute more efficiently.

The lesson learned here is to be judicious in the use of fancy techniques. The coding benefits of using a reachable set iterator are far outweighed by the increase in performance in manually running through the adjacency lists. The latter scheme makes the critical assumption that every node has a "self-edge" to itself in the list of adjacent enodes. This convenient assumption also increases coupling between the `QuotientGraph` class and the descendants of the `MinimumPriorityStrategy` class.

```
void MinimumDegree::prioritize( const QGraph& g, List& l,
                                PriorityQueue& pq )
{
  for( List::iterator it = l.begin(); it != l.end(); ++it ) {
    int i = *it;
    int degree = 0;
    int my_stamp = nextStamp();// get new timestamp
    g.visited[ i ] = my_stamp;    //Mark myself visited
    for( QGraph::enode_iterator e = g.enode_begin(i);
         e != g.enode_end(i); ++e ) {
      int enode = *e;  // for all adjacent enodes
      for( QGraph::node_iterator j = g.node_begin(e);
           j != g.enode_end(e); ++j ) {
        int adj = *j;  // for all adjacent nodes
        if ( visited[ adj ] < my_stamp ) {
          // if not already visited, mark it and add to degree
          visited[ adj ] = my_stamp;
          degree += g.getNodeWeight( adj );
        }
      }
    }
    pq.insert( degree, i );
  }
}
```

Fig. 23. Computing degree without using the reachable set iterators. This piece of code is not as elegant as the equivalent fragment in Figure 22, but in terms of efficiency , this method is the clear winner. In this case, the `visited` array is part of the `MinimumPriorityStrategy` base class along with the `nextStamp()` member function that always returns a number larger than any previous number. If the next stamp is the maximum integer, then the `visited` array is reinitialized to all zeros.

9.3 Conclusions

There is no inherent conflict between object-oriented design and efficient direct solvers in scientific computing. Some features of C++ such as function inlining, templates, encapsulation, and inheritance suffer no performance penalties and can be used aggressively. Other features such as virtual functions, excessive temporary objects that can result from operator overloading, and hidden copy construction can significantly degrade performance if left unchecked. The target of object-oriented design must be the higher layers of the software. With direct solvers they are the most sophisticated, taking most of the code, but just a small fraction of the execution time. There is not much room for abstraction in low level loops, so the focus there must be on performance.

We have learned that good design requires tradeoffs. There is no perfect solution, unless we are talking about artificial problems typically constructed for pedagogical use. Real-life applications have many constraints, many that conflict with each other, and it is difficult to satisfy all of them. The solution is to prioritize the constraints and to satisfy those with high priorities. An example is decoupling, which introduces an overhead since objects have to communicate through well defined interfaces. By decoupling we can localize potential changes in the code, and this is one of the properties that make object-oriented software appealing. To avoid overheads, we have weakened the encapsulation in performance-critical parts of the code.

Well-designed object-oriented software that is implemented in C++ leads to libraries that are easier to use, more flexible, more extensible, safer, and just as efficient as libraries implemented in C or Fortran. Designing and implementing a "good" object-oriented library is also significantly harder. Implementing sparse direct solvers cleanly and efficiently using object-oriented techniques is an area where challenging research issues remain.

Acknowledgments

We thank our colleague, David Hysom, who worked on the integration of our codes into PETSc, provided user feedback, and participated in our spirited debates. We are grateful to Cleve Ashcraft for many engaging technical conversations and Socratic dialogues. Thanks also to the PETSc team: Satish Balay, Lois McInnes, Barry Smith and Bill Gropp, for being so helpful—even with non-PETSc specific issues. Finally, we thank David Keyes for his enthusiasm and support.

This work was supported by U. S. National Science Foundation grants CCR-9412698 and DMS-9807172; by the U. S. Department of Energy by subcontract B347882 from the Lawrence Livermore Laboratory; by a GAANN fellowship from the Department of Education; and by NASA under Contract NAS1-19480 while the third author was in residence at the Institute for Computer Applications in Science and Engineering (ICASE).

References

1. P. Amestoy, T. A. Davis, and I. S. Duff. An approximate minimum degree ordering algorithm. *SIAM J. Matrix Anal. Appl.*, 17(4):886–905, 1996.
2. C. Ashcraft, R. G. Grimes, and J. G. Lewis. Accurate symmetric indefinite linear equation solvers. *SIAM J. Matrix Anal. Appl.*, 20:513–561, 1999.
3. C. Ashcraft and J. W. H. Liu. *SMOOTH: A software package for ordering sparse matrices*, November 1996. www.cs.yorku.ca/ joseph/SMOOTH.html.
4. C. Ashcraft, D. Pierce, D. K. Wah, and J. Wu. *The Reference Manual for SPOOLES, Release 2.2*, February 1999.

5. S. Balay, W. D. Gropp, L. C. McInnes, and B. F. Smith. Efficient management of parallelism in object-oriented numerical software libraries. In *Modern Software Tools for Scientific Computing*, E. Arge, A. M. Bruaset, and H. P. Langtangen, (eds.), Birkhäuser, 1997.
6. A. M. Bruaset and H. P. Langtangen. Object-oriented design of preconditioned iterative methods in DIFFPACK. *ACM Transactions on Mathematical Software*, 23:50–80, 1997.
7. R. Boisvert, R. Pozo, K. Remington, R. Barrett, and J. Dongarra. Matrix Market: a web resource for test matrix collections. In *Numerical Software: Assessment and Enhancement*, R. Boisvert, (ed.), pages 125–137. Chapman and Hall, London, 1997.
8. F. Dobrian, G. Kumfert, and A. Pothen. Object-oriented design of a sparse symmetric solver. In *Computing in Object-oriented Parallel Environments*, Lecture Notes in Computer Science 1505, D. Caromel et al (eds.), pp. 207–214, Springer Verlag, 1998.
9. I. S. Duff, R. G. Grimes, and J. G. Lewis. *Users' Guide for the Harwell-Boeing Sparse Matrix Collection*, Oct 1992.
10. Diffpack homepage: `www.nobjects.com/Diffpack`.
11. I. S. Duff and J. K. Reid. The multifrontal solution of indefinite sparse symmetric linear equations. *ACM Trans. on Math. Software*, 9(3):302–325, 1983.
12. I. S. Duff, A. Erisman, and J. K. Reid. *Direct Methods for Sparse Matrices*. Clarendon Press, Oxford, 1986.
13. E. Gamma, R. Helm, R. Johnson, and J. Vlissides. *Design Patterns: Elements of Reusable Object-Oriented Software*. Professional Computing Series. Addison Wesley Longman, 1995.
14. A. George and J. W. H. Liu. A fast implementation of the minimum degree algorithm using quotient graphs. *ACM Trans. on Math. Software*, 6:337–358, 1980.
15. A. George and J. W. H. Liu. *Computer Solution of Large, Sparse, Positive Definite Systems*. Prentice Hall, 1981.
16. A. George and J. W. H. Liu. The evolution of the minimum degree algorithm. *SIAM Rev.*, 31(1):1–19, March 1989.
17. A. George and J. W. H. Liu. An object-oriented approach to the design of a user interface for a sparse matrix package. Technical Report, Department of Computer Science, York University, Feb. 1997.
18. B. Hendrickson and R. Leland. *The Chaco User's Guide*. Sandia National Laboratories, Albuquerque, NM 87815, Oct 1993.
19. G. Kumfert and A. Pothen. Two improved algorithms for envelope and wavefront reduction. *BIT*, 37:559–590, 1997.
20. G. Kumfert and A. Pothen. An object-oriented collection of minimum degree algorithms: design, implementation, and experiences. In *Computing in Object-oriented Parallel Environments*, Lecture Notes in Computer Science 1505, D. Caromel et al (eds.), pp. 95–106, Springer Verlag, 1998.
21. J. Lakos. *Large-Scale C++ Software Design*. Professional Computing Series. Addison-Wesley, 1996.
22. H. P. Langtangen. *Computational Partial Differential Equations: Numerical Methods and Diffpack Programming*. Springer-Verlag, 1999.
23. J. W. H. Liu. Modification of the minimum-degree algorithm by multiple elimination. *ACM Trans. on Math. Software*, 11:141–153, June 1985.

24. J. W. H. Liu. The role of elimination trees in sparse factorization. *SIAM J. Matrix Anal. Appl.*, 11(1):134–172, 1990.
25. J. W. H. Liu. The multifrontal method for sparse matrix solution: theory and practice. *SIAM Rev.*, 34(1):82–109, 1992.
26. E. G. Y. Ng and P. Raghavan. Performance of greedy ordering heuristics for sparse Cholesky factorization. Technical Report, Computer Science Department, University of Tennessee, Knoxville, 1997.
27. PETSc homepage: `www.mcs.anl.gov/petsc/petsc.html`.
28. E. Rothberg and S. Eisenstat. Node selection strategies for bottom-up sparse matrix ordering. *SIAM J. Matrix Anal. Appl.*, 19(3):682–695, July 1998.
29. E. Rothberg. Ordering sparse matrices using approximate minimum local fill. Preprint, April 1996.
30. SPOOLES homepage:
`www.netlib.org/linalg/spooles/spooles.2.2.html`.

A Sparse Grid PDE Solver; Discretization, Adaptivity, Software Design and Parallelization

Gerhard W. Zumbusch

Institute for Applied Mathematics, University of Bonn, Germany
Email: zumbusch@iam.uni-bonn.de
URL: http://wwwwissrech.iam.uni-bonn.de/people/zumbusch.html

Abstract. Sparse grids are an efficient approximation method for functions, especially in higher dimensions $d \geq 3$. Compared to regular, uniform grids of a mesh parameter h, which contain h^{-d} points in d dimensions, sparse grids require only $h^{-1}|\log h|^{d-1}$ points due to a truncated, tensor-product multi-scale basis representation. The purpose of this paper is to survey some activities for the solution of partial differential equations with method based sparse grids. Furthermore some aspects of sparse grids are discussed such as adaptive grid refinement, parallel computing, a space-time discretization scheme and the structure of a code to implement these methods.

1 Introduction

Quite a lot of phenomena in science and engineering can be modeled by boundary value problems of ordinary differential equation or partial differential equation type. Further assumptions to simplify the model like axis- and radial-symmetries often give rise to a PDE in one or two dimensions ($d = 1, 2$), which can be treated more easily numerically. However, the accurate solution of similar problems in three dimensions, or time-dependent problems in two or three space dimensions requires more computational power. Fortunately, most phenomena in physics live in a three-dimensional space, plus one time dimension. Higher-dimensional problems, which would be extremely expensive or simply impossible to solve numerically, referred to as the 'curse of dimension', do not occur often in the literature. However, there are a lot of higher-dimensional problems $d > 3$ around, but due to their complexity only few approaches address these problems directly. We mention higher-dimensional problems in financial engineering, in quantum physics, in statistical physics and even the four-dimensional problems in general relativity. Hence there is a need for the solution of higher-dimensional problems, but due to the fact that standard methods fail or are extremely expensive, such problems are usually not considered.

2 Sparse Grids

Sparse grids are a multi-dimensional approximation scheme, which is known under several names such as 'hyperbolic crosspoints', 'splitting extrapolation' or as a boolean sum of grids. Probably Smolyak [46] was the historically first reference. Directly related to the boolean construction of the grids was the construction of a multi-dimensional quadrature formula. Both quadrature formulae and the approximation properties of such tensor product spaces were subject to further research, see Temlyakov [47] and others. The curse of dimension was also subject to general research on the theoretical complexity of higher-dimensional problems. For such theoretical reasons, sparse grids play an important role for higher-dimensional problems. Besides the application to quadrature problems, sparse grids are now also used for the solution of PDEs.

Sparse grids were introduced for the solution of elliptic partial differential equations by Zenger [54], where a Galerkin method, adaptive sparse grids and tree data structures were discussed. At the same time a different discretization scheme based on the extrapolation of solutions on several related, regular grids was proposed, see [21].

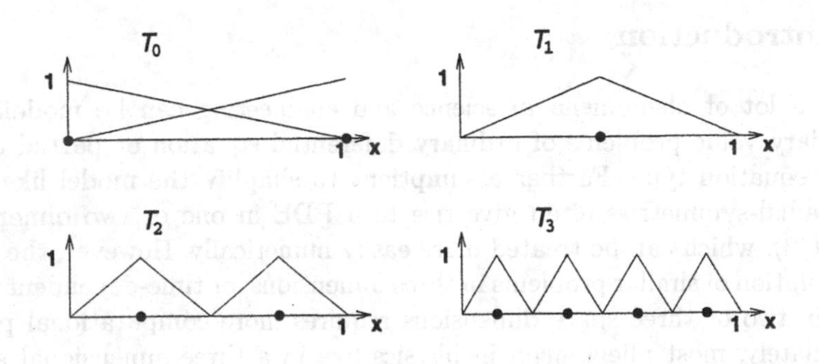

Fig. 1. The hierarchical basis.

The multi-dimensional approximation scheme of sparse grids can be constructed as a subspace of the tensor-products of one-dimensional spaces represented by a hierarchical multi-resolution scheme, such as the hierarchical basis, see Figure 1 and the historical reference [17], or generally any basis system of pre-wavelets or wavelets [24]. Each one-dimensional basis function can be derived from a model function ϕ by a scaling of 2^{-l} (also called level l) and a translation by an integer multiple of 2^{-l}. In the case of the hierarchical basis, the model function ϕ is a hat function.

Let us consider piecewise linear functions. The space of functions on the interval $[0, 1]$ of dyadic level l is based on a mesh with mesh parameter $h =$

2^{-l-1}. There are $2^{l+1}+1$ nodes. In the nodal basis, there are $2^{l+1}-1$ interior hat functions of support $2h$ and two boundary functions of support h, each associated with one node (or grid point). In the hierarchical basis, there are two boundary functions (T_0) and one global function (T_1) of support 1, see Figure 1, two functions of support 1/2 (T_2), four functions of support 1/4 (T_3) and so on. All hierarchical basis functions of support $2h$ and larger span the function space of level l.

Based on the one-dimensional case, we now construct spaces of piecewise linear interpolants on a d-dimensional unit hyper-cube. The space of functions on the regular grid of dyadic level l has a mesh parameter $h = 2^{-l-1}$. It can be represented by the space of all tensor-products (also referred to as direct products) of one-dimensional nodal or hierarchical basis functions of level l, i.e. support larger than 2^{-l-1}. On level l, the two-dimensional space e.g. can be written as

$$T_{ll} = \langle T_i \otimes T_j \rangle_{i,j \le l} .$$

In contrast to the regular grid, the corresponding sparse grid space consists of products of hierarchical basis functions with support larger than a d-dimensional volume of size 2^{-l-1}, see Figure 2. On level l, the two-dimensional space e.g. can be written as

$$T_{ll} = \langle T_i \otimes T_j \rangle_{i+j \le l} .$$

This is a subset of the regular grid space. A regular grid has about $2^{d \cdot l}$ nodes, which is substantially more than the $2^l \cdot l^{d-1}$ nodes of the sparse grid.

The major advantage of sparse grids compared to regular grids is their smaller number of nodes (or grid points) for the same level l and resolution 2^{-l}, which means a smaller number of basis functions (shape functions in FEM) and therefore a smaller dimension of the function space and fewer degrees of freedom in a discretization. This is especially true in higher dimensions $d \gg 1$.

However, the question whether sparse grids have an advantage compared to regular grids does also depend on the discretization accuracy of a solution obtained on a grid. We are interested in a comparison of accuracy versus number of nodes for both types of grids, a regular and a sparse one. We define the storage ε-complexity of an approximation method by the accuracy ε, which can be achieved with a storage of N nodes. The accuracy depends on the smallest mesh parameter h and an approximation order p like $\varepsilon = \mathcal{O}(h^p)$ for smooth data. For regular grids the number of nodes depends on the space dimension d as $N_{storage} = h^{-d}$, which results in $\varepsilon = \mathcal{O}(N^{-p/d})$. In the case of sparse grids, the dependence on the dimension d is much weaker and we denote $\varepsilon = \mathcal{O}(N^{-p+\gamma})$ for every $\gamma > 0$ and an approximation order p. The sparse grid approximation is said to break the curse of dimensionality.

Let us assume that a sparse grid approximation is of first order $p = 1$, which of course depends on the discretization order, the error norm, the smoothness of the solution and the sparse grid approximation itself. Then the sparse grid is competitive to a second-order method in two dimensions and to a third-order method in three dimensions, which is usually much more expensive and harder to construct, see Table 1.

Furthermore the number-of-operations complexity is of interest, because it is an estimate for the computing time a specific algorithm needs. In some cases the work count is proportional to the number of nodes. This is true for a single time-step of a standard explicit finite difference code. We will see that this is also true for the corresponding sparse grid code. However, the work count usually is higher than the number of nodes for implicit discretizations of time-dependent problems and for stationary problems involving the solution

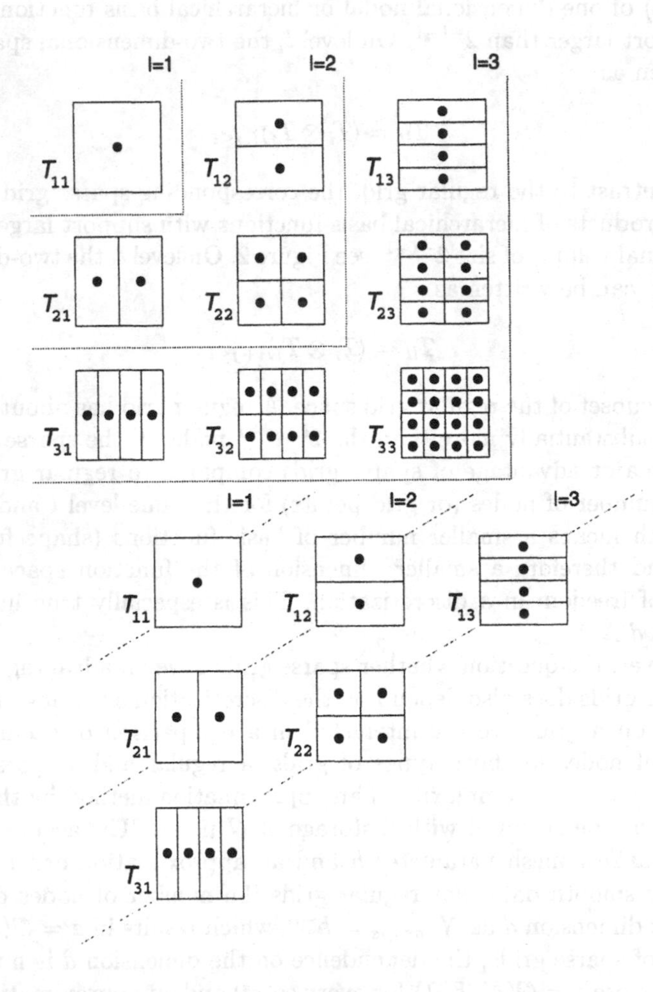

Fig. 2. Tableau of supports of the hierarchical basis functions spanning a two-dimensional regular space (top) and the corresponding sparse grid space. The space T_{ij} is defined by the tensor product $T_i \otimes T_j$.

	$d=1$	$d=2$	$d=3$	$d=1$	$d=2$	$d=3$
first order	$\mathbf{N^{-1}}$	$N^{-1/2}$	$N^{-1/3}$	$\mathbf{N^{-1}}$	$\mathbf{N^{-1+\gamma}}$	$\mathbf{N^{-1+\gamma}}$
second order	N^{-2}	$\mathbf{N^{-1}}$	$N^{-2/3}$	N^{-2}	$N^{-2+\gamma}$	$N^{-2+\gamma}$
third order	N^{-3}	$N^{-3/2}$	$\mathbf{N^{-1}}$	N^{-3}	$N^{-3+\gamma}$	$N^{-3+\gamma}$

Table 1. Storage-complexity of a regular (left) and a sparse (right) grid discretization. Terms printed in bold face indicate linear ε-complexity.

of (non-) linear equation systems, and for time-dependent problems in total. The number of time-steps for an evolution problem of a fixed time interval depends on the spatial resolution h, e.g. due to a CFL stability condition, which leads to a higher work count complexity in N. On regular grids we obtain $N_{work} = \mathcal{O}(h^{-d-1})$, which is equivalent to the storage complexity in $d+1$ space dimensions. This means that any reduction in storage $N_{storage}$, e.g. through sparse grids, may reduce the number of operations even further.

We still have to check the assumption on the approximation order $p = 1$ (or some other constant) of a sparse grid discretization. Up to now such orders had been verified for the extrapolation method, for the interpolation error, for the energy error of the Galerkin method, and for the consistency of the finite difference method. However, either quite strong regularity assumptions

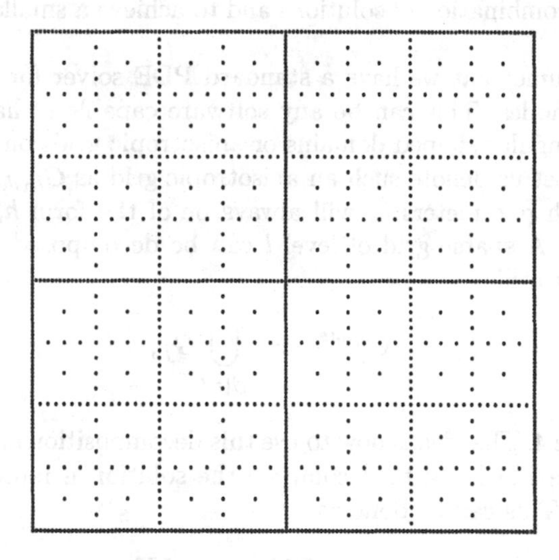

Fig. 3. A sparse grid on the unit square. Depicted are the nodes associated with the hierarchical basis functions.

or model problems were considered. Furthermore, numerical experiments indicate that in more general cases and in connection with adaptive sparse grids the approximation orders obtained for model situations hold.

3 Discretization of a PDE

There are several, completely different ways to discretize a partial differential equation. Many methods provide grid-based discretization schemes and can be generalized to sparse grids. Very popular are the Finite Element Method (FEM) based on the Galerkin approach, Finite Differences, the Finite Volume Method and spectral methods. Sparse grid discretizations have been so far constructed by the Galerkin method [54,16], finite differences [19,43], by extrapolation based on finite differences [21] and finite volumes [25], and by a spectral method [30] based on a fast Fourier transform defined on sparse grids.

3.1 The Extrapolation Method

Probably the simplest way to solve a PDE on a sparse grid is to use the extrapolation method, also called the 'combination'- [21] or 'splitting extrapolation'-technique [32]. The idea is to combine solutions computed on several different regular grids to a more accurate sparse grid solution. The approach uses the extrapolation idea of numerical analysis to cancel out some low order error terms by the combination of solutions and to achieve a smaller discretization error.

Let us assume that we have a standard PDE solver for regular grids of $n_1 \cdot n_2 \cdots n_d$ nodes. This can be any software capable of handling uniform grids on rectangular shaped domains or anisotropic grids on the unit square respectively. Let us denote such an anisotropic grid as $G_{h_1,h_2,\dots,h_d}$. Furthermore the mesh parameters h_i will always be of the form $h_i = 2^{-j_i}$ with a multi-index j. A sparse grid of level l can be decomposed into the sum of several regular grids

$$G_l^{\text{sparse}} = \bigcup_{|j|=l} G_j \,,$$

see also Figure 4. The idea is now to use this decomposition and the numerical solution of the PDE and to decompose the solution u into solutions u_j on regular grids. This can be done as

$$u_l^{\text{sparse}} := \sum_{|j|=l} u_j - \sum_{|j|=l-1} u_j \,, \tag{1}$$

which is an extrapolation formula with weights $+1$ and -1. If we depict the grid points of a grid G_j, each node is associated with a shape function in a

FEM discretization. The solution originally obtained only in the nodes is extended by the shape functions of a linear FEM discretization to a piecewise linear function on the whole domain. Now we can sum up functions from several grids e.g. $G_{2,0}$, $G_{1,1}$ and $G_{0,2}$. The union of the nodes gives the nodes of the corresponding sparse grid, compare Figures 3 and 4. Computationally, the combination of the grids requires interpolation on each grid, or the evaluation of the shape functions on the sparse grid points, because the sparse grid contains always a superset of nodes of each of the regular grids. The remaining nodes are determined by interpolation. For the correct extrapolation we also need to subtract several coarse grids, which can be done in the same way.

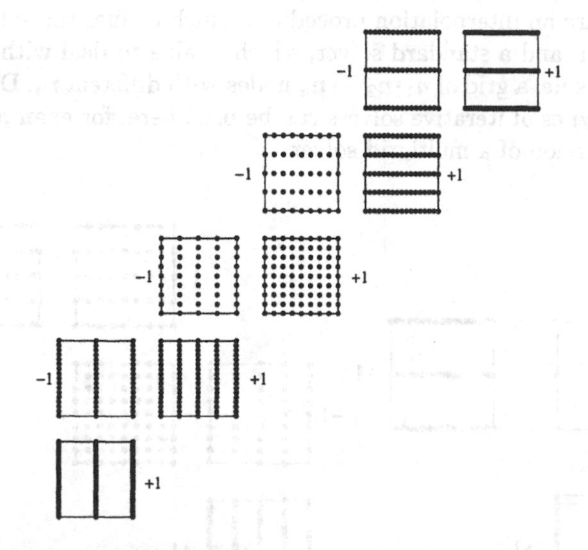

Fig. 4. The sparse grid combination technique. Several solutions of the PDE obtained on different regular grids by standard PDE codes are summed up to form a sparse grid solution of higher accuracy. The weights of $+1$ and -1 are depicted next to the grids. The grids are arranged analogous to the tableau in Figure 2. The union of nodes of all regular grids results in the nodes of the corresponding sparse grid, see Figure 3.

The fact that the extrapolation solution equals the solution of the PDE on the sparse grid can be proven by an error expansion of the regular grid solutions and the cancelation of the lower order error terms, see [21]. The quality of an extrapolation formula, i.e. the sum of the remaining error terms, depends on higher derivatives of the solution u and can be measured in higher order Sobolev norms (e.g. $H^{2d}(\Omega)$). Given sufficient smoothness, it is possible to combine solutions obtained by Finite Elements, Finite Volumes or Finite Differences. The latter two require an appropriate, order-preserving interpolation scheme for the nodal values obtained on single grids. The FEM provides a natural extension of the solution from the nodes to the whole domain through the shape functions used. Error bounds in different norms are available and the error usually compares to the regular grid error, with a logarithmic deterioration $|\log h|$.

Besides its simplicity, there are several advantages for this method: The solution can be computed concurrently, that is on different processors of a parallel computer (almost embarrassingly parallel, see [18]) or one after the other requiring only little computer memory. Multiple instances of a standard PDE code can be used and only little coding is necessary to implement the method. Furthermore such a code may be a very efficient code optimized and tuned for structured grids on a specific computer. The prerequisites of such a code are an interpolation procedure, which defines the solution in the whole domain, and a standard solver, which is able to deal with anisotropic discretizations i.e. a grid of $n_1 \cdot n_2 \cdots n_d$ nodes with different n_i. Direct solvers and several types of iterative solvers can be used here, for example the semi-coarsening version of a multigrid solver.

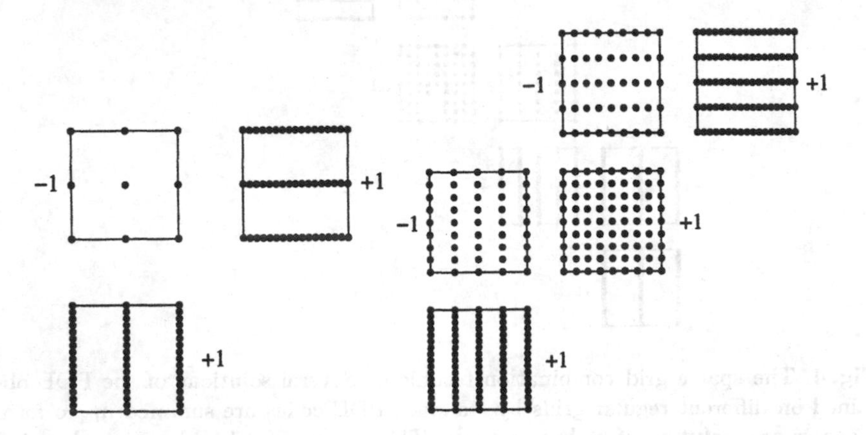

Fig. 5. Sparse grid extrapolation schemes: splitting extrapolation (left), 'semi'-sparse grid (right). A smaller number of solutions than for the sparse grid extrapolation technique are combined to form a higher accuracy solution. Again, each solution on a regular grid can be obtained by any suitable software package or PDE solver.

Here is where the first problems occur. Think of a fluid flow problem or some convection problem discretized on an anisotropic grid with $3 \times 2^{l-1} + 1$ nodes. A discretization may have problems with stability, or at least the numerical linear algebra is difficult. The extrapolation does work as long as there is a suitable error-expansion, which depends on the problem and the discretization. If the solution of extremely anisotropic refined grids is useless for some larger convection terms, the extrapolation does not work properly. A possible fix is to restrict the aspect ratios of the grids under consideration and to remove these from the extrapolation. This can be done by some additional restriction to the index j

$$u_l^{\text{sparse}} := \sum_{|j|=l,\ j_i \leq \gamma l} u_j - \sum_{|j|=l-1,\ \leq \gamma l} u_j,$$

with some parameter $\gamma < 1$. For three-dimensional fluid flow problems this is heavily used by Hemker [25], see Figure 5. Further modifications of the extrapolation method consist of a piecewise constant interpolation for some finite volume discretizations [25] and of the combination of few grids [32]:

$$u_l^{\text{sparse}} := \sum_{k=1}^{d} u_{j+(l-|j|)e_k} - u_j\,.$$

The idea is to refine the grid in each coordinate direction only once. Of course this gives a slightly different solution than the originally proposed method, see Figure 5, but we will see the relation in Section 4.1 on adaptive grids.

We have mentioned the stability problem for anisotropic discretizations. But there are more possible pitfalls: The solution of time-dependent problems and of non-linear problems can be done in several ways. The cheapest way would be to extrapolate the solution of the last time-step or the final non-linear iterate. Under some circumstances this is a legal approach. However, in the presence of phenomena not resolved on some grids, the extrapolation might cause a disaster. The error expansion of the discretization scheme is not accurate enough on such coarse scales and the final result is polluted by higher-order terms. A more expensive version of the extrapolation methods is to combine the solution at every time-step or every non-linear iteration step. This requires more CPU-time, some more changes in the code and does not parallelize so well. However, it is definitely more accurate. Further difficulties may arise with the resolution of coefficients, boundary conditions and the source terms on every grid, with non-linearities, and transport terms and singularities.

Adaptive grids are more difficult to incorporate into the extrapolation scheme, because the grid refinement of the single grids has to match. An adaptive refinement procedure has to use a complicated collaborative grid refinement, which is why this has not yet been implemented.

3.2 The Galerkin Method

The standard approach to discretize an elliptic differential equation on a sparse grid or even on an adapted sparse grid is the Galerkin method (FEM). Given a sparse grid along with the functions on that grid, it is straightforward to apply a Galerkin scheme with test and trial functions ϕ_i. These functions are defined by the hierarchical basis and some other multi-scale resolution, each one related to a node. The equation system is derived from the variational form $a(.,.)$ of the differential equation:

$$\begin{aligned} a(u,v) &= f(v), \qquad \forall v \in H, \\ a_{i,j} &= a(\phi_i, \phi_j), \\ f_j &= f(\phi_j), \\ \sum_i a_{i,j} u_i &= f_j, \qquad \forall j. \end{aligned} \tag{2}$$

This method can be applied to any set of linear independent shape functions in a finite-dimensional setting and furthermore for a complete basis in the case of infinite-dimensional function spaces like H^1. The idea now is to use shape functions of the sparse grid, which are direct products of one-dimensional functions of the hierarchical basis or some other (pre-) wavelet basis. At a coordinate $x = (x_1, \ldots, x_d)$ in space,

$$\phi_i(x) = \prod_{j=1}^{d} \phi_{i_j}(x_j)$$

holds. In the case of the hierarchical basis, the functions ϕ_i are the standard multi-dimensional hat functions also used in FEM. This discretization is symmetric as long as the bilinear form $a(.,.)$ is self-adjoint. The discretization error can be estimated by the interpolation error of the sparse grid. Error bounds for the energy norm of the Laplacian are available and compare to the regular grid error, with a logarithmic deterioration $|\log h|$. However, sharp L_2 norm estimates are not known yet for the Galerkin solution, while pure best-approximation results in L_2 norm do not give the logarithmic deterioration $|\log h|$, see [35]. However, computationally some logarithmic deterioration is observed also in the L_2 norm, see [14].

There are two main drawbacks for a naive implementation of this approach: The stiffness-matrix $a_{i,j}$ is not sparse, unlike the FEM case, and the computation of f_j is quite expensive. As a consequence, the performance of the sparse grid degrades to the performance of a full h^{-d} grid. This is basically due to the fact that a lot of shape functions do have a large support. For the FEM discretization, these supports are small and only a limited number of shape functions might interact in the computation of $a(\phi_i, \phi_j)$.

The fix is not to assemble the stiffness matrix, but to use an algorithm for a matrix multiply or Gauss-Seidel step on the fly. This algorithm can be formulated in terms of tree traversals and has indeed linear complexity [54]. This means that if the number of matrix multiplies and Gauss-Seidel iterations is limited, for example by some suitable preconditioner or accelerator, the equation system can be solved in optimal complexity. Extensions of this algorithm to several dimensions and to certain types of variable coefficients have been developed, see [16] and references therein. The main difficulty with this respect is the symmetry of the discretization. The optimal order algorithms are usually based on the assumption of constant coefficients. The treatment of the variable coefficients case is complicated. In order to maintain optimal complexity, the coefficient function is approximated on the sparse grid. This approximation can cause asymmetries in the operator for a naive implementation. Further difficulties come from jumping coefficients not aligned to a grid axis and from more complicated differential operators.

3.3 The Finite Difference Method

A different way of a sparse grid discretization is a finite difference scheme [19,43], which we will employ further on in this paper. It is simpler to implement and to apply to different types of equations, but there is not that much known analytically.

We define the hierarchical transformation **H** as the hierarchical basis transformation on the regular grid from nodal values to hierarchical values, which are restricted to the sparse grid nodes. Both the nodal basis and the hierarchical basis span the space of piecewise linear functions. Hence any function of the space is uniquely represented as a linear combination of basis functions, and such a representation can be uniquely converted to the representation by another basis. However, computationally such a transformation has to be fast in order for the transformation to be useful. All wavelet-type basis functions provide such fast $\mathcal{O}(n)$ transformation to and from the nodal basis representation. The transformation is especially simple for the one-dimensional hierarchical basis: given the nodal values u_j with $j = 0, 1, \ldots, 2^{l+1}$, the hierarchical representation for interior points can be obtained by

$$\hat{u}_j = u_j - \frac{1}{2}\big(u_{\text{left father}} + u_{\text{right father}}\big), \tag{3}$$

and the boundary nodes u_0 and $u_{2^{l+1}}$ remain unchanged. The nodal values are replaced by their hierarchical excess or deterioration, compared to the value obtained by interpolation on the next coarser level grid. The inverse transformation can be implemented similarly:

$$u_j = \hat{u}_j + \frac{1}{2}\big(u_{\text{left father}} + u_{\text{right father}}\big). \tag{4}$$

However, the coarse nodes have to be computed before the finer grid nodes. Furthermore, the transformation can be implemented in place, without an auxiliary vector. The hierarchical basis transformation **H** is also abbreviated by the stencil [1/2 1 1/2].

Based on the hierarchical basis transformation **H**, we define the action of a one-dimensional finite difference operator for the discretization of a differential operator, see Figure 6. We apply the associated standard difference

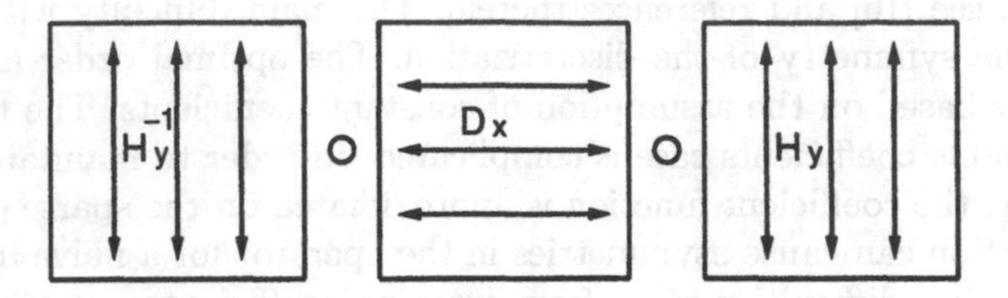

Fig. 6. Scheme for a finite difference operator in x-direction.

stencil $\mathbf{D}_i$ along the x_i-axis to values located on the sparse grid nodes in a specific basis representation. To this end the values are given in nodal basis in direction i and in hierarchical basis representation in all other directions $I \setminus \{i\}$. The associated transformation is denoted by $\mathbf{H}_{I\setminus\{i\}}$. The stencil $\mathbf{D}_i$ for each node itself is chosen as the narrowest finite difference stencil available on the sparse grid. It is equivalent to the corresponding stencil on a regular, anisotropic refined grid. The finite difference stencil can be a 3-point Laplacian $\frac{1}{h^2}[1\ -2\ 1]$, an upwind-stabilized convection term $\frac{\partial}{\partial x_i}$, some variable coefficient operators and so on. In nodal values the finite difference operator reads

$$\frac{\partial}{\partial x_i} u \approx \mathbf{H}_{I\setminus\{i\}}^{-1} \circ \mathbf{D}_i \circ \mathbf{H}_{I\setminus\{i\}} u \,. \tag{5}$$

A general difference operator is then obtained by dimensional splitting. A linear convection-diffusion equation, as a simple example, can be discretized in nodal basis representation as usual as

$$\nabla \cdot u \approx \sum_{i=1}^{d} \mathbf{H}_{I\setminus\{i\}}^{-1} \circ \mathbf{D}_i \circ \mathbf{H}_{I\setminus\{i\}} u \,, \tag{6}$$

where the one-dimensional difference operators $\mathbf{D}_i$ may be chosen as a two-point upwind stencil $c \cdot \frac{1}{x_i - x_{i-1}} \cdot [-1\ 1\ 0]$ (convection term) or plus the three point centered Laplacian $a \cdot \frac{1}{(x_{i+1}-x_{i-1})^2}[1\ -2\ 1]$ (diffusion term). On adaptively refined grids, the nearest neighbor nodes are chosen, which may lead to asymmetric stencils, i.e. non-uniform one-dimensional stencils. Further higher order modifications of the stencils have been tested, too. In the presence of a transport term in the equation, the unsymmetry is believed to be no problem. There are many ways to create discretizations of all kind of equations, e.g. for the Navier-Stokes equations [43] or some hyperbolic conservation laws [22].

There is not that much theory known for the finite difference discretizations. However, the consistency error has been analyzed for some model problems. It behaves like the consistency of regular grids. The second ingredient of a convergence analysis about the stability is still missing, but numerical experiments indicate that the stability deteriorates by a logarithmic factor,

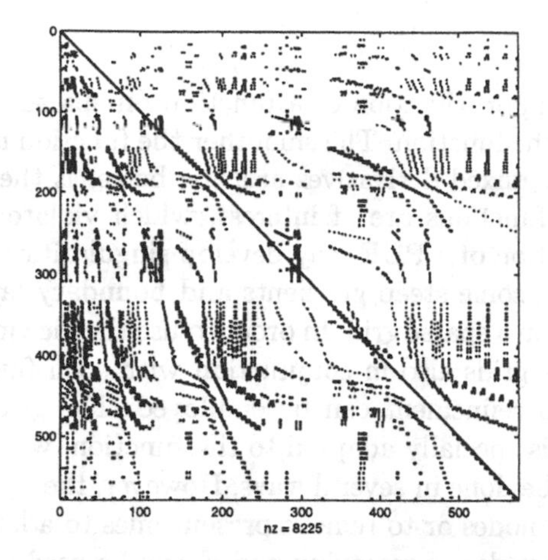

Fig. 7. Sparsity pattern of a 2D finite difference discretization.

which results in similar convergence results as the Galerkin method on sparse grids.

The algorithmic part of the finite difference discretization is much simpler than for the Galerkin method if one accepts the fact that the discretization is unsymmetric. As usual only the matrix multiply is implemented and the matrix is not assembled for complexity reasons. The sparsity pattern, see Figure 7, shows that the matrix is not really sparse and a matrix assembly would deteriorate the overall complexity. Analytically it is known that the average number of non-zero entries per row in the stiffness matrix with $N \approx 2^l \cdot l^{d-1}$ unknowns is roughly $2^l/l$ for the Galerkin method, i.e. the matrix is almost dense, see [5]. The finite difference discretization results in a lower average of l^{d-1} entries, which is still too much to be stored in a matrix. Due to unsymmetric finite difference discretization, any iterative solver for the equation solvers has to deal with the unsymmetry and methods like BiCGstab and GMRES are quite popular.

Basically three steps have to be implemented:

- transform a vector of nodal values to the hierarchical basis: $\mathbf{H}_{I\setminus\{i\}}$
- apply a one-dimensional finite difference stencil along the coordinate axis x_i: $\mathbf{D}_i$
- transform the vector back to the nodal basis $\mathbf{H}^{-1}_{I\setminus\{i\}}$

The basis transformation can be implemented as a sequence of one-dimensional transformations $\mathbf{H}_i$ and $\mathbf{H}_i^{-1}$, each along one coordinate axis x_i. Furthermore, we have to be able to sum up the resulting vectors. A complete PDE solver furthermore includes at least one iterative solver, some output or post-processing features and a treatment of different boundary conditions.

4 Adaptivity

The quality of the approximation of a function on sparse grids depends on the smoothness on the function. The smoother the function is, i.e. the smaller several higher order mixed derivatives are, the better is the sparse grid representation. Often functions are of interest, which violate this smoothness condition. The solution of a PDE may develop singularities at the boundary of the domain. Also some steep gradients and boundary layers are not efficiently represented on a sparse grid. In order to use the nice multi-dimensional properties of sparse grids also in conjunction with such functions and solutions, adaptive grid refinement can be employed. The grid is not longer a regular one, but it is specially adapted to the function, which has to be represented. This can be done in several ways. However, the general approach is to insert additional nodes or to remove present ones to adapt the grid. Both approaches, adding nodes or removing nodes, can be used to define the term of an 'adaptive sparse grid'.

However, such general definitions of sparse grids include all types of regular grids and their adaptively refined versions. A regular grid of level l can be obtained from a sparse grid, either by inserting nodes into a sparse grid of level l, or by removing nodes from a sparse grid of level dl, which contains the regular grid already. A more precise definition of adaptivity can be obtained if one describes the adaptive refinement algorithm instead of the grid.

4.1 A Priori Adapted Spaces

An a priori way of grid adaptivity is to optimize sparse grids towards special classes of functions. Classes of interest are spaces with special smoothness properties. Either a general higher-order smoothness is required, or a smoothness appropriate to the tensor product structure of the grid. This can be done by spaces like H^{2d}, where special smoothness restrictions are imposed on mixed derivatives of the function. Further optimization can be geared towards different error norms. From a best-approximation point of view, the sparse grids are optimal in an L_2 sense. However, another norm like the energy norm or equivalently the H^1 norm gives rise to other sparse grids constructed in [14]. Here emphasis is put on strong anisotropic grids, while some of the isotropic grids can be removed. This results in an overall number of nodes of h^{-1} and breaks the curse of dimension completely. Any other positive and negative Sobolev norm leads to a different sparse grid, which can be constructed systematically, see [20]. An extreme case is the splitting extrapolation grid of [32].

4.2 Adaptive Grid Refinement

A completely different type of adaptivity is the a posteriori approach: The sparse grid is refined locally in space, to resolve jumps, singularities and related features of a specific solution. The grid refinement is done during computation. The general approach to adaptive grid refinement consists of an error indicator or estimator, a selection procedure and a geometric grid refinement procedure. Along with a solution method for the equation system, this forms the ingredients of an iterative refinement cycle. This works very well for other types of grids such as unstructured grids based on triangles or tetrahedra, based on composite grids as an overlapping system of regular grids, and for more complicated unstructured grids with mixed element types. The theoretical foundation of error estimators is well understood in the elliptic case, given some regularity of the solution, see [49]. The grid refinement for a FEM scheme is also sound in the elliptic case.

However, for global approximation schemes like sparse grids, or transport equations, where a small error in one nodal point affects large parts of the domain and many other nodes, this method has to be modified. In the finite element case, an approximation is local and a change in one node does only affect its neighbor nodes immediately. A sparse grid discretization uses hierarchical basis transformations, which link whole lines of nodes. Hence, many more nodes can be affected. The discretization scheme is said to be non-local.

Let us take a closer look why the adaptive refinement cycle usually does work: regular parts of the solution, say H^2, do not need adaptive refinement. Additional refinement, which has been done accidently or due to refinement nearby, does not hurt, on the other hand. So we have to take care of the rough parts of the solution, in the elliptic case singularities of $H^{1+\alpha}$ type. The precise knowledge of the one-dimensional case for x^α singularities, along with their approximation by piecewise polynomials, gives rules for the optimal grading of a mesh. An optimal grid in d dimensions can be derived approximately from the 1D case. A point singularity requires a graded grid radially to the point, while tangentially the grid can be arbitrary. Line singularities require a mesh grading radial to the line, and so on. An adaptive grid refinement procedure will try to mimic this grid grading. This can be achieved by element bisection schemes, which do not refine steep enough compared to an optimal grid, but sufficiently for the efficient resolution of the singularity. An adaptive procedure, which bisects elements with large errors for example, can produce such sub-optimal grids. It is enough to refine all elements next to a node with a large error, because the elements contribute to the error, and a refinement decreases the local error. Accidental further refinement does not spoil the performance, loosely speaking, as long as the number of unnecessary nodes does not exceed the required nodes.

Let us return to sparse grids. They are a non-local approximation scheme. Given a node of the grid with a large error, there are a lot of elements and other neighbor nodes which contribute to the error. Their number is proportional to the grid level and may be large. This is in contrast for the FEM, where the number of neighbors is limited. Hence local refinement is quite expensive. Furthermore, any change in one node does affect large parts of the domain. It is no longer possible to localize the error and to approximate the elliptic differential operator by its diagonal part (i.e. the diagonal entries of the stiffness matrix). As a consequence, we cannot expect that the (expensive) local grid refinement does actually reduce the error of the node under consideration.

One way to improve the situation is to look at error estimators constructed for other global phenomena like transport equations. Here influence factors are computed instead of local errors, which describe the influence of a node onto the (global) error, see [9]. Instead of the local error, a number which indicates the influence of the respective node onto the error is used for refinement. Such indicators can be obtained from the solution of a dual problem based on the current solution of the original problem. An adaptive refinement procedure will refine the grid in the vicinity of nodes with a large influence onto the global error. This coincides with nodes of large errors in the elliptic case with a FEM discretization. However, for transport problems the location where the error occurs and where the error originates differ, because the error is transported along with the solution. Hence it is essential not to refine where the error is measured, but to remove the source of the error.

A similar procedure can be used for global approximation schemes of sparse grid type: given some global error criterion such as the H^1 or L_2 norm of the error, it is possible to compute the influence of each node onto the global error [15]. For efficiency reasons, the dual problem may be solved on a coarser scale than the original problem. However, the problem remains how to refine the sparse grid after selecting a set of nodes with a large error contribution. Due to the global support of many of the shape functions, it is not clear which nodes in the vicinity of the selected nodes should be created. It is even unclear what a vicinity in the presence of global functions means. Hence we have to introduce a further extension of the refinement procedure. The adaptive grid refinement procedure does work for FEM, because the differential operator is spectrally similar to its diagonal part. However, we are able to transform the equation system to a basis where the sparse-grid operator is almost a diagonal operator: the appropriate wavelet basis, where the condition number of the operator is a constant independent from the mesh parameter h and allows one to neglect the global coupling of the operator. Hence the coefficients of the estimate error, represented in this wavelet basis, lead to a better and provable refinement criterion [35]. The question of geometric grid refinement is therefore partially solved. However, the algorithmical side is still open.

Another approach to adaptive sparse grids are the shield grids [27]. The idea is to 'shield' the singularity by several boxes of nodes in order to avoid pollution outside the box. Standard sparse grids can be used outside. The grid is constructed as the union of nodes of a standard sparse grid and the boxes. Each quadratic box is aligned to the grid and contains all nodes on the outline of the square. The nodes are located with a distance $h_{\min}$. The effect is that all nodes, whose shape function support contains the singularity, are shielded, i.e. their next neighbors in all coordinate directions are very close, $h_{\min}$ away.

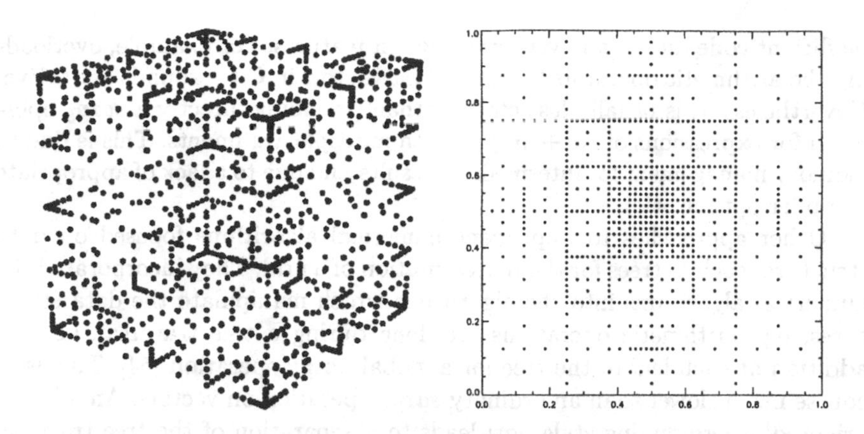

Fig. 8. Adaptive sparse grids. 'Standard' refinement (left) and a 'shield' grid (right).

Numerical experiments indicate the effectiveness of this approach for certain point singularities caused by a singular source term. There are also adaptive refinement procedures to construct such grids. However, the complexity of the grid, that is the number of grid points, is higher than standard sparse grids: The largest box has a side length of 1/2 and a surface of $2d\frac{1}{2}^{d-1}$ (in several dimensions). This means that the nodes required to cover the surface of the box with a mesh of nodes at a distance of h is $h^{-d+1} \log h$ compared to $h \log^{d-1} h$ nodes of a regular sparse grid. As a consequence the curse of dimension is present in this discretization and one of the advantages of sparse grids for higher-dimensional problems is lost. Furthermore, it does cause harm to add further nodes to a shield grid accidently near the singularity, because a point-type singularity might not be shielded by the box any longer. A conceptual difficulty lies in the global parameter $h_{\min}$ of the boxes for a specific grid, because grid refinement is often considered a local procedure. Think of a problem with several areas of refinement, where the grid needs different strength of mesh grading. A global parameter would be adjusted to the strongest refinement and would therefore spend too much effort on weaker refinement regions.

5 Software Abstractions

The goal for the development of our finite difference sparse grid code is to be able to test and to verify different types of discretizations on sparse grids and to tackle different types of partial differential equations. Hence a flexible and modular design is a must.

Techniques such as abstract data types and object-oriented programming provide such a flexibility, see [7,2]. However, they can easily lead to slow and inefficient code due to an over-use of design features. For example, overloading the arithmetic operators '+' and '*' for vectors in C++ sounds attractive. Nevertheless, it is usually less efficient than providing directly a *saxpy* operation for expressions of $\alpha v + w$ type with multiple arguments. This is due to memory management for intermediate results and due to a lack of appropriate compiler optimization.

Other approaches to implement numerical algorithms focused on data structures such as trees for the representation of a grid. They incorporated the numerical algorithms into the algorithms which manipulate the data structures, e.g. arithmetic operations are done during a tree traversal, i.e. one addition at each leaf of the tree for a global *saxpy* operation [54]. This is of course much slower than an ordinary *saxpy* operation on vectors. An 'object-oriented' programming style now leads to a separation of the tree traversal (as an iterator) and the arithmetic operation, resulting in additional overhead, see [39]. Hence, splitting a large code into many small functions and loops may lead to inefficiencies, which cannot be resolved by an optimizing compiler. A very careful code design is needed.

We have based our code on several fundamental higher-level abstractions, which are well separated both in functionality and in implementation. Each implementation of an abstraction itself is coded in a more classical style. This guarantees that we do not loose efficiency in a substantial way due to this separation. We have identified the following building blocks, see Figure 9 (left). They are ordered from low-level, computationally expensive and efficient, to high-level routines, where efficiency is achieved through the call of some efficient subroutines. Similar abstractions can be found in other object-oriented software packages for partial differential equations such as Diffpack, see [13,31].

- *vector*: a large container for real numbers, including *Blas* level 1 arithmetic
- *grid*: geometric description of an adaptively refined sparse grid. It provides I/O, refinement and addressing, and even standard grids
- *field*: a (solution) function on a sparse grid. It provides a mapping between a grid and a vector and gives an interpretation of the vector data as (collocated) scalar or vector field
- *operator*: the finite difference operators that are defined on a sparse grid and operate on vectors, see Figure 9 (right)
- *solver*: different iterative (Krylov) solvers, uses a differential operator

5.1 Numerical Data: Vectors

Large amounts of data are stored efficiently in vectors. We avoid its storage inside some other data structures such as trees or hash maps which will be explained in Section 5.2. The declaration of a vector class, based on a primitive C vector, which allows also for an efficient interface to Fortran subroutines, see [7], reads as follows

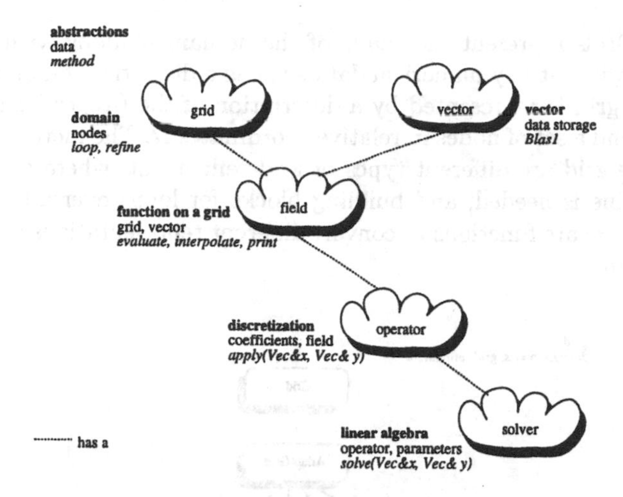

Fig. 9. Abstractions used in the sparse grid code.

```
class Vec {
protected:
  real *r;
  int n;
...
};
```

The member functions implement some management and the functionality of Blas level 1 operations such as the following vector plus vector operation. These member functions can either be optimized by the compiler or used to call an optimized library.

```
void Vec:: add(const Vec& a, const Vec& b)
{
  assert((a.n==n)&&(b.n==n));
  for (int i=0; i<n; i++)
    r[i] = a.r[i] + b.r[i];
}
```

Additionally a short vector *SVec* is defined. It provides a memory efficient way to handle e.g. coordinates. If the code is optimized for three-dimensional applications for example, the underlying data structure is *real[3]*. There are some administration member functions.

5.2 The Sparse Grid

One of the central abstractions is the grid, see Figure 10. Along with standard grids and adaptive girds with hanging nodes [23], there are sparse grids available. Grids represent the shape of the domain Ω along with the set of nodes, but without any numerical data such as a discretization or a function. The sparse grid is represented by a description of the (hyper-) cube shaped domain Ω and a set of nodes in relative coordinates Ω. The methods provided for a sparse grid are different types of grid refinement, where an indication what to refine is needed, and building blocks for loops over all nodes. Furthermore there are functions to convert different representations of nodes and its coordinates.

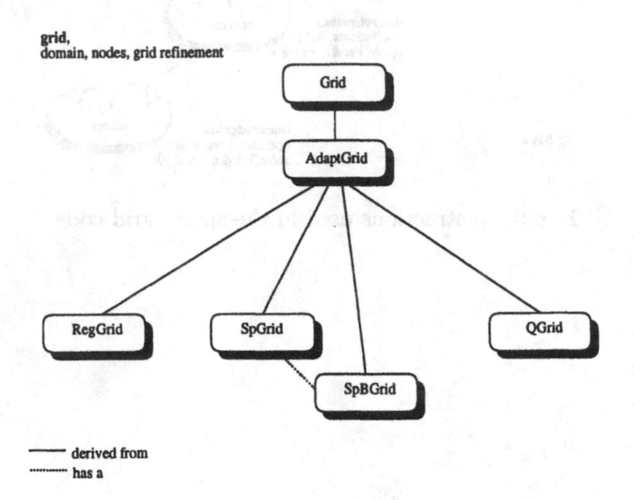

Fig. 10. The abstractions for the geometric properties of a grid.

The approach to store adaptive sparse grids has been so far tree data structures as described in [54]. Each node contains one pointer to its parent and two pointers to its children in each direction. This is in total an amount of $3d$ pointers per node. Using the pointers, different types of tree traversal can be implemented and many numerical algorithms developed for sparse grids are based on these algorithms. The large number of pointers requires a lot of memory and a lot of operations to administer them. For large scale problems however, one is especially interested in a low operation count and low memory consumption implementation. Of course there are modifications of the tree data structure to reduce these requirements slightly, see [16]. Imagine that an algorithm accesses all allocated memory in some random order, then the algorithm will be faster if one is able to reduce the amount of allocated memory. Furthermore, we are interested in the parallelization of such a code. Each pointer poses a potential problem, because the administration of a pointer pointing to data on a distant processor is troublesome.

So there are two goals for us to choose a different data structure: We want to get rid of pointers and we want to separate data storage from the algorithm. The first goal is accomplished by a key-based addressing and for the second one we choose a hash table, but there are of course alternatives, see [42].

Key-based addressing substitutes the memory address stored in a pointer with an integer value uniquely describing the entity. In our case, each node can be characterized by its position in space, that is the local coordinates. Let us begin with a one-dimensional scheme. We number the nodes of a regular grid level by level. At each level the additional nodes are given a number, while the other numbers stay the same. Depicted in Figure 11 are five levels of a one-dimensional grid. Any grid can be represented by the set of its nodes which are stored as keys. The key of a node and its coordinates can be uniquely converted into another. The nodes which are actually present in a grid will later be numbered in the sparse grid. This numbering can be very crude if the grid was created by adaptive refinement and the nodes were numbered in the order of their creation.

```
0                                           -1
0                     1                     -1
0         2           1          3          -1
0    4    2     5     1     6    3     7    -1
0  8 4  9 2 10  5 11  1 12  6 13 3 14  7 15 -1
```

Fig. 11. Key values of the nodes in five levels of a one-dimensional grid.

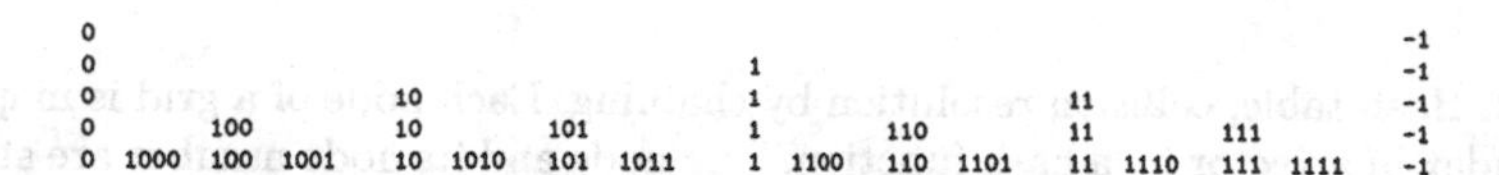

Fig. 12. Key values of the nodes in binary representation.

Looking at the binary representation of the keys in Figure 12, we see some more structure. Hence the level of a node can be computed by its binary logarithm, the relative coordinate is counted from left to right. Transformations from coordinate to key and backwards are available. In the multi-dimensional case, a coordinate vector can be transformed into a vector of numbers. The enumeration scheme, along with many others, can be used to name nodes uniquely.

Furthermore we need some storage mechanism, where nodes can be accessed by their respective keys. We choose a hash map, where the key is mapped to an address by a hash function. Hashing is a very general storage concept in computer science, see [29], often used in data base systems, but also in compilers or e.g. the Unix shell. It is used to store and retrieve large amounts of data without relying on any special structure or distribution of the data. A very general universe of all possible key values is mapped to a finite number of cells in an ordinary vector in memory by a hash function. The key and its related data is stored in that vector cell, given that no other key occupies it, which is called a collision. In that case some collision resolution mechanism assigns another storage location to the key. The hash implementation of SGI's version of the STL [4] we use is based on a collision resolution scheme by chaining, see also Figure 13. Keys mapped to the same address are stored in a linked list at that address. The overall performance of the hash storage scheme depends on the data and the hash function. A nearly statistically random distribution of the nodes to the hash table entries is the ideal case. Many heuristic hash functions work very well in this respect.

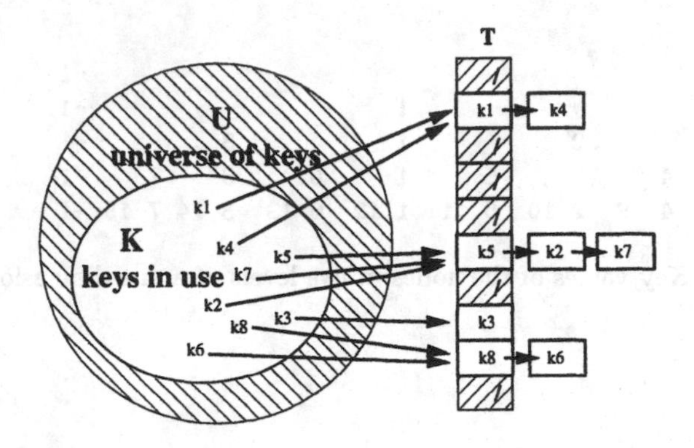

Fig. 13. Hash table, collision resolution by chaining. Each node of a grid is mapped to an index in a vector by a hash function. The node and its node number are stored in the vector.

We have mentioned PDE solvers where the numerical data are stored in a tree data structure. There has also been an approach to store the numerical data directly in a hash table, see Figure 14 (top) and [43]. This means that all numerical algorithms are tied to the hash table. Even operations which are independent of any geometric information of the grid or the nodes such as the saxpy operation have to loop trough the hash table. This means unnecessary overhead and a performance penalty for this type of operation. Furthermore, for software design reasons one is interested to separate the geometric information from the algebraic data, i.e. the sparse grid from the vectors. A fixed amount of storage related to each hash table entry limits the implementation of numerical algorithms such as a BiCGstab solver or the solution of systems of PDEs.

Assume that we want to run a conjugate gradient iteration. In addition to the solution vector and the right hand side, five auxiliary vectors are required. We have to reserve five scalar variables in each hash table entry for this purpose, if we actually store data in the hash table. Assume furthermore that there are auxiliary fields for grid refinement, for the differential operator, for coefficients and so on. It is quite difficult to know in advance how much auxiliary memory is needed in each hash table entry. Hence standard vectors separated from the hash table, which can be allocated and deleted easily, increase the flexibility of the overall code, see Figure 14. The hash table maps the node to a number, which serves as an index for all vectors. Hence floating point vectors, bit-fields and integer vectors can be addressed in the same way, without the danger of memory alignment problems.

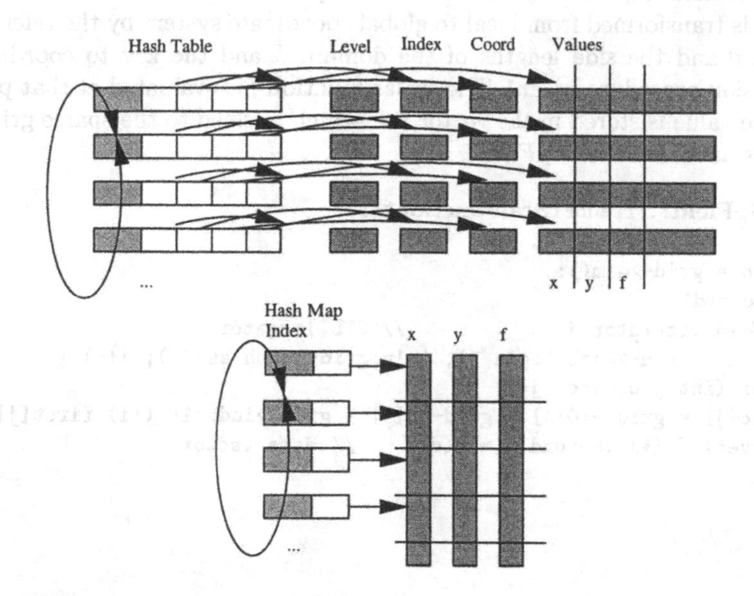

Fig. 14. Numerical data storage within a hash table (upper half) and indexing of a hash table into standard vectors of different types located elsewhere. In the first case all numerical algorithms have to access the hash table in order to manipulate the data. In the second case however, numerical data is separate and some algorithms such as **saxpy** do not have to touch the hash table. Geometric (grid) and algebraic data (vectors) are separate.

5.3 Data on a Grid: Field

The solution of the partial differential equation, the right hand side and other functions represented on the sparse grid are stored as *fields*. The numerical data is contained in a *vector*, while the geometry is stored in a *grid*. The *field* abstraction glues both together and gives an interpretation of the vector as a continuous function. There are scalar fields and vector fields, both aligned to the grid (collocated) and in a staggered grid configuration, see Figure 15. The *field* abstraction contains an interpolation procedure and it can perform I/O. Several output formats for different graphics packages are implemented for several types of fields. An input procedure is given below, which evaluates a scalar function on all nodes of a grid and stores the values at the appropriate locations in the numerical vector. The sparse grid is represented as a hash map, where each key is mapped to its node number. The STL implementation provides an iterator (*SpGrid::iterator*) to loop through all nodes in the hash table. The iterator points to a tuple of values, the coordinate vector and the node number, which can be accessed as *first* and *second*. The coordinate vector is transformed from local to global coordinate system by the reference point *x0* and the side lengths of the domain *h* and the key to coordinate conversion procedure *index1*. The scalar function *f* is evaluated at that point and the value is stored in the vector *vec*, which is glued to the sparse grid by this instance of a field *SpField*.

```
void SpField:: readSF(ScalFunction& f)
{
  int d = grid->dim();
  SVec c(d);
  SpGrid::iterator i;                    // STL iterator
  for (i = grid->hash.begin (); i != grid->hash.end (); i++) {
    for (int j=0; j<d; j++)
      c[j] = grid->x0[j] + grid->h[j] * grid->index1( (*i).first[j] );
    (*vec) [ (*i).second ] = f(c);     // data vector
  }
}
```

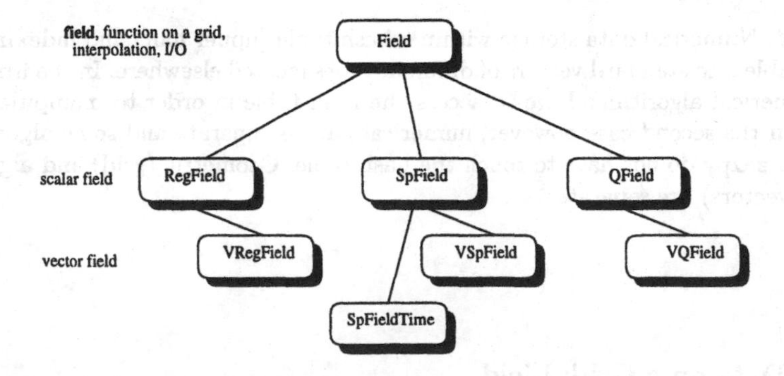

Fig. 15. The abstractions for the functions defined on a grid.

The finite difference discretization on sparse grids is defined by hierarchical basis transformations $\mathbf{H}$ of a vector, one-dimensional finite difference stencils $\mathbf{D_i}$ and back-transforms $\mathbf{H}^{-1}$, see formula (5). The vectors are stored as instances of the class *Vec*, the transformations defined on a sparse grid are implemented in the *field* class and the difference stencils are located in the *operator* classes. The hierarchical basis transformation $\mathbf{H}$ and its inverse $\mathbf{H}^{-1}$ are coded within the *field* classes, because they depend heavily on the type of function representation encoded there. Furthermore, they are used in other parts of the code besides the finite difference operators. The transformation $\mathbf{H}$ is heavily used for the definition of finite difference operators. The implementation of the one-dimensional transformation of the nodal to the hierarchical basis representation is given as an example and follows formula (3). The vector $\mathbf{x}$ contains the nodal representation and the hierarchical representation is stored in $\mathbf{y}$. The transformation is done along coordinate axis *dir*.

```
void SpField:: node2hierDir(const Vec& x, Vec& y, int dir) const
{
  assert(x.dim() == y.dim());
  for (SpGrid::iterator i = grid->begin(); i != grid->end(); i++)
    y[(*i).second] = x[(*i).second] - interpolDir(x, dir, (*i).first);
}

real SpField:: interpolDir(const Vec& x, int n, SIVec p) const
{
  if ((p[n] != 0)&&(p[n] != -1)) {     // boundary node?
      int il, ir;
      grid->neighbour(p[n], il, ir);  // compute neighbour keys
      p[n] = il;
      SpGrid::const_iterator l = grid->find(p);  // STL lookup
      p[n] = ir;
      SpGrid::const_iterator r = grid->find(p);
      return .5 * ( x[ (*l).second ] + x[ (*r).second ] );
      // interpolation
  }
  return 0.;
}
```

5.4 Differential Operators

The heart of the sparse grid research code is the partial differential operators. The straightforward representation of a discretized operator is a matrix, which is one implementation (a dense and a sparse matrix). Of course this matrix has to be initialized (or assembled). However, the number of non-zero entries for large sparse grids is prohibitively large and mainly only the *apply* method of the operator is needed (i.e. matrix multiply), as in conjugate gradient-like solvers. Hence, another implementation of the operator is the finite difference sparse grid operator which implements '*apply*' on the fly. The sparse grid finite differences are implemented as defined in formula (5), i.e. by the hierarchical basis transformation $\mathbf{H}$ and back-transform $\mathbf{H}^{-1}$ coded within the *SpField* classes and the one-dimensional finite difference stencils $\mathbf{D}_i$ which are implemented in the various operator classes. Further implementations of operators include finite differences and finite elements on regular grids.

The *operator* classes are based on a certain type of function representations, given as a *field* class. Furthermore the hierarchical basis transform of the field is used. The *operator* abstraction contains some parameter data, such as the coefficient functions and references to a *field* and the underlying sparse *grid*. The class provides a method for applying the operator to some vector. This matrix multiplication is implemented here, calling the necessary basis transforms and applying finite difference stencils. Each PDE results in its own operator class.

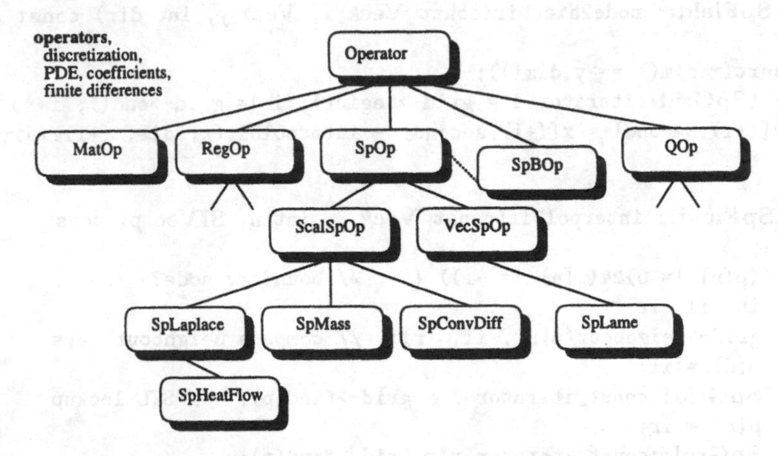

Fig. 16. The discretized differential operators. The abstract *Operator* class is implemented as a matrix based operator, a finite element discretization on a regular grid, the sparse grid finite difference methods *SpOp*, a block grid version of sparse grids *SpBOp*, and a general finite difference discretization on adaptive grids *QOp*. There are scalar equations *ScalSpOp* and systems of equations *VecSpOp* and model problems like the Laplace equation, the convection-diffusion equation or the equations of linear elasticity.

Our operator works on vectors instead of fields. There are several reasons for this choice. We are using the operator abstraction in a pure linear algebra context in a linear solver, where a *field* does not make sense. Furthermore, operating directly on vectors improves the performance of the code and we do not want to implement operators where the source and target field are represented on different types of grids. However, it would be possible to create a grid which is independent of the operator class and which operates on arbitrary fields. This would lead to more general concepts of algorithms.

5.5 Linear Algebra

The goal of the linear algebra abstraction of an iterative Krylov solver is to provide a collection of algorithms, see Figure 17. The abstraction is based on *vectors* and *operators*.

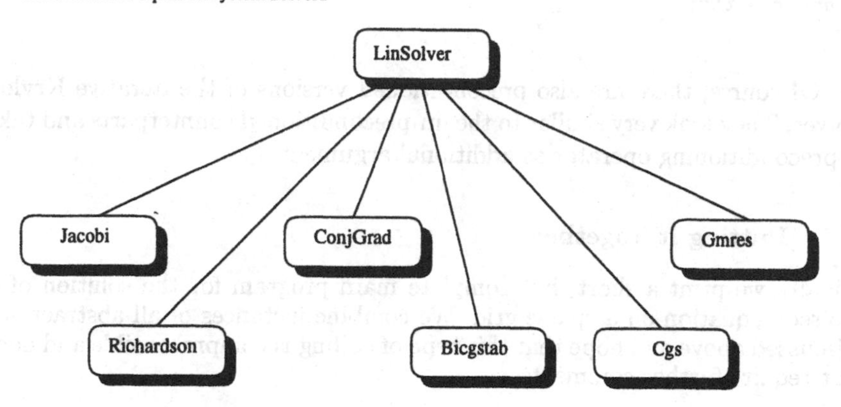

Fig. 17. The iterative equation solvers.

We present the implementation of a conjugate gradient method to demonstrate that the code looks pretty much like a pseudo-code from a numerical analysis text book. A reference to the operator is stored as *op* and the vectors are fed as arguments *b* and *x*.

```
void ConjGrad:: solve(const Vec& b, Vec& x)
{
  assert(b.dim()==x.dim());
  int n = x.dim();
  Vec r(n), p(n), ap(n);
  iter = 0;
  real beta, alpha, rtr, rtrold;
  op->apply(x, ap);              // ap = A * x
  p.sub(b, ap);                  // p  = b - ap
  r.copy(p);                     // r  = p
  rtr = r.prod(r);               // rtr = <r,r>

  while (rtr>tol) {
    op->apply(p, ap);            // ap = A * p
```

```
    alpha = rtr / p.prod(ap);    // alpha = rtr / <p, ap>
    x.add(x,  alpha, p);         // x = x + alpha * p
    r.add(r, -alpha, ap);        // r = r - alpha * ap
    rtrold = rtr;
    rtr = r.prod(r);             // rtr = <r,r>
    beta = rtr / rtrold;
    p.add(r, beta, p);           // p = r + beta * p
    iter++;
    printIter("ConjGrad it ", " res= ", rtr);
    if ((iter>=maxiter)||(error>=HUGE)) break;
  }
  error = rtr;
}
```

Of course, there are also preconditioned versions of the iterative Krylov solver. They look very similar to the un-preconditioned counterparts and take a preconditioning operator as additional argument.

5.6 Putting it Together

Finally we print a short, but complete main program for the solution of a Poisson equation on a sparse grid. We combine instances of all abstractions discussed above and hope that this type of coding is comprehensible and does not require further comments.

```
main() {
  int dim = 2;
  SVec x0(dim), x1(dim);          // lower left and upper right corner
  x0.set(-1.0);
  x1.set(1.0);
  SpGrid grid(dim, x0, x1);       // sparse grid on the square
  grid.refineAll();
  uint n = grid.nodes();
  Vec x(n), f(n);
  x.set(0.0);
  f.set(1.0);
  SpField field(x, grid);         // scalar solution field
  SpLaplace lap(field, 1.0);      // differential operator, coefficient
  Bicgstab j;                     // iterative solver
  j.attach(lap);
  j.setTol(1e-8);
  j.setMaxIter(10000);
  j.setVerbose(1);
  j.solve(f, x);                  // solve equation system
  cout<<j<<endl;                  // write statistics
  ofstream of("lap.vtk");
  field.print(of, Grid::vtk);     // dump solution in Vtk format
}
```

6 Parallel Algorithms

The parallelization of an adaptive code usually is non-trivial and requires a substantial amount of code for the parallelization only. In this respect the parallelization of the sparse grid extrapolation method has a big advantage. If we are interested in a parallel version for adaptive sparse grids, however, we have to consider a more complicated approach to be described now.

Hierarchies of refined grids, where neighbor elements may reside on different processors, have to be managed [12]. That is, appropriate ghost nodes and elements have to be created and updated, when the parallel algorithm performs a communication operation. This happens both in the numerical part, where an equation system is set up and solved, and in the non-numerical part, where grids are refined and partitioned, see also [8,28].

The key point of any dynamic data partition method is efficiency. We look for a cheap, linear time heuristic for the solution of the partition problem. Furthermore the heuristic should parallelize well. Here, parallel graph coarsening is popular. It results in a coarser graph on which then a more expensive heuristic on a single processor can be employed. However, graph coarsening is at least a linear time algorithm itself and lowers the quality of the heuristic further. This is why we look for even cheaper partition methods. They are provided by the concept of *space-filling curves.*

6.1 Space-Filling Curves

First we have to define curves. The term curve shall denote the image of a continuous mapping of the unit interval to the $\mathbb{R}^d$. Mathematically, a curve is space-filling if and only if the image of the mapping does have a classical positive d-dimensional measure. The curve fills up a whole domain. For reasons of simplicity we restrict our attention to a simple domain, namely the unit square. We are interested in a mapping

$$f \ : \ [0,1] =: I \ \mapsto Q := [0,1]^2, \qquad f \text{ continuous and surjective.}$$

One of the oldest and most prominent space-filling curve, the Hilbert curve, can be defined geometrically [26], see also [40]. If the interval I can be mapped to Q by a space-filling curve, then this must be true also for the mapping of four quarters of I to the four quadrants of Q, see Figure 18. Iterating this subdivision process, while maintaining the neighborhood relationships between the intervals leads to the Hilbert curve in the limit case. The mapping is defined by the recursive sub-division of the interval I and the square Q.

Often some curves as intermediate results of the iterative construction procedure are more interesting than the final Hilbert curve itself. The construction begins with a generator, which defines the order in which the four quadrants are visited. The generator is applied in every quadrant and their sub-quadrants. By affine mappings and connections between the loose ends of the pieces of curves, the Hilbert curve is obtained.

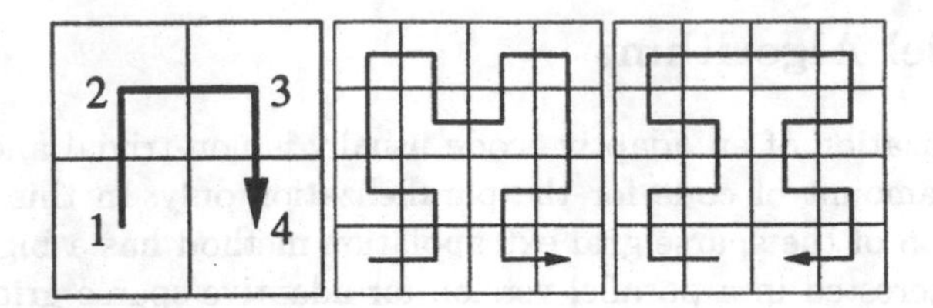

Fig. 18. Construction of a Hilbert space-filling curve. The open and the closed curve.

There are basically two different Hilbert curves for the square, modulo symmetries, see Figure 18. An open and a closed space-filling curve can be constructed by the Hilbert curve generator maintaining both neighborhood relations and the sub-division procedure. The resulting curve is indeed continuous and space-filling. Although each curve of the iterative construction is injective and does not cross itself, the resulting Hilbert curve is not injective. This can be demonstrated easily by looking at the point $(1/2, 1/2) \in Q$, which is contained in the image of all four quadrants. Hence several points on I are mapped to this point of Q. In three dimensions, there are 1536 versions of Hilbert curves for the unit cube [1].

The Hilbert curve may be the most prominent curve. However it is not the oldest one. The discovery of 'topological monsters', as they were called initially, started when Cantor in 1878 proved the existence of a bijective mapping between two smooth, arbitrary manifolds of arbitrary, but finite dimension. A year later Netto proved that such mappings are non-continuous. This means that they are not curves according to our previous definition. However, without the property of an injective mapping, Peano constructed a space-filling curve in 1890, see Figure 19. Later on several other curves were constructed by Hilbert, Moore, Lebesgue and Sierpinski and many others.

The Peano space-filling curve can be constructed by a sub-division scheme of the square Q into nine squares of side length $1/3$. The generator can be given by the order of the nine squares. In contrast to the Hilbert curve there are 273 different curves, which maintain neighborhood relations. The original construction of the Peano curve was based on a triangulation of the domain and a bisection scheme for triangles, which defines the order in which the sub-triangles are visited. We will use this scheme later on.

There are two basic differences of the Lebesgue space-filling curve compared to the previous curves: The curve is differentiable almost everywhere, while the previously mentioned curves are not differentiable anywhere. The Hilbert and Peano curves are self-similar, a feature they share with fractals: sub-intervals of the unit interval I are mapped to curves of similar structure than the original curve. However, the Lebesgue curve is not self-similar. It can be defined on the Cantor set C. This set is defined by the remainder of the interval I, after successively removing one third $(1/3, 2/3]$. Any element

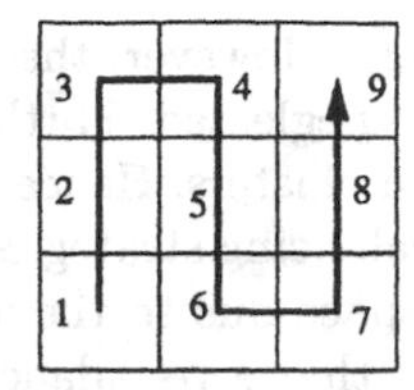

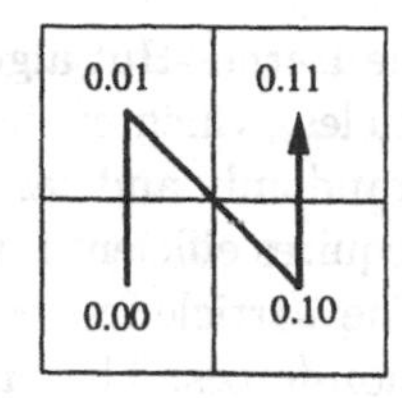

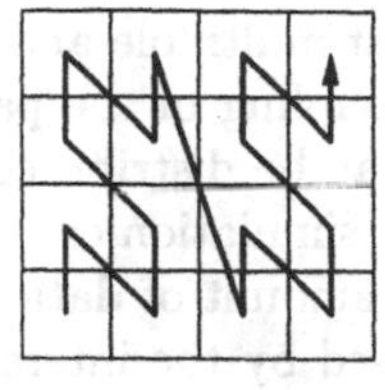

Fig. 19. The construction of a Peano (left) and Lebesgue curve (middle and right).

$x \in C$ of the Cantor set can be represented as a number in base-3 expansion, $0_3.t_1t_2t_3\ldots$ where all digits t_i are zeros and twos $t_i \in \{0,2\}$, because the ones have been removed by the construction of the Cantor set. The mapping of the Lebesgue curve is defined by

$$f\left(0_3.(2d_1)(2d_2)(2d_3)\ldots\right) := \begin{pmatrix} 0_2.d_1d_3d_5\ldots \\ 0_2.d_2d_4d_6\ldots \end{pmatrix} \text{ with binary digits } d_i \in \{0,1\} .$$

The function f defined on the Cantor set C is extended to the unit interval I by linear interpolation. The generator of the Lebesgue curve looks like in Figure 19. Although constructed by digit shuffling, the curve is continuous and a space-filling curve. The order of the quadrants imposed by the generator of the curve can be found in the depth-first traversal of oct-trees and related algorithms.

6.2 Applications of Space-Filling Curves

Space-filling curves were originally created for purely mathematical purposes. However, nowadays there are numerous applications of space-filling curves. Basically, multi-dimensional data are mapped to a one-dimensional sequence. This mapping is useful for load-balancing of parallel simulations on a computer, for data locality of memory or disc accesses inside a computer in geographic information systems [3], for finding shortest paths in optimization [6,11] and ordering data in computer graphics [37,52,50], and in other applications [10].

Another interesting application of space-filling curves are particle or n-body problems which are defined by the interaction of n entities by some interaction forces. This model describes different phenomena like the movement of planets or dust under gravitational forces in astrophysics and the dynamics of atoms or groups of atoms in molecular dynamics. The number of particles of interest easily reaches the range of $10^6 - 10^9$. The model leads to a system of n ordinary differential equations. The right-hand side of each equation consists of the $n-1$ forces of particles, which interact with the particle under consideration. Usual model forces decay with the distance of the particles, which can be exploited by efficient approximation algorithms like the fast-multipole and the Barnes-Hut algorithm. However, there is still a global coupling of the particles, which cannot be neglected. Furthermore, particles can be distributed randomly and can form clusters. Hence a parallel particle simulation code requires efficient load-balancing strategies for the enormous amount of data. The particle moves in space due to the acceleration imposed by the interaction-forces. This means that a re-balancing or a dynamic load-balancing is needed for parallel computing, which can be done by space-filling curves [51,41,36].

A slightly different situation can be found in adaptive discretizations of partial differential equations. Now a grid consisting of n nodes and elements or volumes has to be distributed to a parallel computer. The nodes and elements can be found at arbitrary positions (completely unstructured grids) or at fixed positions, which are a priori known (structured grids) or at least computable by grid-refinement rules from a coarse grid (adapted grids). The degrees of freedom are coupled locally, usually between neighboring nodes. However, algorithms for the solution of the resulting equation systems couple all degrees of freedom together, which imposes difficulties on the parallelization of the solver. For the solution of stationary problems, no nodes are moved in general. However, due to adaptive grid refinement, new nodes are created during the computation. This requires dynamic load-balancing, which can also be done by space-filling curves [34,36,23,38].

6.3 Parallel Sparse Grid Algorithms

First of all we want to convert the sequential finite difference sparse grid code into a parallel code. There will be some extensions and slight modifications, but the general design and most of the lines of code are retained. The parallel version, run on a single processor, will only introduce very little overhead. However, the parallel version run on a larger distributed memory parallel computer with p processors should ran ideally at p times the speed of the sequential code, at least for problems large enough, where sufficient computational load can be distributed to the processors. We are interested in a high number of processors. Hence we consider a message passing programming model, where each processor can access its own memory only. Smaller shared memory parallel computers offer more convenient programming models and have been used for a similar code based on automatic parallelization by loop-parallelism, see [43].

In the previous section we have introduced a key-based node addressing scheme, where each node is stored in a hash map. The advantage in the sequential version was simplicity and little administration overhead. The parallel version is based on the distribution of nodes to the processors. Each processor owns a subset of the sparse grid. Each node is present on exactly one processor. The appropriate process stores the node in a local hash map. The difference to the sequential code is that the hash map does not contain all nodes any longer, but the nodes owned by the processor. Additionally, at some stages of the algorithm there are ghost nodes present, which contain the values of nodes belonging to other processors, but whose values are required by the algorithm. The values of the ghost nodes are updated or filled in a communication step prior to the actual computation. During the creation of such ghost nodes, processors have to communicate which nodes are required: Processor 1 determines which ghost nodes it needs. It finds out that some of the nodes originally belong to processor 2 and asks for them. Hence processor 2 knows that during the communication step, it has to send this data to processor 1. These negotiations would be quite complicated in a code based on pointers, because a reference to local memory does not make sense on another processor's memory. However, using keys, i.e. a unique id for each node derived from its coordinates, there is no problem at all. The global key is understood by all processors and can be used for all kind of requests. Furthermore, the space-filling curve, which provides a unique mapping of nodes to processors, immediately reveals which processor to ask for a node. Other grid partitioning heuristics in contrast would require a substantial bookkeeping effort to decide where a node belongs to.

One detail is still missing: In an adaptive sparse grid it is not exactly clear where to look for a neighbor node, while father and son nodes are determined. Hence there has to be some searching procedure for the appropriate neighbor node. Several requests for potential nodes may have to be raised. The node might or might not exist and might or might not be the nearest node in a certain direction. Note that these requests might be addressed to different processors, because each requested node can belong to another processor.

We list some hints on the parallel version of the stages of the sparse grid discretization and equation solver:

- Krylov iteration: Conjugate gradient-type Krylov methods like BiCG, BiCGstab and CGS can be parallelized such that the parallel version of an iterative Krylov method looks exactly like the sequential one. Each processor operates on the unknowns related to its own nodes. The Blas level one operations of *saxpy* type do not require communication. The communication of a Blas level one scalar product can be implemented by local summation and a standard reduce operation over all processors. This communication library call is hidden in the vector classes scalar product in our implementation. The main source of trouble is the application of the operator (matrix multiply) and a preconditioner, which is implemented separately.
- Finite difference operator: The operator

$$\sum_{i=1}^{d}\Big(\prod_{j=1,\ j\neq i}^{d}\mathbf{H}_j^{-1}\Big)\circ\mathbf{D}_i\circ\Big(\prod_{j=1,\ j\neq i}^{d}\mathbf{H}_j\Big) \tag{7}$$

  is composed of three basic operations, the transform to hierarchical basis $\mathbf{H}_j$, the one-dimensional finite difference stencil $\mathbf{D}_i$ and the transform back to nodal basis $\mathbf{H}_j^{-1}$, which are implemented separately, see also Section 3.3.

- Transform to hierarchical basis $\mathbf{H}_j$: each processor computes the values related to its own nodes. Prior to the computation, in a communication step the required ghost nodes are filled. The ghost nodes for this operation are determined by the direct parent nodes of nodes on the processor.
- Transform to nodal basis $\mathbf{H}_j^{-1}$: this operation can be done in place and requires more communication than the previous one. The sequential implementation cycles through a tree top down, so that the parent nodes are processed before their children. A straightforward parallelization would be to insert a communication step before each tree level is traversed. However, this results in a number of communication steps (= communication latencies) proportional to the maximum number of levels which is unacceptable for large sparse grids.
  We propose an alternative implementation here which is based on a single communication step before the computation: along with the parents of a node, the whole tree of their grand-parents and so on are required as ghost nodes on a processor. When the ghost nodes are filled, the computation can be done top down, such that the values on all nodes owned by the processor and additionally their parents, grand-parents and so on are computed. Hence, this implementation requires a larger amount of computation and a larger volume of communication than the straightforward version. However, the overall execution time is smaller because the number of communication steps is reduced to one.
- Finite difference operator $\mathbf{D}_i$: first the appropriate ghost nodes for the difference stencils are filled and afterwards the stencils are applied to all nodes, which belong to the processor. The main point here is the searching procedure for the neighbor nodes that are necessary for adaptive refined sparse grids. We create the necessary ghost nodes so that the sequential search algorithm can be re-used in this situation.

– Adaptive grid refinement: Following some refinement rules, new nodes are created. This can be done also in parallel. Afterwards, a repartitioning has to take place, which is also responsible for the elimination of multiple instances of a node that might have been created during the grid refinement. The partitioning of nodes is done by the space-filling curve heuristic and can be implemented as cutting a sorted list of all nodes into equal sized pieces. Of course, one has to avoid storing all nodes on a single processor. That is why the sorting is also executed in parallel by a bucket sort algorithm. The old partitioning serves as the buckets, each one mapped to one processor. Afterwards the new partitioning is computed. This procedure can be run completely in parallel and scales very well. The execution time of this repartitioning step usually is so low that it is below .01 of the execution time spent in the numerical algorithms, see also [55].

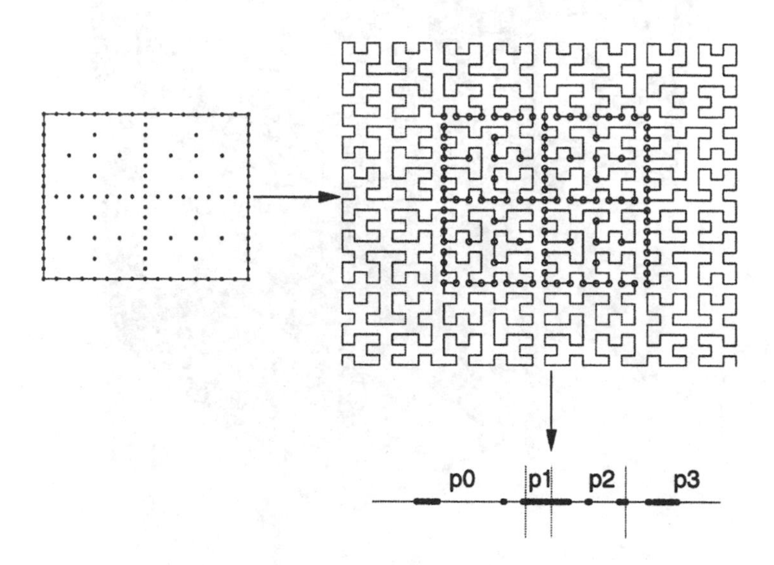

Fig. 20. An example of a sparse grid partitioned and mapped to four processors. The sparse grid (left) can be represented by its nodes in coordinate space. A Hilbert type space-filling curve, which fills the domain, is laid over the grid (right). Each node lies on the curve. Now we straighten the space-filling curve with the nodes fixed to the curve (bottom). The interval is cut into four sub-intervals assigned to one processor, each containing the same number of nodes.

All numbers reported are scaled CPU times measured on our parallel computing cluster 'Parnass2'. It consists of dual processor Pentium II 400MHz BX boards with at least 256 Mbytes of main memory, interconnected by a Myrinet network in a fat-tree configuration. The MPI message passing protocol implementation Mpich-PM showed a bandwidth between each two boards of 850 Mbit/s, see also [45].

In the first test we consider the solution of a three-dimensional convection-diffusion equation discretized on sparse grids with up-winding (standard refinement). Table 2 shows wall clock times for the solution of the equation system on a sparse grid of different levels using different numbers of processors.

For a fixed number of processors, we observe a scaling of a factor slightly above 2 from one level to the next finer level, which corresponds to a similar factor of increase in the amount of unknowns on that level. Furthermore, for a fixed level the measured times scale roughly with $1/p$ of the number of

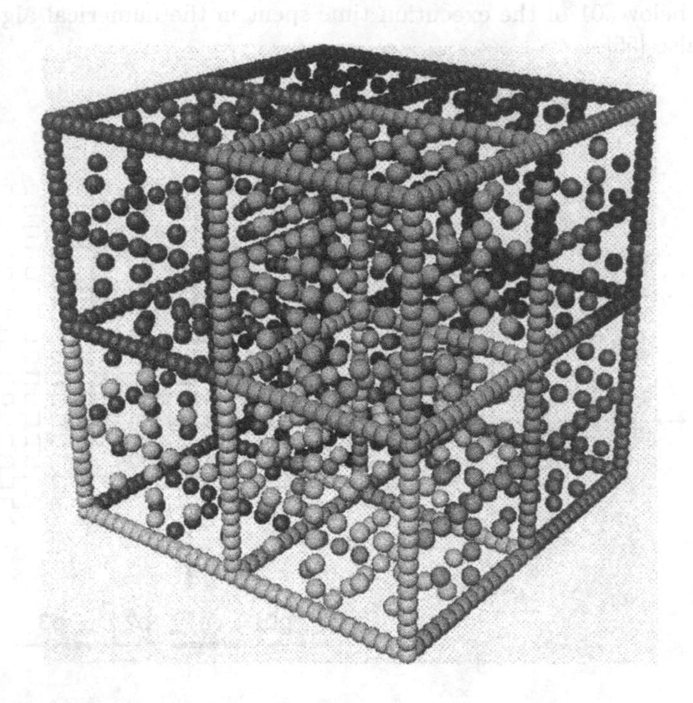

Fig. 21. A three-dimensional sparse grid, partitioned to eight processors (color coded).

time		processors						
nodes	1/h	1	2	4	8	16	32	64
81	4	0.03	0.03	0.07	0.08	0.11		
225	8	0.12	0.09	0.09	0.11	0.16	0.20	0.20
593	16	0.63	0.41	0.33	0.32	0.38	0.44	0.53
1505	32	3.78	2.29	1.60	1.34	1.26	1.33	1.53
3713	64	22.1	13.3	8.79	6.39	5.17	4.47	4.47
8961	128	68.1	40.7	24.8	16.2	11.9	8.89	7.56
21249	256	201	119	66.1	40.1	28.0	18.6	13.5
49665	512	575	379	169	106	71.6		28.0
114689	1024	1630			275	179		62.6

Table 2. Parallel execution times for a 3D convection-diffusion problem on sparse grids. Solution of the equation system in seconds on Parnass2.

processors p up to a parallel efficiency of 0.4 for 64 processors. However, the 32 and 64 processors perform efficiently only for sufficiently large problems, i.e. for problems with more than some thousands degrees of freedom. If we fix the amount of work, that is the number of nodes per processor, we obtain the scale-up. Comparing a time at one level l and a number of processors p with the time of one finer level $(l+1)$ and $2p$ processors, we obtain very good scaling of the method. Note that in this case of uniform grid refinement, some a priori partition schemes would be superior to our dynamic partitioning scheme. However, our dynamic load balancing scheme performs well and introduces only little overhead and results in good partitions, see Figure 21.

In the next test we consider adaptively refined sparse grids for a problem with singularities, where the sparse grids are refined towards a singularity located in the lower left corner, see also Figure 8. Table 3 depicts times in the adaptive case. These numbers give the wall clock times for the solution of the equation system again, now on different levels of adaptive grids and on different numbers of processors. Due to the solution-dependent adaptive refinement criterion, the single processor version contained slightly more nodes, indicated by *. For the same reason, the equation systems have been solved up to rounding error instead of the weaker discretization error condition in the uniform sparse grid experiment.

time		processors					
nodes	1/h	1	2	4	8	16	32
81	4	0.03	0.04	0.05	0.07	0.11	
201	8	0.07	0.05	0.05	0.07	0.08	0.09
411	16	0.21	0.13	0.12	0.13	0.17	0.20
711	32	0.78	0.48	0.38	0.36	0.41	0.51
1143	64	2.60	1.49	1.06	0.93	0.92	1.14
1921	128	8.69	5.99	3.70	2.88	2.70	2.83
3299	256	39.3*	20.7	13.8	9.62	7.79	7.32
6041	512	177*	91.0	56.8	39.5	28.6	22.0
11787	1024	949*	525	271	177	138	88.2
22911	2048			1280	761	660	

Table 3. Parallel execution times for adaptive sparse grids. A 3D convection-diffusion problem is solved and the solution times in seconds on Parnass2 are given.

We obtain a good scaling, both in the number of unknowns and in the number of processors, i.e. the times are proportional to the number of unknowns for a fixed number of processors and are inverse proportional to the number of processors. Increasing the number of processors speeds up the computation accordingly. The parallel efficiencies are somewhat smaller than for the uniform refinement case, due to the imbalance in the tree of nodes. This is also the case for other parallel adaptive methods. Hence this parallelization approach does perform very well, even in the range of higher number of processors 16 and 32, where a number of other strategies are not competitive.

7 Application to Time-Dependent Problems

Sparse grid discretizations have been applied so far to stationary problems and to time-dependent, transient problems. Besides methods based on the extrapolation method, where standard time-dependent codes can be re-used, there are also attempts to solve parabolic problems [5] and Navier-Stokes equations [43] by native sparse grid discretizations. We will discuss some features of a sparse grid finite difference discretization of space and time for scalar parabolic and hyperbolic equations.

We are interested in the numerical solution of

$$u_t + \nabla \cdot f(u) = q(u) \quad \text{for } u(x,t), \quad x \in \Omega \subset \mathbb{R}^d, \quad t \in [0, t_0], \tag{8}$$

written as an initial-boundary value problem. The standard procedure for the numerical solution of (8) is to discretize the space Ω and the initial value $u^0 = u(x,0)$ on Ω and to step forward in time. The solution u^{t+1} at time step $t+1$ is computed from u^t and the boundary conditions. This *'time stepping'* scheme is iterated until $t = t_0$ is reached.

7.1 Space-Time Discretization

An alternative solution algorithm uses a discretization of (8) in the *'space-time'* domain $\Omega \times [0, t_0]$, see Figure 22. The transient problem can be re-written as a boundary value problem

$$\nabla \cdot F(u) = q(u) \quad \text{for } u(x), \quad x \in \Omega \times [0, t_0] \subset \mathbb{R}^{d+1}, \tag{9}$$

with F given by the components $F_0(u) = u$ and $F_{1,\dots,d}(u) = f(u)$. A standard finite difference or finite element discretization, implicit in time, on a regular grid leads to a sequence of equation systems, for one time step each. However, the corresponding sparse grid discretization, where different step sizes in time and in space are coupled, requires the numerical solution of a single large (non-) linear equation system and returns an approximation of u at all time steps at once. The hierarchical basis transformation glues all time slices and all location in space together. Together with an iterative equation solver on the global equation system, this approach is related to the waveform relaxation [48]. It has been used for standard grids by many authors, see e.g. [33] and with the Galerkin method for periodic parabolic equations on sparse grids in [5].

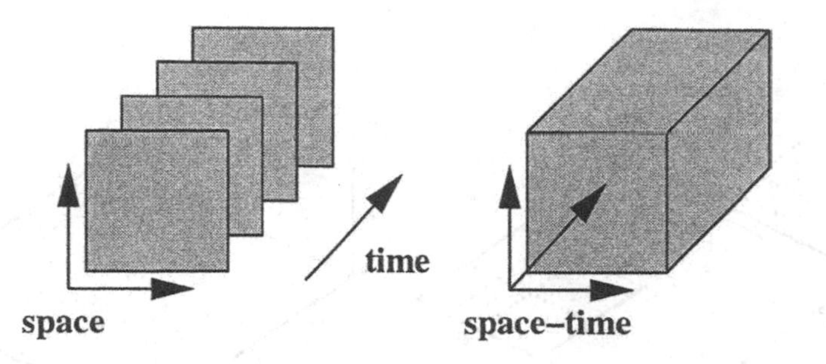

Fig. 22. Time stepping vs. space-time discretization.

We present two test cases of finite difference sparse grid discretizations for problems in space-time. The heat conduction equation (parabolic, $u_t = \Delta u$) and the linear advection equation (hyperbolic, $u_t + u_x = 0$). We prescribe some simple initial values and use a zero source term $q \equiv 0$. The boundary conditions of the heat conduction as well as the inflow of the advection are set to zero. The solutions are depicted in Figures 23 and 24, resp. 25. Errors and convergence rates are given in Table 4. The ratio of number of unknowns to global error is much better on sparse grids than on standard grids. We obtain a weaker dependence on the dimension for sparse grid discretizations, as predicted. However, the actual performance depends on the smoothness of the solution in space-time, which is present for the heat conduction problem, but is missing for typical solutions of hyperbolic equations. That is why we consider adaptive grid refinement next. For details and further considerations and numerical schemes for the solution of hyperbolic equations on sparse grids we refer to [22].

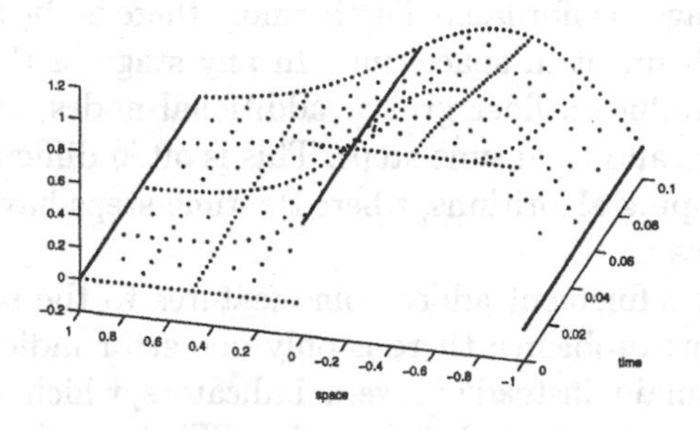

Fig. 23. Sparse grid solution of the parabolic heat conduction equation in 2D space-time.

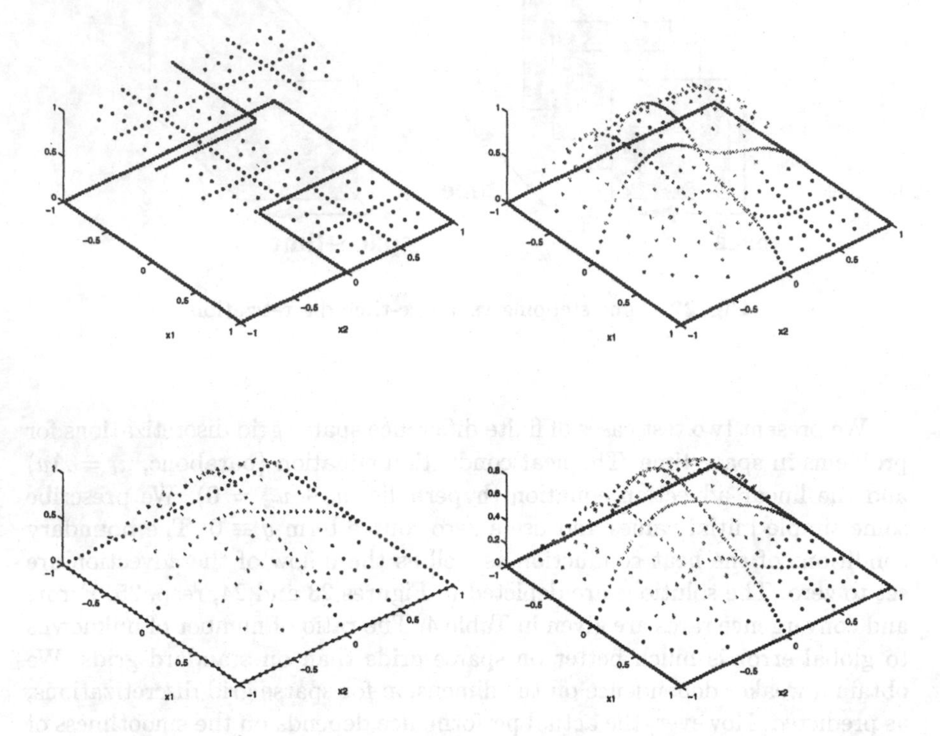

Fig. 24. Sparse grid solution of the parabolic heat conduction equation in 3D space-time. Snapshots at different times t.

7.2 Adaptive Grid Refinement and Macro Time Stepping

The storage requirements of the space-time formulation are often considered as prohibitively high. However, by the sparse grid technique the additional dimension in storage is affordable. Furthermore there is the advantage of easy adaptive grid refinement in space-time. In any stage of the computation it is possible to introduce a finer grid or additional nodes, which gives better resolution in space and local time steps. This is often difficult or even impossible for time stepping algorithms, where the time-steps have to be computed again at a finer scale.

Adaptive grid refinement added some features to the solution algorithm for time-dependent problems: there is only one error indicator operating in the space-time domain, instead of several indicators, which operate separately in space and time and are coupled through a CFL type of stability condition. Grid refinement now enhances local space resolution and at the same time introduces local time stepping. However, due to the discretization implicit in time, there is no temporal stability condition.

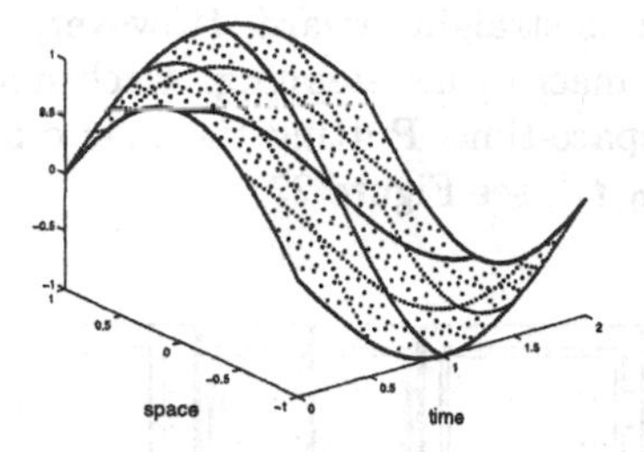

Fig. 25. Sparse grid solution of the hyperbolic linear advection equation in 2D space-time with a sine wave as initial data.

$2/h$	L_1	L_2	L_∞	L_1	L_2	L_∞
32	1.8973	1.8713	1.7746	1.4347	1.6189	1.6752
64	1.9348	1.9194	1.8696	1.5539	1.7186	1.7834
128	1.9610	1.9534	1.9297	1.6518	1.7985	1.8652
256				1.7247	1.8558	1.9235
512				1.7746	1.8932	1.9606

Table 4. Convergence rates on regular (left) and on sparse grids, sine wave initial data for the linear advection equation.

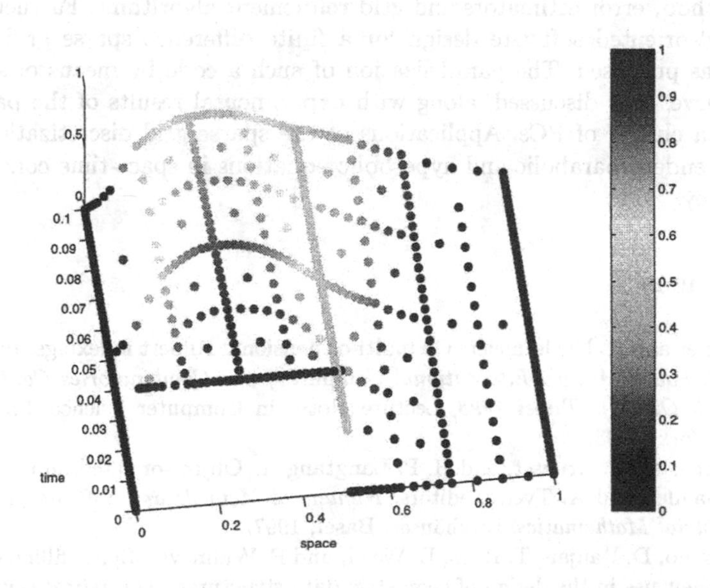

Fig. 26. Adaptive sparse grid solution in space-time.

Initial value problems, where the final time t_1 is unknown, cannot be discretized in space-time straightforward. However, variable time steps can be introduced with a macro time stepping. Each macro step is based on a single sparse grid in space-time. Putting the macro steps together gives the whole time interval $[t_0, t_1]$, see Figure 27.

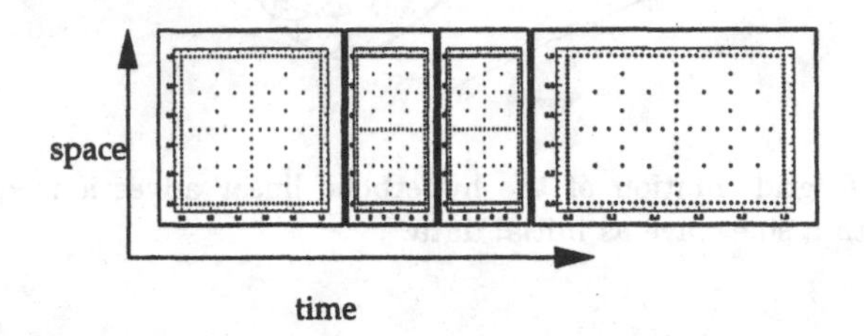

Fig. 27. Macro time stepping in a space-time scheme.

8 Conclusion

We have introduced sparse grids, several ways to discretize partial differential equations on sparse grids and strategies to create adaptive sparse grids. This includes the extrapolation method, the Galerkin method and the finite difference method, error estimators and grid refinement algorithms. Furthermore an object-oriented software design for a finite difference sparse grid PDE solver was proposed. The parallelization of such a code by means of space-filling curves was discussed, along with experimental results of the parallel code on a cluster of PCs. Applications of the sparse grid discretizations to time-dependent parabolic and hyperbolic equations in space-time concluded this survey.

References

1. J. Alber and R. Niedermeier. On multi-dimensional Hilbert indexings. In *Proc. of the Fourth Annual International Computing and Combinatorics Conference (COCOON'98), Taipei 1998*, Lecture Notes in Computer Science. Springer, New York, 1998.
2. E. Arge, A. M. Bruaset, and H. P. Langtangen. Object-oriented numerics. In M. Dæhlen and A. Tveito, editors, *Numerical Methods and Software Tools in Industrial Mathematics.* Birkhäuser, Basel, 1997.
3. T. Asano, D. Ranjan, T. Roos, E. Welzl, and P. Widmayer. Space filling curves and their use in the design of geometric data structures. *Theoretical Computer Science*, 181:3–15, 1997.
4. M. H. Austern. *Generic Programming and the STL.* Addison-Wesley, 1999.

5. R. Balder. *Adaptive Verfahren für elliptische und parabolische Differentialgleichungen auf dünnen Gittern.* PhD thesis, TU München, Inst. für Informatik, 1994.
6. J. Bartholdi and L. K. Platzman. Heuristics based on space-filling curves for combinatorial optimization problems. *Management Science*, 34:291–305, 1988.
7. J. J. Barton and L. R. Nackman. *Scientific and Engineering C++ – An Introduction with Advanced Techniques and Examples.* Addison-Wesley, Reading, MA., 1994.
8. P. Bastian. Load balancing for adaptive multigrid methods. *SIAM J. Sci. Comput.*, 19(4):1303–1321, 1998.
9. R. Becker and R. Rannacher. Weighted a posteriori error control in FE methods. In *Proc. ENUMATH 95*, 1995.
10. I. Beichl and F. Sullivan. Interleave in peace, or interleave in pieces. *IEEE Computational Science & Engineering*, 5(2):92–96, 1998.
11. D. Bertsimas and M. Grigni. On the spacefilling curve heuristic for the Euclidean traveling salesman problem. Technical report, Massachusetts Institute of Technology, Cambridge, MA, 1988.
12. K. Birken and C. Helf. A dynamic data model for parallel adaptive PDE solvers. In B. Hertzberger and G. Serazzi, editors, *Proceedings of HPCN Europe 1995*, volume 919 of *Lecture Notes in Computer Science*, Milan, Italy, 1995. Springer.
13. A. M. Bruaset and H. P. Langtangen. A comprehensive set of tools for solving partial differential equations; Diffpack. In M. Dæhlen and A. Tveito, editors, *Numerical Methods and Software Tools in Industrial Mathematics.* Birkhäuser, Basel, 1997.
14. H.-J. Bungartz. *Dünne Gitter und deren Anwendung bei der adaptiven Lösung der dreidimensionalen Poisson-Gleichung.* PhD thesis, TU München, Inst. für Informatik, 1992.
15. H.-J. Bungartz. *Finite Elements of Higher order on Sparse Grids.* PhD thesis, TU München, Inst. für Informatik (Habilitation), 1998.
16. H.-J. Bungartz, T. Dornseifer, and C. Zenger. Tensor product approximation spaces for the efficient numerical solution of partial differential equations. In *Proc. Int. Workshop on Scientific Computations*, Konya, 1996. Nova Science Publishers, Inc. to appear.
17. G. Faber. Über stetige Funktionen. *Mathematische Annalen*, 66:81–94, 1909.
18. M. Griebel. The combination technique for the sparse grid solution of PDEs on multiprocessor machines. *Parallel Processing Letters*, 2:61–70, 1992.
19. M. Griebel. Adaptive sparse grid multilevel methods for elliptic PDEs based on finite differences. In *Proc. Large Scale Scientific Computations, Varna, Bulgaria.* Vieweg, 1998.
20. M. Griebel and S. Knapek. Optimized approximation spaces for operator equations. Technical Report 568, University Bonn, SFB 256, 1998.
21. M. Griebel, M. Schneider, and C. Zenger. A combination technique for the solution of sparse grid problems. In P. de Groen and R. Beauwens, editors, *Iterative Methods in Linear Algebra*, pages 263–281. IMACS, Elsevier, 1992.
22. M. Griebel and G. Zumbusch. Adaptive sparse grids for hyperbolic conservation laws. In *Proceedings of Seventh International Conference on Hyperbolic Problems, Zurich.* Birkhäuser, 1998.
23. M. Griebel and G. Zumbusch. Hash-storage techniques for adaptive multilevel solvers and their domain decomposition parallelization. In J. Mandel,

C. Farhat, and X.-C. Cai, editors, *Proc. Domain Decomposition Methods 10*, volume 218 of *Contemporary Mathematics*, pages 279–286, Providence, Rhode Island, 1998. AMS.
24. A. Harten. Multi-resolution representation of data: A general framework. *SIAM J. Numer. Anal.*, 33:1205–1256, 1995.
25. P. W. Hemker. Sparse-grid finite-volume multigrid for 3D-problems. *Advances in Computational Mathematics*, 4:83–110, 1995.
26. D. Hilbert. Über die stetige Abbildung einer Linie auf ein Flächenstück. *Mathematische Annalen*, 38:459–460, 1891.
27. S. Hilgenfeldt, S. Balder, and C. Zenger. Sparse grids: Applications to multi-dimensional Schrödinger problems. Technical Report 342/05/95, TU München, Inst. für Informatik, SFB 342, 1995.
28. M. T. Jones and P. E. Plassmann. Parallel algorithms for adaptive mesh refinement. *SIAM J. Scientific Computing*, 18(3):686–708, 1997.
29. D. E. Knuth. *The Art of Computer Programming*, volume 3. Addison-Wesley, 1975.
30. F. Kupka. *Sparse Grid Spectral Methods for the Numerical Solution of Partial Differential Equations with Periodic Boundary Conditions.* PhD thesis, Universität Wien,Inst. für Math., 1997.
31. H. P. Langtangen. *Computational Partial Differential Equations – Numerical Methods and Diffpack Programming.* Springer-Verlag, 1999.
32. C. B. Liem, T. Lu, and T. M. Shih. *The Splitting Extrapolation Method: A New Technique in Numerical Solution of Multidimensional Problems*, volume 7 of *Applied Mathematics.* World Scientific, 1995.
33. S. F. McCormick. *Multilevel Adaptive Methods for Partial Differential Equations.* Frontiers in Applied Mathemtics. SIAM, 1989.
34. J. T. Oden, A. Patra, and Y. Feng. Domain decomposition for adaptive hp finite element methods. In *Proc. Domain Decomposition 7*, volume 180 of *Contemporary Mathematics*, pages 295–301. AMS, 1994.
35. P. Oswald. Best N-term approximation of singularity functions in two Haar bases. Technical report, Bell Labs, Lucent Technologies, 1998.
36. M. Parashar and J. C. Browne. On partitioning dynamic adaptive grid hierarchies. In *Proceedings of the 29th Annual Hawais International Conference on System Sciences*, 1996.
37. A. Pérez, S. Kamata, and E. Kawaguchi. Peano scanning of arbitrary size images. In *Proc. Int. Conf. Pattern Recognition*, pages 565–568, 1992.
38. S. Roberts, S. Kalyanasundaram, M. Cardew-Hall, and W. Clarke. A key based parallel adaptive refinement technique for finite element methods. In *Proc. Computational Techniques and Applications: CTAC '97*. World Scientific, 1998. to appear.
39. U. Rüde. Data structures for multilevel adaptive methods and iterative solvers. Technical Report I-9217, TU München, Inst. für Informatik, 1992.
40. H. Sagan. *Space-Filling Curves.* Springer, New York, 1994.
41. J. K. Salmon, M. S. Warren, and G.S. Winckelmans. Fast parallel tree codes for gravitational and fluid dynamical n-body problems. *International Journal of Supercomputer Applications*, 8(2).
42. H. Samet. *The Design and Analysis of Spatial Data Structures.* Addison-Wesley, 1990.

43. T. Schiekofer. *Die Methode der Finiten Differenzen auf dünnen Gittern zur Lösung elliptischer und parabolischer partieller Differentialgleichungen.* PhD thesis, Universität Bonn, Inst. für Angew. Math., 1998. to appear.
44. T. Schiekofer and G. Zumbusch. Software concepts of a sparse grid finite difference code. In W. Hackbusch and G. Wittum, editors, *Proceedings of the 14th GAMM-Seminar Kiel on Concepts of Numerical Software*, Notes on Numerical Fluid Mechanics. Vieweg, 1998.
45. M. A. Schweitzer, G. Zumbusch, and M. Griebel. Parnass2: A cluster of dual-processor PCs. In W. Rehm and T. Ungerer, editors, *Proceedings of the 2nd Workshop Cluster-Computing*, number CSR-99-02 in Informatik Berichte. University Karlsruhe, TU Chemnitz, 1999.
46. S. A. Smolyak. Quadrature and interpolation formulas for tensor products of certain classes of functions. *Dokl. Akad. Nauk SSSR*, 4:240–243, 1963.
47. V. N. Temlyakov. Approximation of functions with bounded mixed derivative. *Proc. of the Steklov Institute of Mathematics*, 1, 1989.
48. S. Vandewalle. *Parallel Multigrid Waveform Relaxation for Parabolic Problems.* Teubner, Stuttgart, 1992.
49. R. Verfürth. *A Review of A Posteriori Error Estimation and Adaptive Mesh-Refinement Techniques.* J. Wiley & Teubner, 1996.
50. D. Voorhies. Space-filling curves and a measure of coherence. In J. Arvo, editor, *Graphics Gems II*, pages 26–30. Academic Press, 1994.
51. M. Warren and J. Salmon. A portable parallel particle program. *Comput. Phys. Comm.*, 87:266–290, 1995.
52. R. E. Webber and Y. Zhang. Space diffusion: An improved parallel halfoning technique using space-filling curves. In *Proc. ACM Comput. Graphics Ann. Conf. Series*, page 305ff, 1993.
53. H. Yserentant. On the multilevel splitting of finite element spaces. *Numer. Math.*, 49:379–412, 1986.
54. C. Zenger. Sparse grids. In W. Hackbusch, editor, *Proc. 6th GAMM Seminar*, Kiel, 1991. Vieweg.
55. G. Zumbusch. Dynamic loadbalancing in a lightweight adaptive parallel multigrid PDE solver. In B. Hendrickson, K. Yelick, C. Bischof, I. Duff, A. Edelman, G. Geist, M. Heath, M. Heroux, C. Koelbel, R. Schrieber, R. Sinovec, and M. Wheeler, editors, *Proceedings of 9th SIAM Conference on Parallel Processing for Scientific Computing (PP 99)*, San Antonio, T., 1999. SIAM.

Java as an Environment for Scientific Computing

John K. Arthur

Telenor Research and Development
Applied Media Technology, Kjeller, Norway
Email: john.arthur@telenor.com

Abstract. Java is an object-oriented programming language that has attracted some interest in the software community due, in part, to its syntactic simplicity, robustness, platform independence and relative ease of programming. A key question with respect to the suitability of Java in scientific computing is its performance in numerically intensive tasks, especially when compared to other languages such as C++. Also of interest is the potential of Java as a replacement environment for scripting languages such as Perl and Tcl. Java's potential in these areas is explored through feature comparisons and a case study in the use of Java for geometric modelling.

1 Introduction

Java is an object-oriented programming language that has attracted much attention in the information technology world, partly due to its pre-eminent use on the Internet. Since the information technology industry is constantly devising new tools, such as programming languages and scripting languages, finding the tool appropriate for a given task is made difficult by the bewildering choice. One often uses the tools which are familiar, and with which one is most comfortable. However, these may not be the best tools for the job. Though if it is an unfamiliar tool a balance must be struck between the time taken to learn the new tool, and using the tool to achieve the task. In addition the type of task to be performed, whether it is a one-off or once-a-day task, is also of importance. Object-orientation (in analysis, design and programming) has proven its worth in recent years, allowing greater complexity, robustness, extendibility and re-usability of software. Moreover network computing (LAN, Internet and Intranet etc.), is now beginning to play a role in supporting collaborative scientific and engineering computing environments [7,2]. Java can be seen as a programming language built on the cornerstones of object-orientation and network computing. In this sense it has the potential to be the key enabling technology for heterogeneous scientific problem solving environments [7], providing the *glue* that seamlessly allows information to flow between heterogeneous systems.

2 Historical Background

Java was created around 1991 by engineers at Sun Microsystems in the USA. The chief architect was James Gosling [3], and the original goal of the project was to create a programming language that was compact enough to fit on consumer electronic devices such as cable-TV switch boxes and hi-fi systems. Since these devices would be very limited in power and memory, the code would have to be very compact. In addition different devices and manufacturers would tend to use different CPUs, thus the language would have to be architecture neutral. The latter feature was implemented through the concept of a *virtual machine*, which interpreted the compiled code called *byte code*, translating it into native code on-the-fly, (see Section 3.2). Thus any platform with a correct implementation of the virtual machine could run the compiled code. Unfortunately neither SUN nor consumer electronics companies were interested in the system, and by 1994 SUN stopped trying to market the product.

During this time, the Internet and more specifically the World Wide Web was beginning to gain momentum, and the Java engineers at Sun realised that the features such as architecture neutrality and compactness were ideal for building a web browser. Thus the first 100% Java Web browser — *Hot-Java* was born. This browser demonstrated the power of Java applets (small interpreted programs) and in 1995 Netscape incorporated Java technology in the next release, allowing web authors to embed dynamic Java applets in their pages. It is through Netscape's support that Java became widespread over the Internet.

3 The Java Computing Environment

3.1 Java's Main Features

There follows a brief list of the main features of the Java language, a more detailed overview of the language can be found in [3] and [12].

- *simple*: The designers wanted to build a system that could be programmed easily without the need for highly specialised training, thus the syntax and grammar of Java resembles that of the more widely used C and C++ languages, though is less complex.
- *object-oriented:* Since Java is *strongly* object-oriented (unlike C++), no Java statements exist in the absence of a defined class, even the main program, be it application or applet, is a class. This forces the developer to always think in terms of objects and classes, which has advantages and disadvantages depending on the programming task.
- *distributed:* Java contains extensive libraries of routines with network, Internet protocol and RMI (Remote Method Invocation) functionality for facilitating object message passing in distributed environments. Java's

ability to dynamically load remote libraries, classes, objects and methods can allow truly distributed applications to be built.

- *robust:* In Java there is a lot of emphasis on early error checking, as well as run time checking. Java has a pointer model that ensures that memory cannot be overwritten. Note that the developer does not have access to Java pointers.
- *secure:* Since Java is intended to be used in distributed/networked environments, some emphasis has been placed on security. However there is no system that is 100% secure and 'holes' have been found in the security model. These holes, though, are very difficult to find, and Java's security system is generally well regarded.
- *architecture neutral:* The Java compiler generates bytecode, which is interpreted by the Java run-time system (known as the Java Virtual Machine or JVM). Java bytecode will run on any platform with the standard implementation of the JVM.
- *portable:* There are no implementation dependent aspects of the Java specification. The libraries defining the system define portable interfaces.
- *interpreted:* As stated above, the JVM interprets Java bytecodes. Unfortunately the fact that Java is interpreted, makes for lower performance compared to native code. However technologies such as Just-In-Time compilation (compiling Java bytecode to native code once, at run time) improves Java's performance significantly. Newer technologies to be found in Sun's HotSpot [11] virtual machine should, according to Sun, improve Java's performance by orders of magnitude.
- *multithreaded:* This feature gives a Java program the ability to perform more than one task at the same time. This is intended to give better interactive response and real-time behaviour.
- *dynamic:* Java was designed to adapt easily to changing environments and allows type information to be acquired at run time.

3.2 Compilation and Execution Environment

As stated, Java source code is compiled into byte code, an intermediate code interpreted by the JVM, which translates it to native code (see Figure 1). In addition to simply supplying the native CPU with translated Java code, the JVM also performs all memory management through automated *garbage collection.* This feature means that the memory used by the JVM is periodically freed of areas allocated to unreferenced objects. In this way the developer does not need to worry about de-allocating memory for objects, arrays etc. However there is a performance penalty for this simplification. Clearly, the JVM must spend some execution time searching for unreferenced objects, and this will inevitably degrade performance with respect to a non-garbage collecting language such as C++. The advantage is that memory leaks, segmentation/protection faults and core dumps are virtually unheard of in Java, and Java is therefore much more robust than C and C++.

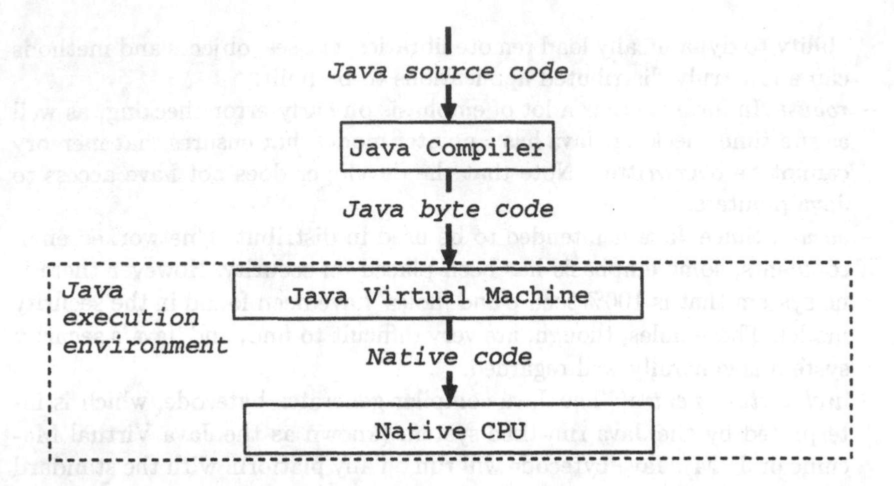

Fig. 1. Compilation and execution of Java code.

4 Java Functionality Comparisons

There are two main possibilities when considering the use of Java for scientific computing:

1. Java as a replacement for numerical code (typically written in Fortran or C/C++)
2. Java as a gluing framework linking together modules possibly written in other languages, lying on different computers etc., performing the role of a scripting language such as Perl, Python and Tcl.

Clearly some combination of the above two functions is also a possibility. However, it would be beneficial to consider the two separately, to see where Java's strengths and weaknesses lie in these respects. When choosing a programming language for any task or application, the features of the language, the existence of in-house/third party components and libraries, integration with legacy code and experience of the developers are all important, inter-related factors.

4.1 Java as a Programming Language for Numerical Computing

Since Java was not designed for scientific computing, the features that make it attractive for such applications are, in general, features that make it good for programming other types of applications. Its deficiencies with regard to scientific computing stem from attempts to simplify the language and reduce the risk of introducing bugs (especially in relation to C++) while maintaining much of the same syntax. A comprehensive survey and analysis of the use of

Java in numerical computing can be found in [5]. The following two sections outline some of the advantages and disadvantages of using Java for numerical computing.

Advantages for Numerical Computing.

- *platform independence:* encapsulating both the architecture neutrality and portability features of Java, this essentially means that Java byte code should run and behave identically on different platforms, as long as the JVM is correctly implemented on that platform — the so-called *write once, run anywhere* feature. In practice different platforms implement the JVM in different ways, which can lead to inconsistencies in the behaviour of code. In general, though, porting Java code is usually trivial and significantly less labour intensive than porting programs in other languages.
- *networking functionality:* Java has an extensive and comprehensive library of network classes based on TCP/IP protocols and higher level protocols such as HTTP and FTP. This makes the tasks of building network functionality into a Java program much simpler than doing so with C or C++. Moreover Java's platform independence means that no platform specific calls or procedures need to be learned by the developer. This is a major advantage considering the esoteric nature of networking software development.
- *database functionality:* The standard Java Development Kit (JDK) also comes with a complete set of classes for interfacing with standard databases, and communicating with SQL.
- *multithreading functionality:* As stated, threads allow a single program to execute more than one task simultaneously. In addition to better user interactivity, there is also the possibility to take advantage of multiprocessor systems, though this depends on the implementation of the JVM.
- *dynamic:* Java possesses a number of features which allow it to adapt to a dynamic IT environment. Since the language is interpreted and linking takes place at run-time, classes and libraries do not need to be re-compiled when changes occur in other modules in the software system. For example classes can add new methods and variables without requiring dependent classes to be re-compiled.
- *distributed computing environment:* Java objects can be dynamically loaded, transferred and controlled over a network, allowing for a distributed computing environment to be created with Java as the foundation.
- *robustness:* Java does not give the developer direct access to memory through pointers as in C. This reduces the risk of segmentation faults and core dumps. Java also has run-time exception handling built into the language.

– *ease of programming:* As mentioned, Java's syntax is designed to make programming easier, the lack of pointers, operator overloading and multiple inheritance, all make for a simpler language compared to C++. In addition the programming infrastructure of Java is much simpler than that of C/C++ requiring no makefiles, header files or static linking at compilation time. Interestingly, Java is beginning to become the programming language of choice, within educational institutions, for learning to design and develop object-oriented software, and many CASE (*Computer Aided Software Engineering*) tools now support Java code. This will give rise to a generation of IT experts who have learnt object-oriented programming through Java, making the language all the more attractive for development projects.

Disadvantages for Numerical Computing. Java's simplicity and robustness do not come without a price. They result in a number of deficiencies with regard to scientific computing.

– *relatively low performance:* An interpreted language will always be slower than one running compiled, native code. Add to this performance degrading features such as automatic garbage collection, automatic array bounds checking, and exception handling and you have the potential for a language that performs like a brick in a bucket of cement. The JVM, though, has several features (such as code optimisations, and Just-In-Time compilation) that improve performance such that Java code runs at *acceptable* levels for most applications.
– *no pre-processing:* The C/C++ preprocessor allows for features such as macros and conditional compilation. Although there are Java preprocessors available developed by third parties, they are not a standard part of the language. Despite the possibility for introducing errors through macros, pre-processor use in conditional compilation is a very useful tool for outputting debug information, without degrading performance in the final version.
– *no templates:* Templates have only recently become a standard feature of most C++ implementations, and their late arrival has discouraged their use, especially in projects to be delivered across different platforms. In the context of numerical computing, though, templates are an invaluable way to save development time by eliminating the need to write different functions for different data types.
– *no operator overloading:* This is another feature found in C++, that has been left out of Java to keep the language simple — to write and to read. Although open to abuse, operator overloading in mathematical contexts is useful for keeping the length of a given expression to a minimum, and maintaining a close mapping from the mathematical expression to the source code implementation.

- *copying of sub-arrays:* Since there are no pointers in Java, it does not have the level of flexibility when dealing with arrays that C has. Most critical is the inability to specify a sub-array without copying from the original array. In numerical computing this is a serious flaw, which not only degrades performance but also creates data redundancy.
- *non-contiguous multidimensional arrays:* The specification for the JVM does not stipulate that multidimensional arrays should be contiguous. Thus when iterating from one row of elements to the next in a multidimensional array there is a risk of having to jump to a completely different area of memory and thus waste time. C, C++ and Fortran allocate multidimensional arrays in contiguous areas of memory.
- *platform specific floating point hardware cannot be exploited:* Since Java requires that floating point calculations produce bitwise the same results on different architectures, it inhibits the use of efficient floating point hardware that may be available on that platform. For example the extended precision registers on Intel platforms cannot be used by the JVM.
- *no standard complex data type:* A standard complex data type is a must for scientific computing, however in the absence of a standard type or class it should be possible to define such an entity, and use it as though it was a pre-defined type. Unfortunately the inability to overload operators for classes makes it a clumsy and inelegant solution to define one's own complex type in Java, and use it in mathematical expressions.

The above list of disadvantages is among the issues currently being addressed by the *Java Grande Forum* [10]. This group of numerical computing experts see Java's potential in this field and aim to bring about changes to the language such that it becomes the programming environment of choice for high performance computing.

The feature comparison in Table 1 indicates that Java has a balance of advantages and disadvantages with respect to numerical software development in relation to the more commonly used C, C++ and Fortran 90. As stated, the appropriateness of the tool depends on the type of application, among other factors, and so it is impossible to generalise with a global statement regarding Java's general suitability. However it is clear that if performance is a critical factor, then Java is not an appropriate tool, though if robustness, distributed execution, GUI functionality, and/or ease of development are more important than performance Java appears very attractive.

Existing Numerical Libraries. A number of numerical libraries have been developed for use in the Java programming environment. An up-to-date and comprehensive list can be found at the Java Numerics Web site [13]. These libraries can be divided into two types: original Java language versions and conversions from other languages.

	Language			
Feature	Java	C	C++	F90
Object-oriented	✓	×	✓	×
Pointers	×	✓	✓	✓
Templates	×	×	✓	×
Long label names	✓	✓	✓	✓
Sub-arrays w/o copying	×	✓	✓	✓
Automatic array bounds check	✓	×	×	×
Built-in exception handling	✓	×	✓	×
Operator overloading	×	–	✓	✓
Built-in multi-threading	✓	×	×	×
Standard GUI library	✓	×	×	×
Dynamic object loading	✓	–	×	–
Pre-processing	×	✓	✓	✓
Ease of programming rank	2	3	4	1
Platform independence	✓	×	×	×
Interpreted(I)/Compiled(C)	I	C	C	C
Contiguous multidimensional arrays	×	✓	✓	✓
Performance rank	4	2	3	1
Robustness rank	1	4	2	3

Table 1. Feature table for numerical programming languages.

JAMA. Of note in the first category is JAMA [8] — the JAva MAtrix package, developed by the National Institute for Standards in Technology (NIST) and the MathWorks in the USA. This library provides basic Matrix and Linear algebra functionality and is being proposed as the standard Matrix class for Java, by the two parties involved in its development.

The package currently supports the following Matrix operations:

- Elementary operations (addition, transpose etc.)
- Decompositions (Cholesky, LU, QR, SVD, symmetric/non symmetric eigenvalue)
- Equation solvers (non-singular systems, least squares)
- Derived quantities (condition number, determinant, rank, inverse, pseudo inverse)

The following functionality is planned for a later stage:

- support for special matrices (e.g. sparse, banded)
- complex matrices

Note that this is the package used for interpolation in the geometric modelling suite developed in the case study discussed in Section 5.

4.2 Java as a Scripting Language

Scripting languages provide the *glue* that links tasks, procedures, and applications into homogeneous interfaces (sometimes graphical interfaces). In addition, since they are generally interpreted, they allow for rapid application development. Since Java is interpreted and dynamic, it would be natural to use it as a scripting language, and to compare it with other scripting languages. The main scripting languages used in the scientific community are Perl, Tcl and Python, usually combined with Tk and its extensions as GUI library. Each of these languages have their own features, advantages and disadvantages depending on the application in question. The following two sections outline Java's advantages and disadvantages as a scripting language.

Advantages as a Scripting Language.

- *interpreted:* This is almost a pre-requisite for a scripting language, since it encourages platform independence, and facilitates dynamic linking, and operation on distributed systems.
- *standard GUI library:* Java's cross-platform GUI library is a big advantage in heterogeneous environments.
- *built-in multithreading:* The JVM's multithreading implementation allows for multithreading functionality to be implemented in a platform independent way. Though most scripting languages can take advantage of multithreading embedded in a given operating system (OS), if the functionality is not in the OS, the scripting language cannot use it. Java has no such limitation.
- *built-in exception handling:* Makes for naturally more robust programs.
- *object-oriented:* Promotes code re-usability, extendibility, robustness and modularity.
- *standard network library:* This facilitates the development of network centric, distributed application environments for which a scripting glue is ideal.
- *Good CASE tool support*: CASE tools can greatly simplify and speed-up application development. Java support for such tools is generally good, whereas for scripting languages it is often poor.

Most of these advantages are present in Perl, Tcl and Python as well.

Disadvantages as a Scripting Language.

- *strongly object-oriented:* Unlike C++ and Python, Java is a *strongly* object-oriented language, which means that it is not possible to write programs that do not conform to the object-oriented paradigm. This is a disadvantage for a scripting language, since there will be an object overhead in creating even the simplest of programs. In addition a script may not fit into an object model, forcing the developer to *think objects* when there are no natural objects to work with.

- *difficult to interface with native code:* Though Java native interfaces exist for calling native procedures (e.g. written in C/C++ or Fortran) from Java, implementing these native calls is by no means an easy task and compared to other scripting languages it is somewhat labour intensive. This is especially the case when trying to interface with C++ objects that are part of a large class hierarchy (say more than 40 classes) where data must be transferred between Java and C++ objects. Perl, Tcl and Python are easy to combine with C, C++, F77 or Java code [1].
- *more primitive text processing tools:* Text manipulation via regular expressions is often a major part of scripts, but Java lacks the extensive support for text processing as found in Tcl, Perl and Python.
- *less support for build-in OS operations:* Especially Perl and Python come with comprehensive libraries that offer a unified, cross-platform interface to operating systems operations. Similar functionality in Java is more primitive, which makes it harder to avoid direct OS calls, with the result of decreased portability.

When considering Java as a scripting language the size and complexity of the project is of crucial importance, considering the overhead involved. Unless a developer is already familiar with C++ or Java, it would not seem prudent to choose Java as the scripting medium. However when the application as a whole is fairly complex, integrates stand-alone native programs or Java code, and/or is distributed, then Java could provide a good solution. However, the optimal solution in such cases is perhaps provided by JPython which offers a seemless integration of Java and Python.

5 Case Study: Building a Java-based Geometric Modelling Suite

5.1 Objectives

A set of packages was developed to model basic curves and surfaces in Java in order to assess Java's suitability for computational geometry and computational science and engineering in general. The package would allow a number of parametric geometric entities to be generated and visualised in Java. The core mathematical classes of the suite would be written in Java, and the visualisation part would use the standard Java3D library [9].

5.2 Mathematical Foundation

The *Parametric Geometry* package, see Figure 2, contained high-order geometric objects (typically used in Computer Aided Design applications). The geometry of these entities is described by parameters (hence the term parametric) in a different space from the 3D Euclidean space where they are displayed. For a more rigorous analysis of the mathematics described in this section see [6] and [4]. Classes were created for the entities described in the following sections.

Bézier Curve. A Bézier curve is a parametric curve defined as a linear combination of Bernstein polynomial functions (defined below). This was the type of curve used in the performance tests carried out in Section 5.4. To aid the performance comparison with C++ code, the Bézier curve class was kept separate from the Parametric Geometry package seen in Figure 2.

This family of curves can be thought of as a subset of the set of B-spline curves, discussed in the next section. The curve is defined as follows:

$$s(u) = \sum_{i=0}^{n} B_{i,n}(u)\mathbf{b}_i,$$

where $s(u)$ is the value of the function at the parameter value and n is the degree of a given Bernstein polynomial function:

$$B_{i,n}(u) = \binom{n}{i} u^i(1-u)^{n-i},$$

$$\binom{n}{i} = \begin{cases} \frac{n!}{i!(n-i)!} & \text{if } 0 \leq i \leq n \\ 0 & \text{else} \end{cases}$$

where $\{b_0, \ldots, b_n\}$ is the coefficient set for the curve.

The algorithm for evaluating a Bézier curve, and that used in the performance tests, is the recursive *de Casteljau algorithm.*

Polynomial coefficients are generated using this formula:

$$\mathbf{b}_i^r(u) = (1-u)\mathbf{b}_i^{r-1}(u) + u\mathbf{b}_{i+1}^{r-1}(u) \begin{cases} r = 1, \ldots, n \\ i = 0, \ldots, n-r. \end{cases}$$

The value of the function at the parameter value u is:

$$s(u) = \mathbf{b}_0^n(u).$$

Entities in the Parametric Geometry Package. A B-spline curve and three different types of surface were created in the *Parametric Geometry* package:

- *A B-spline curve* (see Figure 3) is a parametric curve whose geometry is described through a linear combination of piecewise polynomial basis functions. This allows for a very flexible, free-form geometry, and forms the building-block for the parametric surfaces described later. The mathematical formulation for a B-spline function is the following:

$$s(u) = \sum_{i=0}^{n} P_{i,k}(u)\mathbf{d}_i.$$

- *A tensor product B-spline surface* can be viewed as a linear combination of B-spline functions in the u and v parameter directions. The *control net* (grid of control points) defines the coefficients of the basis functions. This method of formulating a spline-based parametric surface allows for the surface to be visualised by evaluating a sequence of curves defined by the characteristic coefficients of the surface. See Figure 4 for the visualisation of a tensor product B-spline surface, whose formula is below:

$$\mathbf{S}(u,v) = \sum_{i=0}^{p}\sum_{j=0}^{q} P_{i,k}(u)P_{j,k}(v)\mathbf{d}_{i,j}.$$

- *A surface of revolution* can be thought of as the surface produced by rotating a plane (2D) curve defined about an axis which does not intersect the curve. See Figure 5 for the visualisation of a surface of revolution, the formula for which is the following:

$$\mathbf{S}(u,v) = \{f(v)\cos u, f(v)\sin u, g(v)\}.$$

- *A lofted tensor product B-spline surface* is a surface generated by interpolating a set of input curves. The control points of the given curves are interpolated to give the control net of the surface. For every row of control points in the curve, the following linear system must be solved:

$$\begin{pmatrix} -3 & 3 & & & & \\ f_{-1} & h_{-1} & g_{-1} & & & \\ & & \ddots & & & \\ & & & f_{n-4} & h_{n-4} & g_{n-4} \\ & & & & -3 & 3 \end{pmatrix} \begin{pmatrix} \mathbf{d}_0 \\ \mathbf{d}_1 \\ \vdots \\ \mathbf{d}_{n-2} \\ \mathbf{d}_{n-1} \end{pmatrix} = \begin{pmatrix} \Delta_0 m_0 \\ y_0 \\ \vdots \\ y_1 \\ \Delta_{n-1} m_1 \end{pmatrix}.$$

The linear solver found in the JAMA (see Section 4.1) package, was used for solving the linear system for spline interpolation. See Figure 4 for the visualisation of a lofted surface.

5.3 Packages in the Modeling Suite

The system was divided into a set of libraries responsible for different functions. Figure 2 shows the individual libraries (known in Java as *packages*) used in the system. The *Computational Geometry* package contained classes for a basic 3D object, a point and vector. The *Parametric Geometry* package contains classes for the geometric entities mentioned in Section 5.2. The *Visual Geometry* package contains classes derived from standard Java3D classes for visualising curves and surfaces.

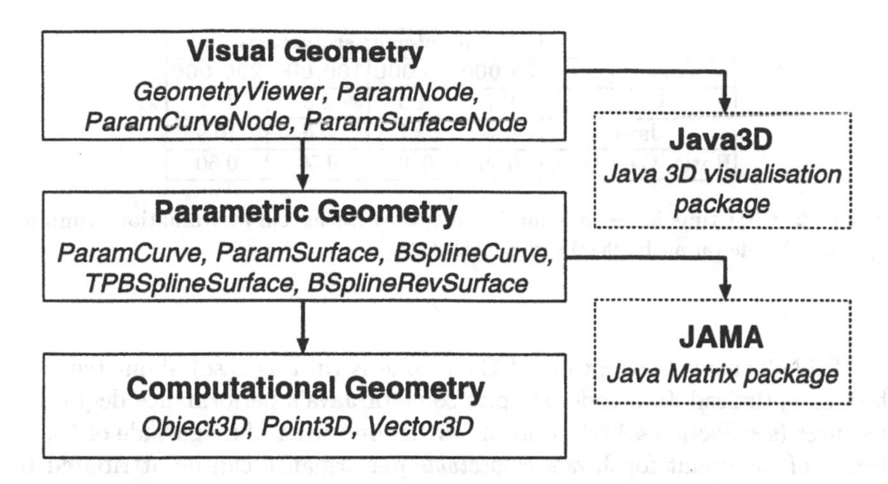

Fig. 2. Packages in the Geometric Modelling suite.

5.4 Results

Prototyping and Debugging with Java. As a programming language Java is very approachable, and for those with C++ experience it is an excellent environment for quickly building prototypes and testing solutions to problems. Compared to C++, setting up a working software system from scratch takes considerably less time. Since Java is interpreted, compiling, linking and testing took very little time, and this greatly decreased overall development time. However there was one feature of C++ that could be useful to include in Java for testing — namely pre-processing. This would allow debug information to be easily switched on or off without affecting performance (i.e. using conditional compilation).

The core classes in the modelling suite were developed using both UNIX (HP-UX) and IBM PC Java environments, i.e. were developed on heterogeneous platforms. Porting Java source and byte code proved no problem whatsoever and this is a clear advantage of the language, particluarly in more complex development projects which might involve different platforms and operating systems in different locations.

Performance. To assess the performance of Java code with respect to C++ code, experiments were carried out using Java and C++ implementations of a Bézier curve (see Section 5.2), evaluated using the classic de Casteljau algorithm [6]. The simple class hierarchy implemented in Java and C++, consisted of a Bézier curve class and point class. The tests were conducted using a 2D Bézier curve of degree 3, on an HP9000 series workstation with JDK version HP-UXC.01.16.00 installed.

	number of evaluations			
language	**25 000**	**50 000**	**100 000**	**200 000**
C++	0.7	1.3	2.8	5.5
Java	1.4	2.7	5.6	10.9
Ratio, C++/Java	0.50	0.48	0.50	0.50

Table 2. CPU time in seconds for 2D, degree 3 Bézier curve evaluations running optimised code on an HP9000 series workstation.

Table 2 shows that optimised C++ code is (in this case) about twice as fast as optimised Java code. Despite some of Java's performance degrading features (see Section 4.1) it performs within one order of magnitude of C++. Much of the credit for Java's *respectable* performance can be attributed to the Just In Time (JIT) compiler.

JIT execution means compiling Java byte code into native machine code the first time on-the-fly, thereafter the native code is executed. Specifically when a class is loaded by the JVM, its code is compiled into machine code. The penalty to be paid is that there is a delay when a class is loaded. Thus if a program involves more class loading and unloading than execution, the JIT will not pay dividends. In the case of the Bézier test program, all classes are loaded when the program first runs and tests showed the JIT improves performance by a factor of about 3.

Visualisation. The Java3D [9] library is intended to be used as a layer above a low-level graphics API such as OpenGL [16]. It incorporates a *Scene Graph* data model, namely a directed acyclic graph that reflects the spatial and topological relations between the objects to be rendered. The library is also designed to support distributed virtual environments and virtual reality equipment such as head mounted displays and data gloves. These features combined with its free availability give it some advantage over alternatives such as Mesa [15], the Visualisation Toolkit [14] and OpenInventor [17].

The beta version of Java3D used was run on an IBM compatible PC using OpenGL as the low-level library. Future versions of Java3D should be able to use other low-level graphics APIs such as Direct3D. The use of the OpenGL API means that once curves/surfaces have been generated they can be manipulated (rotated, translated, zoomed) in real-time on modest graphics hardware. The visual quality of curves and surfaces was also high, due mostly to the underlying functionality of OpenGL, used in this PC implementation of Java3D. The availability of a standard, quality and fully functional graphics API is a key advantage for the Java environment, especially with regard to visualisation in science and engineering.

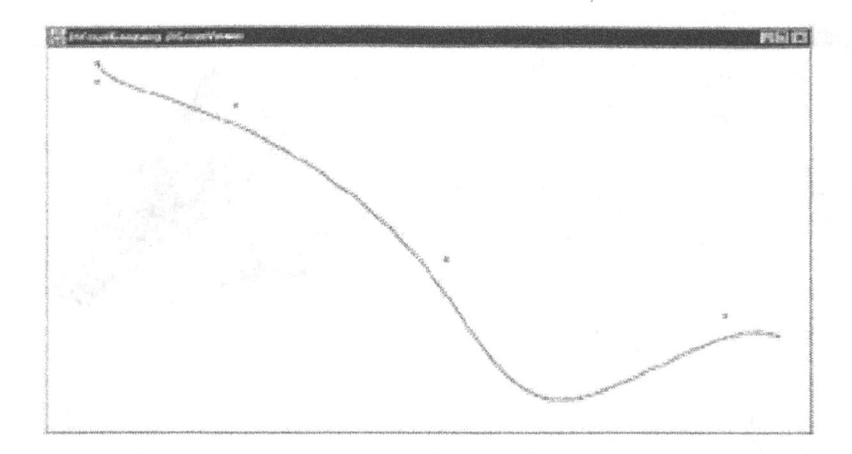

Fig. 3. A cubic B-spline curve shown with control points.

5.5 Lessons Learned

Although the Bézier curve evaluation comparison was a useful measure of comparative performance, it would have been more useful to extend the comparison to surfaces, to assess how well 2D arrays are handled by the JVM. Since there exist tools for converting C/C++ to Java, it would have been helpful to convert existing C/C++ code to Java, and then compare performance; this would also be a good test of the conversion tool. There also exist tools and enhancements to Java, that allow the use of pre-processor directives, this would have simplified the task of debugging without affecting performance in the non-debug state, though this is not a standard feature of the Java environment. When developing in Java it is important to stay abreast of developments especially where numerical computing is concerned, the Java Numerics Web [13] site is an excellent resource for information tools and news for Java users in this respect, and contains information and links to the resources mentioned here.

6 Conclusion

6.1 Java for Numerical Computing

Java is seen to have much potential in the scientific computing community, however its deficiencies (summarised in Section 4.1) prohibit its wide spread adoption for numerical software. If the Java Grande Forum [10] can achieve the desired changes in the language, then Java could easily become the language of choice in this context.

The Java programming environment as a whole is well rounded, well integrated and the set of standard libraries for networking, GUIs, databases and

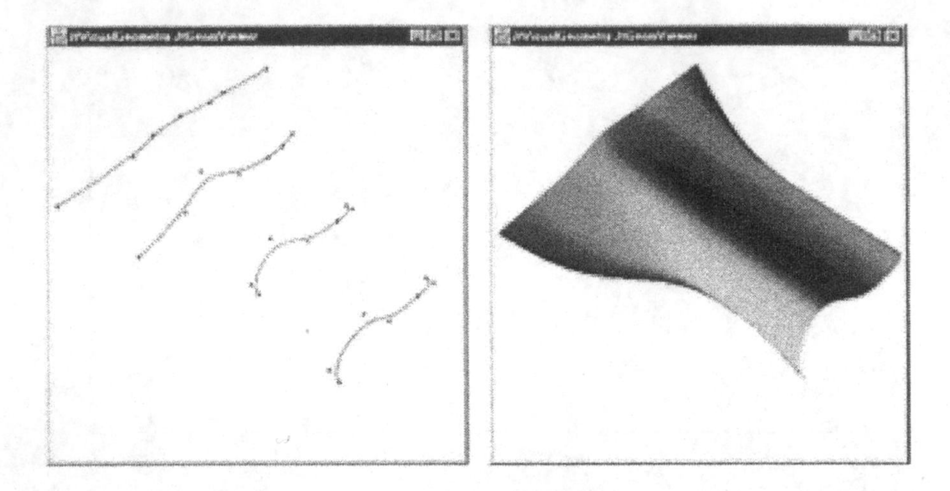

Fig. 4. Input curves and resulting lofted surface: bi-cubic Tensor Product B-spline surface with 6x7 control net.

visualisation (among others) allows for quite complex and extensive applications to be built rapidly. In addition features such as automatic array bounds checking, the lack of pointers, and exception handling make for very robust code. At the same time, this robustness comes at the price of performance and this combined with the fact that the language is interpreted, results in relatively (compared to C/C++ or Fortran) low performance for optimised numerical code. However the JIT compiler *does* improve performance significantly, and advances such as the HotSpot [11] JVM are designed to resolve such performance issues.

Many of Java's advantages over C/C++ and Fortran for scientific computing can also be obtained by using a scripting language like Python for program administration and executing heavy numerics in native C, C++ or F77 code, see e.g. [1].

6.2 Java for Script Writing

Java's glue-like properties lend themselves well to its use as a scripting environment. Platform independence and the functionality for operating in distributed networked environments is a clear advantage. However, since Java is meant to be a fully fledged programming language, there is some overhead attached to using it for this purpose that does not come with other scripting languages such as Perl and Tcl/Tk. Moreover, the somewhat arduous task of integrating Java with native C++ or C code is a handicap. The use of Java as a script depends chiefly upon the complexity of the tasks and the nature of the other applications/modules involved. With better native integration facilities, and the ability to side-step the object-oriented model (as in

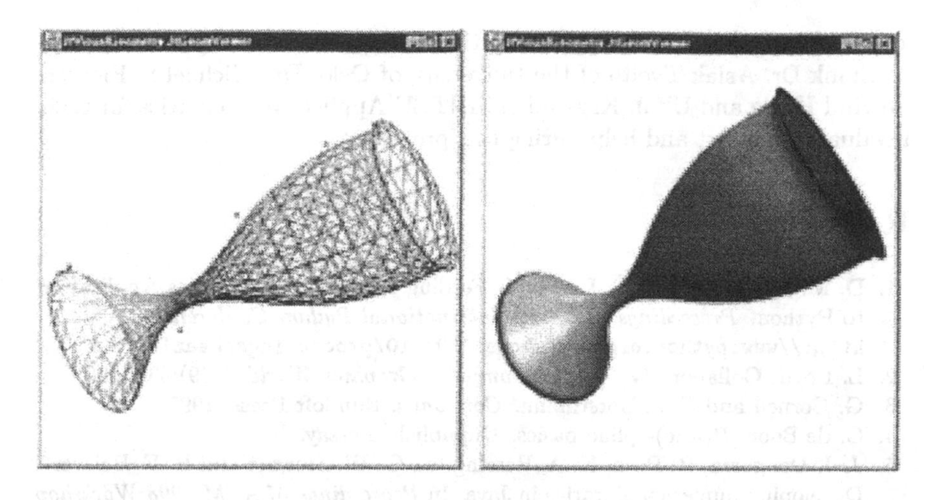

Fig. 5. Surface of revolution: cubic B-spline generating curve with 7 control points.

Python), Java could be a viable alternative for scripting. However, it seems unlikely that such changes will come about in the foreseeable future. Java is not really designed for script writing and as such, it will usually have better alternatives.

6.3 Java for Collaborative Distributed Environments

The most potential for Java in the fields of science and engineering lies in its use in collaborative networked environments [7,2] where people in geographically different locations are required to work together on one project, usually over the Internet. Endeavours such as collaborative computer aided design and scientific visualisation are just two that benefit from new networking technologies, increases in bandwidth and improved local processing power. Java is an ideal technology to use in such distributed environments as the enabling glue linking platforms, users and systems. Although its drawbacks are still apparent in this context, these are more than outweighed by its advantages. This is where Java has the potential to make the path for computational science and engineering into the network-centric distributed environments much easier. It remains to be seen whether Java can fulfill its potential and whether computational scientists are willing take full advantage of the language.

Acknowledgments

This work has been supported by the Research Council of Norway through the *Numerical Computations in Applied Mathematics* (NAM) program, number

110673/420, at SINTEF Applied Mathematics, Oslo, Norway. I would like to thank Dr. Aslak Tveito of the University of Oslo, Dr. Michael S. Floater, Øyvind Hjelle and Ulf J. Krystad of SINTEF Applied Mathematics for their invaluable support and help during this project.

References

1. D. M. Beazley and P. S. Lomdahl. Feeding a Large-scale Physics Application to Python. *Proceedings of the 6th International Python Conference.* `http://www.python.org/workshops/1997-10/proceedings/beazley.html`
2. L. Boyd. Collaborative efforts. *Computer Graphics World*, 21(9):33, 1998.
3. G. Cornell and C. S. Horstmann. *Core Java.* SunSoft Press, 1997.
4. C. de Boor. B(asic)-spline basics. Unpublished essay.
5. J. J. Dongarra, R. Pozo, K. A. Remington, G. W. Stewart and R. F. Boisvert. Developing numerical libraries in Java. In *Proceedings of ACM 1998 Workshop on Java for High-Performance Network Computing*, 1998.
6. G. Farin. *Curves and Surfaces for CAGD, A Practical Guide.* Academic Press, London, 1993.
7. E. Gallopoulos, R Bramley, J R. Rice and E. N. Houstas. Problem-solving environments for computational science. *IEEE Computational Science and Engineering*, 4(3):18–21, 1997.
8. The JAMA home page. URL: `http://math.nist.gov/javanumerics/jama/`
9. Java3D home page. URL: `http://java.sun.com/products/java-media/3D/`
10. The Java Grande Project. URL: `http://www.npac.syr.edu/projects/javaforcse/javagrande/`
11. Java HotSpot virtual machine. URL: `http://java.sun.com/products/hotspot/index.html`
12. The Java Language: An Overview. URL: `http://java.sun.com/docs/overviews/java/java-overview-1.html`
13. The Java Numerics web site. URL: `http://math.nist.gov/javanumerics/`
14. K. Martin, W. Shroeder and B. Lorensen. *The Visualization Toolkit.* Prentice Hall, Upper Saddle River, New Jersey, 1997.
15. The Mesa home page. URL: `http://www.ssec.wisc.edu/~brianp/Mesa.html`
16. The OpenGL home page. URL: `http://www.opengl.org`
17. J. Wernecke. *The Inventor Mentor: Programming Object Oriented 3D Graphics with Open Inventor, Release 2.* Addison-Wesley Longman Inc, Reading, Massachusetts, 1994.

ODE Software that Computes Guaranteed Bounds on the Solution

Nedialko S. Nedialkov and Kenneth R. Jackson

Department of Computer Science, University of Toronto, Canada
Email: {ned,krj}@cs.toronto.edu.

Abstract. Validated methods for initial value problems (IVPs) for ordinary differential equations (ODEs) produce bounds that are guaranteed to enclose the true solution of a problem. In this chapter, we briefly survey validated methods for IVPs for ODEs, discuss software issues related to the implementation of a validated ODE solver, and describe the structure of a package for computing rigorous bounds on the solution of an IVP for an ODE.

1 Introduction

Standard numerical methods aim at computing approximations that satisfy a user-specified tolerance. These methods are robust and reliable for most applications, but it is possible to find examples for which they return inaccurate results. However, there are situations when guaranteed bounds on the mathematically correct result are desired or needed. For example, such bounds can be used to prove a theorem [45]. Also, some calculations may be critical to the safety or reliability of a system. Therefore, it may be necessary to ensure that the true result is within computed bounds. Furthermore, methods that produce guaranteed bounds may be used to check a sample calculation to be performed by a standard floating-point scheme, to give the user an indication of the accuracy and reliability of the floating-point scheme.

Validated (also called interval) methods produce bounds that are guaranteed to contain a mathematically correct result. These methods use interval arithmetic [33] to enclose rounding errors, truncation errors, and possibly even uncertainties in the model, assuming these can be quantified.

A key application of interval methods is to compute rigorous enclosures for the range of a function. In addition to bounding errors in the computation, such enclosures can be used, for example, to prove the existence of a solution to a nonlinear system of equations and to compute interval vectors that are guaranteed to contain the true solution of the system [37]. One of the most successful applications of interval computations is in global optimization where bounds on the range of a function are used to reject regions that cannot contain global optima [20]. We can also obtain tight bounds on the true solution of a linear system of equations [26].

Interval techniques can also be used to bound the error of a numerical approximation. For example, as explained below, we can enclose the truncation error of a Taylor series expansion by generating Taylor coefficients in interval arithmetic. Such enclosures can be used to compute guaranteed bounds either for an integral [9] or for an initial value problem (IVP) for an ordinary differential equation (ODE) (see [36] for a survey). An extensive summary of applications of and resources for interval computations can be found in [21].

Interval methods have not been popular in the past for several reasons. Chief amongst these are:

- a validated method frequently requires much more computational work and storage than an approximate method for the same problem;
- the bounds produced by many early validated methods were not tight;
- it is difficult to develop good, robust, portable software for validated methods in older languages such as Fortran or C.

However, many of these difficulties are not as serious now as they were several decades ago when interval methods were first proposed.

It remains true that a validated method may run several orders of magnitude slower than an approximate method for the same problem. (We hope to quantify the extent to which this is true in another paper.) Consequently, a validated method may not be appropriate for large computationally intensive problems or ones that are time critical, such as real-time control problems. However, when interval methods are criticized for being more expensive than traditional methods, we should remember that the former compute rigorous enclosures for the mathematically correct result, not just approximations as the latter do. Moreover, with the speed of processors increasing and the cost of memory decreasing, many IVPs for ODEs are not expensive to solve by modern standards. For such problems, we believe the time is now ripe to shift the responsibility for determining the reliability of numerical results from the user to the computer. If a problem is not computationally expensive to solve by standard methods, but accuracy is important, it may be much more cost-effective to let the computer spend a few more minutes computing a validated solution with a guaranteed error bound, rather than having the user attempt to verify the reliability of a numerical solution computed by approximate methods. Note that the reliability issue does not arise for validated methods, since, if implemented correctly, software based on a validated method always produces reliable results.

Also note that in some cases an interval computation is not more work than a standard floating-point calculation. For example, interval methods can be faster than standard floating-point schemes for finding all solutions or global optima of a nonlinear equation [15] (see [21] for references).

The bounds produced by many early interval methods were not tight. One reason is that a straightforward replacement of real operations by interval arithmetic operations in a point algorithm often leads to unacceptably wide bounds. For example, to bound the solution of a linear system, we can apply Gaussian elimination with floating-point numbers replaced by intervals. This, though, often produces large overestimations of the error. A good interval linear equation solver is significantly different from the corresponding floating-point algorithm [37]. Similarly, early interval methods for ODEs suffered from the *wrapping effect.* As a result, they often produced very poor bounds. However, as described below, more sophisticated modern methods cope quite effectively with this difficulty and produce tight bounds for a wide range of problems.

Until recently, there has been a scarcity of interval-arithmetic software. In part, this is because it is difficult to write good, robust, portable interval-arithmetic software in older languages such as Fortran or C. This task is greatly facilitated by modern programming languages such as C++ and Fortran 90, which allow user-defined types and operator overloading. As a result, we can implement an interval data type and write interval arithmetic operations without explicit function calls. An important step towards wider use of interval computations is the development of a Fortran compiler that supports intervals as an intrinsic data type [43].

This chapter focuses on interval methods for IVPs for ODEs: we give a brief overview of these methods, discuss software issues related to the implementation of a validated ODE solver, and describe the structure of the VNODE (Validated Numerical ODE) package that we are developing.

Sections 2 to 4 present the mathematical background necessary for understanding the rest of this chapter. Section 2 discusses interval arithmetic operations and computing ranges of functions. Section 3 states in more detail the IVP that is a subject of this work. Section 4 reviews validated methods for computing guaranteed bounds on the solution of an IVP for an ODE. We consider one step of these methods, automatic generation of Taylor coefficients, validating existence and uniqueness of the solution, explain the wrapping effect, and describe Lohner's method for reducing it. We also discuss an Hermite-Obreschkoff scheme for computing bounds on the solution of an IVP for an ODE.

Sections 5 to 10 are devoted to software issues in validated ODE solving. First, we discuss the goals that we have set to achieve with the design of VNODE (§5), available software for computing validated solutions of IVPs for ODEs (§6), object-oriented concepts (§7), and the choice of C++ to implement VNODE (§8). Then, in §9, we describe the structure of VNODE and illustrate how it can be used. Section 10 discusses problems that we have encountered during the development of VNODE.

2 Interval Arithmetic and Ranges of Functions

An interval $[a] = [\underline{a}, \bar{a}]$ is the set of real numbers

$$[a] = [\underline{a}, \bar{a}] = \{x \mid \underline{a} \le x \le \bar{a}, \quad \underline{a}, \bar{a} \in \mathbb{R}\}.$$

If $[a]$ and $[b]$ are intervals and $\circ \in \{+,-,*,/\}$, then the interval-arithmetic operations are defined by

$$[a] \circ [b] = \{x \circ y \mid x \in [a],\ y \in [b]\}, \quad 0 \notin [b] \text{ when } \circ = / \tag{1}$$

[33]. Definition (1) can be written in the following equivalent form (we omit $*$ in the notation):

$$[a] + [b] = [\underline{a} + \underline{b}, \bar{a} + \bar{b}], \tag{2}$$

$$[a] - [b] = [\underline{a} - \bar{b}, \bar{a} - \underline{b}], \tag{3}$$

$$[a]\,[b] = [\min\{\underline{a}\underline{b}, \underline{a}\bar{b}, \bar{a}\underline{b}, \bar{a}\bar{b}\},\ \max\{\underline{a}\underline{b}, \underline{a}\bar{b}, \bar{a}\underline{b}, \bar{a}\bar{b}\}], \tag{4}$$

$$[a] / [b] = [\underline{a}, \bar{a}]\,[1/\bar{b}, 1/\underline{b}], \quad 0 \notin [b]. \tag{5}$$

The formulas[1] (2–5) allow us to express the interval arithmetic operations using the endpoints of the interval operands, thus obtaining formulas for real interval arithmetic.

The strength of interval arithmetic when implemented on a computer is in computing rigorous enclosures of real operations by including rounding errors in the computed bounds. To include such errors, we round the real intervals in (2–5) outwards, thus obtaining machine interval arithmetic. For example, when adding intervals, we round $\underline{a} + \underline{b}$ down and round $\bar{a} + \bar{b}$ up. For simplicity of the discussion, we assume real interval arithmetic in this chapter. Because of the outward roundings, intervals computed in machine interval arithmetic always contain the corresponding real intervals.

Definition (1) and formulas (2–5) can be extended to vectors and matrices. If their components are intervals, we obtain interval vectors and matrices. The arithmetic operations involving interval vectors and matrices are defined by the same formulas as in the scalar case, except that scalars are replaced by intervals.

An important property of interval-arithmetic operations is their monotonicity: if $[a]$, $[a_1]$, $[b]$, and $[b_1]$ are such that $[a] \subseteq [a_1]$ and $[b] \subseteq [b_1]$, then

$$[a] \circ [b] \subseteq [a_1] \circ [b_1], \quad \circ \in \{+,-,*,/\}. \tag{6}$$

Using (6), we can *rigorously* enclose the range of a function.

Let $f : \mathbb{R}^n \to \mathbb{R}$ be a function on $\mathcal{D} \subseteq \mathbb{R}^n$. The interval-arithmetic evaluation of f on $[a] \subseteq \mathcal{D}$, which we denote by $f([a])$, is obtained by replacing each occurrence of a real variable with a corresponding interval, by replacing the standard functions with enclosures of their ranges, and by performing interval-arithmetic operations instead of the real operations. From (6), the range of f, $\{f(x) \mid x \in [a]\}$, is always contained in $f([a])$.

Although $f([a])$ is not unique, since expressions that are mathematically equivalent for scalars, such as $x(y+z)$ and $xy+xz$, may have different values

[1] Addition and subtraction are nearly always expressed as in (2) and (3), but different sources may give different formulas for multiplication and division.

if x, y, and z are intervals, a value for $f([a])$ can be determined from the code list, or computational graph, for f.

One reason for obtaining different values for $f([a])$ is that the distributive law does not hold in general in interval arithmetic [1, pp. 3–5]. However, for any three intervals $[a]$, $[b]$, and $[c]$, the subdistributive law

$$[a]([b]+[c]) \subseteq [a][b]+[a][c]$$

holds. This implies that if we rearrange an interval expression, we may obtain tighter bounds. An important case when the distributive law does hold is when $[a]$ is a degenerate interval ($\underline{a} = \bar{a}$).

Addition of interval matrices is associative, but multiplication of interval matrices is not associative in general [37, pp. 80–81]. Also, the distributive law does not hold in general for interval matrices [37, p. 79].

If $f : \mathbb{R}^n \to \mathbb{R}$ is continuously differentiable on $\mathcal{D} \subseteq \mathbb{R}^n$ and $[a] \subseteq \mathcal{D}$, then, for any y and $b \in [a]$, $f(y) = f(b) + f'(\eta)(y - b)$ for some $\eta \in [a]$ by the mean-value theorem. Thus,

$$f(y) \in f_M([a], b) = f(b) + f'([a])([a] - b) \tag{7}$$

[33, p. 47]. The mean-value form, $f_M([a], b)$, is popular in interval methods since it often gives better enclosures for the range of f than the straightforward interval-arithmetic evaluation of f itself. Like f, f_M is not uniquely defined, but a value for it can be determined from the code lists for f and f'.

3 The Initial-Value Problem

We consider the set of autonomous IVPs

$$y'(t) = f(y) \tag{8}$$

$$y(t_0) \in [y_0], \tag{9}$$

where $t \in [t_0, T]$ for some $T > t_0$. Here $t_0, T \in \mathbb{R}$, $f \in C^{k-1}(\mathcal{D})$, $\mathcal{D} \subseteq \mathbb{R}^n$ is open, $f : \mathcal{D} \to \mathbb{R}^n$, and $[y_0] \subseteq \mathcal{D}$. The condition (9) permits the initial value $y(t_0)$ to be in an interval, rather than specifying a particular value.

We assume that the representation of f contains only a finite number of constants, variables, elementary operations, and standard functions. Since we assume $f \in C^{k-1}(\mathcal{D})$, we exclude functions that contain, for example, branches, abs, or min. For expositional convenience, we consider only autonomous systems, but this is not a restriction of consequence since a nonautonomous system of ODEs can be converted into an autonomous system, and our methods can be extended easily to nonautonomous systems.

We consider a grid $t_0 < t_1 < \cdots < t_m = T$, which is not necessarily equally spaced, and denote the stepsize from t_j to t_{j+1} by h_j ($h_j = t_{j+1} - t_j$). The step from t_j to t_{j+1} is referred to as the $(j+1)$st step. We denote the solution of (8) with an initial condition y_j at t_j by $y(t; t_j, y_j)$. For an interval, or an interval vector in general, $[y_j]$, we denote by $y(t; t_j, [y_j])$ the set of solutions $\{y(t; t_j, y_j) \mid y_j \in [y_j]\}$.

Our goal is to compute intervals (or interval vectors) $[y_j]$, $j = 1, 2, \ldots, m$, that are guaranteed to contain the solution of (8–9) at $t_1, t_2, \ldots, t_m$. That is,

$$y(t_j; t_0, [y_0]) \subseteq [y_j], \quad \text{for } j = 1, 2, \ldots, m.$$

4 An Overview of Validated Methods for IVPs for ODEs

Most validated methods for IVPs for ODEs are based on Taylor series [33], [27], [12], [30], [40], [6]. The Taylor series approach has been popular because of the simple form of the error term, which can be readily bounded using interval arithmetic. In addition, the Taylor series coefficients can be efficiently generated by automatic differentiation, the order of the method can be changed easily by adding or deleting Taylor series terms, and the stepsize can be changed without doing extra work to recompute Taylor series coefficients.

4.1 One Step of a Validated Method

The $(j+1)$st step $(j \geq 0)$ in most Taylor series methods consists of two phases [36]:

ALGORITHM I: Compute a stepsize h_j and an a priori enclosure $[\tilde{y}_j]$ of the solution such that $y(t; t_j, y_j)$ is guaranteed to exist for all $t \in [t_j, t_{j+1}]$ and all $y_j \in [y_j]$, and $y(t; t_j, [y_j]) \subseteq [\tilde{y}_j]$ for all $t \in [t_j, t_{j+1}]$.

ALGORITHM II: Compute a tighter enclosure $[y_{j+1}]$ for $y(t_{j+1}; t_0, [y_0])$.

In this section, we briefly review methods for implementing these two algorithms. A detailed survey of Taylor series methods can be found in [36].

4.2 Automatic Generation of Taylor Coefficients

Since the interval methods for IVPs for ODEs considered here use Taylor series, we outline the recursive generation of Taylor coefficients.

If we know the Taylor coefficients $u^{(i)}/i!$ and $v^{(i)}/i!$ for $i = 0, \ldots, p$ for two functions u and v, then we can compute the pth Taylor coefficient of $u \pm v$, uv, and u/v by standard calculus formulas [33, pp. 107–130].

We introduce the sequence of functions

$$f^{[0]}(y) = y, \tag{10}$$

$$f^{[i]}(y) = \frac{1}{i}\left(\frac{\partial f^{[i-1]}}{\partial y} f\right)(y), \quad \text{for } i \geq 1. \tag{11}$$

For the IVP $y' = f(y)$, $y(t_j) = y_j$, the Taylor coefficient of $y(t)$ at t_j satisfies

$$(y_j)_i = f^{[i]}(y_j) = \frac{1}{i}\big(f(y_j)\big)_{i-1}, \quad i \geq 1, \tag{12}$$

where $\big(f(y_j)\big)_{i-1}$ is the $(i-1)$st coefficient of f evaluated at y_j. Using (12) and formulas for the Taylor coefficients of sums, products, quotients, and the standard functions, we can recursively evaluate $(y_j)_i$, for $i \geq 1$. It can be shown that to generate k coefficients, we need at most $O(k^2)$ operations [33, pp. 111–112]. Note that this is far more efficient than the standard symbolic generation of Taylor coefficients.

If we have a procedure to compute the point Taylor coefficients of $y(t)$ and perform the computations in interval arithmetic with $[y_j]$ instead of y_j, we obtain a procedure to compute the interval Taylor coefficients of $y(t)$.

4.3 Algorithm I: Validating Existence and Uniqueness of the Solution

Using the Picard-Lindelöf operator and the Banach fixed-point theorem, one can show that if h_j and $[\tilde{y}_j^0] \supseteq [y_j]$ satisfy

$$[\tilde{y}_j] = [y_j] + [0, h_j]\, f\left([\tilde{y}_j^0]\right) \subseteq [\tilde{y}_j^0], \tag{13}$$

then (8) with $y(t_j) = y_j$ has a unique solution $y(t; t_j, y_j)$ that satisfies $y(t; t_j, y_j) \in [\tilde{y}_j]$ for all $t \in [t_j, t_{j+1}]$ and all $y_j \in [y_j]$, [12, pp. 59–67].

We refer to a method based on (13) as a constant enclosure method. Such a method can be easily implemented, but a serious disadvantage of this approach is that the stepsize is restricted to Euler steps. Usually, Algorithm II allows larger stepsizes, but they are often limited by Algorithm I.

One can obtain methods that enable larger stepsizes by using polynomial enclosures [31] or more Taylor series terms in the sum in (13), thus obtaining a Taylor series enclosure method [33, pp. 100–103], [10], [34].

4.4 Algorithm II: Computing a Tight Enclosure

Consider the Taylor series expansion

$$y_{j+1} = y_j + \sum_{i=1}^{k-1} h_j^i f^{[i]}(y_j) + h_j^k f^{[k]}(y; t_j, t_{j+1}), \tag{14}$$

where $y_j \in [y_j]$ and $f^{[k]}(y; t_j, t_{j+1})$ denotes $f^{[k]}$ with its lth component ($l = 1, \ldots, n$) evaluated at $y(\xi_{jl})$, for some $\xi_{jl} \in [t_j, t_{j+1}]$. Using the a priori bounds $[\tilde{y}_j]$ on $[t_j, t_{j+1}]$, we can enclose the local truncation error of the Taylor series (14) on $[t_j, t_{j+1}]$ for any $y_j \in [y_j]$ by $h_j^k f^{[k]}([\tilde{y}_j])$. We can also bound with $f^{[i]}([y_j])$ the ith Taylor coefficient at t_j. Therefore,

$$[y_{j+1}] = [y_j] + \sum_{i=1}^{k-1} h_j^i f^{[i]}([y_j]) + h_j^k f^{[k]}([\tilde{y}_j]) \tag{15}$$

contains $y(t_{j+1}; t_0, [y_0])$. However, as explained below, (15) illustrates how replacing real numbers by intervals often leads to large overestimations.

Denote by $w([a]) = \bar{a} - \underline{a}$ the width of an interval $[a]$; the width of an interval vector or interval matrix is defined componentwise. Since $w([a] + [b]) = w([a]) + w([b])$,

$$w([y_{j+1}]) = w([y_j]) + \sum_{i=1}^{k-1} h_j^i w(f^{[i]}([y_j])) + h_j^k w(f^{[k]}([\tilde{y}_j])) \geq w([y_j]),$$

which implies that the width of $[y_j]$ almost always increases[2] with j, even if the true solution contracts.

A better approach is to apply the mean-value theorem to $f^{[i]}$, $i = 1, \dots, (k-1)$, in (14), obtaining that, for any $\hat{y}_j \in [y_j]$,

$$\begin{aligned} y_{j+1} = \hat{y}_j + &\sum_{i=1}^{k-1} h_j^i f^{[i]}(\hat{y}_j) + h_j^k f^{[k]}(y; t_j, t_{j+1}) \\ &+ \left\{ I + \sum_{i=1}^{k-1} h_j^i J\left(f^{[i]}; y_j, \hat{y}_j\right) \right\} (y_j - \hat{y}_j), \end{aligned} \tag{16}$$

where $J\left(f^{[i]}; y_j, \hat{y}_j\right)$ is the Jacobian of $f^{[i]}$ with its lth row evaluated at $y_j + \theta_{il}(\hat{y}_j - y_j)$ for some $\theta_{il} \in [0, 1]$ $(l = 1, \dots, n)$. This formula is the basis of the interval Taylor series (ITS) methods of Moore [33], Eijgenraam [12], Lohner [30], and Rihm [40] (see also [36]).

Let

$$[S_j] = I + \sum_{i=1}^{k-1} h_j^i J\left(f^{[i]}; [y_j]\right) \quad \text{and} \quad [z_{j+1}] = h_j^k f^{[k]}([\tilde{y}_j]). \tag{17}$$

Using the notation (17), we can bound (16) in interval arithmetic by

$$[y_{j+1}] = \hat{y}_j + \sum_{i=0}^{k-1} h_j^i f^{[i]}(\hat{y}_j) + [z_{j+1}] + [S_j]([y_j] - \hat{y}_j). \tag{18}$$

The ith Jacobian $(i = 1, \dots, k-1)$ in (17) can be computed by generating the Taylor coefficient $f^{[i]}([y_j])$ and then differentiating it [3], [4]. Alternatively, these Jacobians can be computed by generating Taylor coefficients for the associated variational equation [30].

If we compute enclosures with (18), the widths of the computed intervals may decrease. However, this approach may work poorly in some cases, because the interval vector $[S_j]([y_j] - \hat{y}_j)$ may significantly overestimate the set $\{S_j(y_j - \hat{y}_j) \mid S_j \in [S_j],\ y_j \in [y_j]\}$. This is called the wrapping effect.

[2] An equality is possible only in the trivial cases $h_j = 0$ or $w(f^{[i]}([y_j])) = 0$, $i = 1, \dots k-1$, and $w(f^{[k]}([\tilde{y}_j])) = 0$.

The Wrapping Effect. The wrapping effect is clearly illustrated by Moore's example [33],

$$\begin{aligned} y_1' &= y_2 \\ y_2' &= -y_1, \\ y(0) &\in [y_0]. \end{aligned} \tag{19}$$

The interval vector $[y_0]$ can be viewed as a rectangle in the (y_1, y_2) plane. At $t_1 > 0$, the true solution of (19) is the rotated rectangle shown in Figure 1. If we want to enclose this rectangle in an interval vector, we have to wrap it

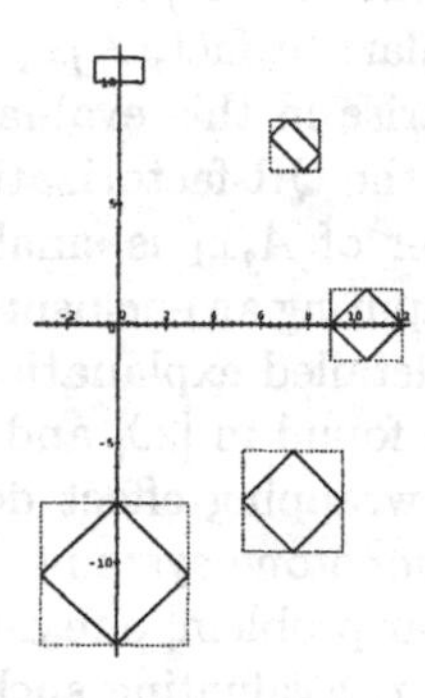

Fig. 1. The rotated rectangle is wrapped at $t = \frac{\pi}{4}n$, $n = 1, 2, 3, 4$.

by another rectangle with sides parallel to the y_1 and y_2 axes. On the next step, this larger rectangle is rotated and so must be enclosed in a still larger one. Thus, at each step, the enclosing rectangles become larger and larger, but the solution set remains a rectangle of the same size. Moore showed that at $t = 2\pi$, the interval inclusion is inflated by a factor of $e^{2\pi} \approx 535$ as the stepsize approaches zero [33, p. 134].

Lohner's Method. Here, we describe Lohner's QR-factorization method [29], which is one of the most successful, general-purpose methods for reducing the wrapping effect.

Denote by $m([a]) = (\bar{a} + \underline{a})/2$ the midpoint of an interval $[a]$; the midpoint of an interval vector or interval matrix is defined componentwise. Let

$$A_0 = I, \; \hat{y}_0 = m([y_0]), \; [r_0] = [y_0] - \hat{y}_0, \quad \text{and} \tag{20}$$

$$\hat{y}_j = \hat{y}_{j-1} + \sum_{i=1}^{k-1} h_{j-1}^i f^{[i]}(\hat{y}_{j-1}) + m([z_j]) \quad (j \geq 1) \tag{21}$$

(I is the identity matrix). From (16–17) and (20–21), we compute

$$[y_{j+1}] = \hat{y}_j + \sum_{i=1}^{k-1} h_j^i f^{[i]}(\hat{y}_j) + [z_{j+1}] + \big([S_j]\,A_j\big)[r_j] \tag{22}$$

and propagate for the next step the vector

$$[r_{j+1}] = \big(A_{j+1}^{-1}([S_j]\,A_j)\big)[r_j] + A_{j+1}^{-1}\big([z_{j+1}] - m\,([z_{j+1}])\big). \tag{23}$$

Here $A_j \in \mathbb{R}^{n\times n}$ is a nonsingular matrix. If $A_{j+1} \in [S_j]\,A_j$, $w([S_j])$ is small, and A_j is well conditioned, then $A_{j+1}^{-1}([S_j]\,A_j)$ is close to the identity matrix. Since we can also keep $w([z_{j+1}])$ small, there is not a large overestimation in forming $[r_{j+1}]$. However, this choice of A_{j+1} does not guarantee that it is well conditioned or even nonsingular. In fact, A_{j+1} may be ill conditioned, and a large overestimation may arise in this evaluation. Usually, a good choice for A_{j+1} is the Q-factor from the QR-factorization of $\widehat{A}_{j+1} = m\,([S_j]A_j)$. As a result, the condition number of A_{j+1} is small and the wrapping effect is reduced (on each step) by computing an enclosure in an orthogonal coordinate induced by this Q-factor. A detailed explanation of the reasons for this and other choices for A_{j+1} can be found in [29] and [36].

We should note that the wrapping effect does not occur in some problems, for example, in quasi-monotone systems (see [36]) or one-dimensional problems. Also, for a particular problem, one anti-wrapping strategy may be more appropriate than another. Evaluating such strategies is not trivial and will not be discussed in this chapter.

An Interval-Hermite Obreschkoff Method. For more than 30 years, Taylor series has been the only effective approach for computing rigorous bounds on the solution of an IVP for an ODE. Recently, we developed a new scheme, an interval Hermite-Obreschkoff (IHO) method [35], [34]. Here, we outline the method and its potential.

Let

$$c_i^{q,p} = \frac{q!}{(p+q)!}\,\frac{(q+p-i)!}{(q-i)!}$$

(q, p, and $i \geq 0$), $y_j = y(t_j; t_0, y_0)$, and $y_{j+1} = y(t_{j+1}; t_0, y_0)$. It can be shown that

$$\begin{aligned}\sum_{i=0}^{q}(-1)^i c_i^{q,p} h_j^i f^{[i]}(y_{j+1}) &= \sum_{i=0}^{p} c_i^{p,q} h_j^i f^{[i]}(y_j) \\ &\quad + (-1)^q \frac{q!p!}{(p+q)!} h_j^{p+q+1} f^{[k]}(y; t_j, t_{j+1})\end{aligned} \tag{24}$$

[51], [34]. This formula is the basis of our IHO method. The method we propose in [34] consists of two phases, which can be considered as a predictor

and a corrector. The predictor computes an enclosure $[y_{j+1}^{(0)}]$ of the solution at t_{j+1}, and using this enclosure, the corrector computes a tighter enclosure $[y_{j+1}] \subseteq [y_{j+1}^{(0)}]$ at t_{j+1}. Since $q > 0$, (24) is an implicit scheme. The corrector applies a Newton-like step to tighten $[y_{j+1}^{(0)}]$.

We have shown in [35] and [34] that for the same order and stepsize, our IHO method has smaller local error, better stability, and requires fewer Jacobian evaluations than an ITS method. The extra cost of the Newton step is one matrix inversion and a few matrix multiplications.

5 Objectives

Our primary goal is to provide a software package that will assist researchers in the numerical study and comparison of schemes and heuristics used in computing validated solutions of IVPs for ODEs. The VNODE (Validated Numerical ODE) package that we are developing is intended to be *a uniform implementation of a generic validated solver for IVPs for ODEs.* Uniform means that the design and implementation of VNODE follow well-defined patterns. As a result, implementing, modifying, and using methods can be done systematically. Generic means that the user can construct solvers by choosing appropriate methods from sets of methods. This property enables us to isolate and compare methods implementing the same part of a solver. For example, we can assemble two solvers that differ only in the module implementing Algorithm II. Then, the difference in the numerical results, obtained by executing the two solvers, will indicate the difference in the performance of Algorithm II (at least in the context of the other methods and heuristics used in VNODE). Since we would like to investigate algorithms, being able to isolate them is an important feature of such an environment.

We list and briefly explain some of the goals we have tried to achieve with the design of VNODE. Provided that a validated method for IVPs for ODEs is implemented correctly, the reliability issue does not arise: if a validated solver returns an enclosure of the solution, then the solution is guaranteed to exist within the computed bounds.

Modularity. The solver should be organized as a set of modules with well-defined interfaces. The implementation of each module should be hidden, but if necessary, the user should be able to modify the implementation.

Flexibility. Since we require well-defined interfaces, we should be able to replace a method, inside a solver, without affecting the rest of the solver. Furthermore, we should be able to add methods following the established structure, without modifying the existing code.

Efficiency. The methods incorporated in VNODE do not have theoretical limits on the size of the ODE system or order of a method. However, these methods require the computation of high-order Taylor coefficients and Jacobians of Taylor coefficients. As a result, the efficiency of a validated solver is determined mainly by the efficiency of the underlying automatic differentiation package. Other factors that contribute to the performance are: the efficiency of the interval-arithmetic package, the programming language, and the actual implementation of the methods. One difficulty is that to achieve flexibility, we may need to repeat the same calculations in two parts of a solver. For example, to separate Algorithm I and Algorithm II, we may need to generate the same Taylor coefficients in both algorithms. However, the repetition of such computations should be avoided, where possible.

Since VNODE is to be used for comparing and assessing methods, it should contain the most promising ones. Moreover, VNODE should support rapid prototyping.

6 Background

The area of computing validated solutions of IVPs for ODEs is not as developed as the area of computing approximate solutions [36]. For example, we still do not have well-studied stepsize and order control strategies nor methods suitable for stiff problems. Also, reducing the wrapping effect in Algorithm II and taking larger stepsizes in Algorithm I are still open problems in validated ODE solving. With respect to the tools involved, a validated solver is inherently more complex than a classical ODE solver. In addition to an interval-arithmetic package, a major component of a validated solver is the module for automatic generation of interval Taylor coefficients (see §8.1).

Currently, there are three available packages for computing guaranteed bounds on the solution of an IVP for an ODE: AWA [30], ADIODES [46] and COSY INFINITY [5]. We briefly summarize each in turn.

AWA is an implementation of Lohner's method (§4.4) and the constant enclosure method (§4.3) for validating existence and uniqueness of the solution. This package is written in Pascal-XSC [24], an extension of Pascal for scientific computing.

ADIODES is a C++ implementation of a solver using the constant enclosure method in Algorithm I and Lohner's method in Algorithm II. The stepsize in both ADIODES and AWA is often restricted to Euler steps by Algorithm I.

COSY INFINITY is a Fortran-based code to study and to design beam physics systems. The method for verified integration of ODEs uses high-order Taylor polynomials with respect to time and the initial conditions. The wrapping effect is reduced by establishing functional dependency between initial and final conditions (see [6]). For that purpose, the computations are carried out with Taylor polynomials with real floating-point coefficients and a guaranteed error bound for the remainder term. Thus, the arithmetic operations and standard functions are executed with such Taylor polynomials as operands. Although the approach described in [6] reduces the wrapping effect substantially, working with polynomials is significantly more expensive than working with intervals.

7 Object-Oriented Concepts

We have chosen an object-oriented approach in designing VNODE and C++ [13] to implement it. This is not the first object-oriented design of an ODE solver. The Godess [38] and TIDE [17] (a successor of CODE++ [16]) packages offer generic ODE solvers that implement traditional methods for IVPs for ODEs. Another successful package is Diffpack [28], which is devised for solving partial differential equations. In [28], there is also an example of how to construct an object-oriented ODE solver.

In this section, we briefly review some object-oriented concepts supported in C++. We use them in the discussion of the choice of programming language (Section 8) and in the description of VNODE (Section 9). A reader familiar with object-oriented terms can skip the current section (7).

A good discussion of object-oriented concepts, analysis, and design can be found in [7]. An excellent book on advanced C++ styles and idioms is [8]. A study of nonprocedural paradigms for numerical analysis, including object-oriented ones, is presented in [47].

Data Abstraction. In the object model, a software system can be viewed as a collection of objects that interact with each other to achieve a desired functionality. An object is an instance of a class, which defines the structure and behavior of its objects. By grouping data and methods inside a class and specifying its interface, we achieve encapsulation, separating the interface from the implementation. Hence, the user can change the data representation and the implementation of a method[3] (or methods) of a class without modifying the software that uses it. By encapsulating data, we can avoid function calls with long parameter lists, which are intrinsic to procedural languages like Fortran 77. A class can encapsulate data or algorithms, or both.

Inheritance and Polymorphism. Inheritance and polymorphism are powerful features of object-oriented languages. Inheritance allows code reuse: a derived class can reuse data and functions of its base class(es). The role of polymorphism is to apply a given function to different types of objects. Often polymorphism and inheritance are used with abstract classes. An abstract class defines abstract functions, which are implemented in its subclasses; it has no instances and an object of such a class cannot be created.

[3] We use *method* in two different contexts: to denote a member function of a class or a method in VNODE.

Operator Overloading. Operator overloading allows the operators of the language to be overloaded for user defined types. To program interval operations without explicit function calls, we have to use a language that supports operator overloading. Without it, programming interval-arithmetic expressions is cumbersome, because we have to code interval-arithmetic operations by using function calls. Both C++ and Fortran 90 provide operator overloading. This feature is used to build interval-arithmetic libraries like PROFIL/BIAS [25] (C++) and INTLIB (Fortran 90) [22], as discussed in Section 8.2.

8 Choice of Language: C++ versus Fortran 90

We considered C++ and Fortran 90 [32] as candidates to implement VNODE, but we preferred C++. Procedural languages like C or Fortran 77 can be used to implement an object-oriented design [2]. However, using a language that supports object-oriented programming usually reduces the effort for implementing object-oriented software.

Our choice was determined by the following considerations, listed in order of importance:

1. availability of software for automatic generation of interval Taylor coefficients;
2. performance and built-in functions of the available interval-arithmetic packages;
3. support of object-oriented concepts; and
4. efficiency.

In this section, we address each of these considerations in turn.

8.1 Software for Automatic Generation of Interval Taylor Coefficients

Although packages for automatic differentiation (AD) are available (see for example [19] and [52]), to date, only two free packages for automatic generation of interval Taylor coefficients for the solution of an ODE and the Jacobians of these coefficients are known to the authors. These are the FADBAD/TADIFF [3], [4] and IADOL-C [18] packages. They are written in C++ and implement AD through operator overloading.

TADIFF and FADBAD are two different packages. TADIFF can generate Taylor coefficients with respect to time. Then, FADBAD can be used to compute Jacobians of Taylor coefficients by applying the forward mode of AD [39] to these coefficients. FADBAD and TADIFF are not optimized to handle large and sparse systems. Also, they perform all the work in main memory.

The IADOL-C package is an extension of ADOL-C [14] that allows generic data types. ADOL-C can compute Taylor coefficients by using the forward mode and their Jacobians by applying the reverse mode [44] to these coefficients. The basic data type of ADOL-C is double. To use a new data type in IADOL-C, the user has to overload the arithmetic and comparison operations and the standard functions for that data type. Then, using IADOL-C is essentially the same as using ADOL-C. Since IADOL-C replaces only the double data type of ADOL-C, IADOL-C inherits all the functionality of ADOL-C. However, this overloading in IADOL-C causes it to run about three times slower than ADOL-C. This appears to be due largely to the C++ compilers rather than the AD package [18].

The ADOL-C package records the computation graph on a so-called tape. This tape is stored in the main memory, but, when necessary, is paged to disk. When generating Jacobians of Taylor coefficients, ADOL-C can exploit the sparsity structure of the Jacobian of the function for computing the right side. Since optimization techniques are used in ADOL-C, we expect the interval version, IADOL-C, to perform better than FADBAD/TADIFF on large and complex problems. On the other hand, FADBAD/TADIFF should perform well on small to medium-sized problems.

Currently, VNODE is configured with FADBAD/TADIFF, but we have also used IADOL-C. VNODE with these AD packages is based on the INTERVAL data type from the PROFIL/BIAS package, which we discuss in §8.2 and §8.4.

8.2 Interval Arithmetic Packages

The most popular and free interval-arithmetic packages are PROFIL/BIAS [25], written in C++, and INTLIB [23], written in Fortran 77 and available with a Fortran 90 interface [22]. The Fortran 90 version of INTLIB uses operator overloading. For references and comments on other available packages, see for example [22] or [25]. Recently, an extension of the Gnu Fortran compiler which supports intervals as an intrinsic data type was reported [43].

PROFIL/BIAS seems to be the fastest interval package. In comparison with other such packages, including INTLIB, PROFIL/BIAS is about one order of magnitude faster [25]. For efficiency, it uses directional rounding facilities of the processor on the machines on which it is installed. Portability is provided by isolating the machine dependent code in small assembler files, which are distributed with the package. Another reason for the efficiency of PROFIL/BIAS is that the number of rounding mode switches and sign tests are minimized in vector and matrix operations. In modern RISC architectures, sign tests and rounding mode switches cost nearly as much or even more than floating-point operations [25], [43]. In addition, PROFIL/BIAS is easy to use, and provides matrix and vector operations and essential routines, such as guaranteed linear equation solvers and optimization methods.

INTLIB is intended to be a totally portable Fortran 77 package. It does not assume IEEE arithmetic and does not assume anything about the accuracy of the standard functions of the underlying Fortran compiler. For portability, directed roundings are simulated by multiplying the computed results by $(1 + \epsilon)$, for an appropriate epsilon. Significant additional execution time is required for these multiplications.

8.3 Support of Object-Oriented Concepts

C++ is a fully object-oriented language, while Fortran 90 is not, because it does not support inheritance and polymorphism. The features of C++ (e.g., data abstraction, operator overloading, inheritance, and polymorphism) allow the goals in §5 to be achieved in a relatively simple way. Inheritance and polymorphism can be simulated in Fortran 90 [11], but this is cumbersome.

8.4 Efficiency

Compared to Fortran, C++ has been criticized for its poor performance for scientific computing. Here, we discuss an important performance problem: the pairwise evaluation of arithmetic expression with arguments of array types (e.g., matrices and vectors). More detailed treatment of this and other problems can be found in [41], [49], and [50].

In C++, executing overloaded arithmetic operations with array data types creates temporaries, which can introduce a significant overhead, particularly for small objects. For example, if `A`, `B`, `C`, and `D` are vectors, the evaluation of the expression

```
D = A + B + C
```

creates two temporaries: one to hold the result of `A + B`, and another to hold the result of `(A + B) + C`. Furthermore, this execution introduces three loops. Clearly, it would be better to compute this sum in one loop without temporaries. In Fortran 90, mathematical arrays are represented as elementary types and optimization is possible at the compiler level.

Because of better optimizing compilers and template techniques [48], [50], C++ is becoming more competitive for scientific computing. A good technique for reducing the overhead in the pairwise evaluation of expressions involving arrays is to use expression templates [48]. The expression template technique is based on performing compile-time transformation of the code using templates. With this technique, expressions containing vectors and matrices can be computed in a single pass without allocating temporaries. For example, with expression templates, it is possible to achieve a loop fusion [48], allowing the above sum to be evaluated in a single loop:

```
for ( int i = 1; i <= N; i++ )
      D(i) = A(i) + B(i) + C(i);
```

However, executing this loop in interval arithmetic may not be the best solution for the following reason. Each interval addition in this loop involves two changes of the rounding mode. As we pointed out, the approach of PROFIL/BIAS is to minimize the number of rounding mode switches. Suppose that we want to compute in PROFIL/BIAS

```
C = A + B,
```

where `A`, `B`, and `C` are vectors of the same dimensions. If we denote the components of `A`, `B`, and `C` by a_i, b_i, and c_i, respectively, PROFIL/BIAS changes the rounding mode downwards and computes $\underline{c}_i = \underline{a}_i + \underline{b}_i$, for $i = 1, 2, \ldots, n$. Then, this package changes the rounding mode upwards and computes $\bar{c}_i = \bar{a}_i + \bar{b}_i$, for $i = 1, 2, \ldots n$. Thus, the result of `A + B` is computed with two rounding mode switches. However, PROFIL/BIAS still creates temporaries.

9 The VNODE package

9.1 Structure

From an object-oriented perspective, it is useful to think of a numerical problem as an object containing all the information necessary to compute its solution. Also, we can think of a particular method, or a solver, as an object containing the necessary data and functions to perform the integration of a given problem. Then, we can compute a solution by "applying" a method object to a problem object. Most functions in VNODE have such objects as parameters. The description of the numerical problem and the methods in VNODE are implemented as classes in C++.

The problem classes are shown in Figure 2, and the method classes are shown in Figure 3. A box in Figures 2 and 3 denotes a class; the rounded, filled boxes denote abstract classes. Each of them declares one or more virtual functions, which are not defined in the corresponding abstract class, but must be defined in the derived classes.

The lines with $\triangle$ indicate an *is-a* relationship, which here denotes inheritance or a specialization of a base class; the lines with $\diamondsuit$ indicate a *has-a* relationship. It is realized either by a complete containment of an object `Y` within another object `X` or by a pointer from `X` to `Y`. The notation in these figures is similar to that suggested in [42].

In the next two subsections, we list the problem and method classes and provide brief explanations. Here, we do not discuss the classes for generating Taylor coefficients in VNODE. A description of VNODE will be given in the documentation of the code at `http://www.cs.toronto.edu/NA`.

Problem Classes. Class `ODE_PROBLEM` specifies the mathematical problem, that is, t_0, $[y_0]$, T, and a pointer to a function to compute the right side of

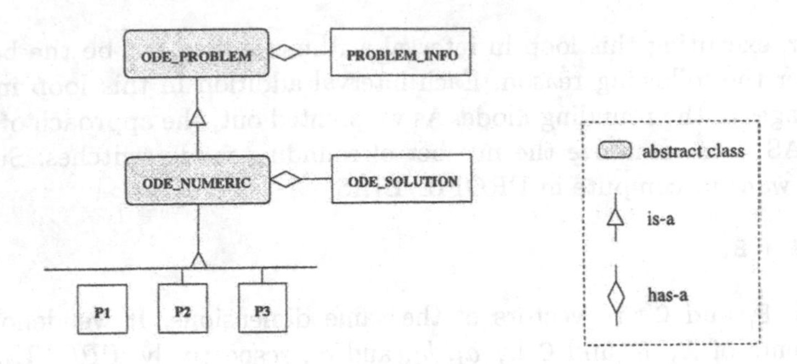

Fig. 2. Problem classes.

the ODE. It also contains a pointer to a class `PROBLEM_INFO`. It indicates, for example, if the problem is constant coefficient, scalar, has a closed form solution, or has a point initial condition. Such information is useful since the solver can determine from it which part of the code to execute.

`ODE_NUMERIC` specifies the numerical problem. This class contains data such as absolute and relative error tolerances,[4] and a pointer to a class `ODE_NUMERIC` representing a solution. The user-defined problems, `P1`, `P2`, and `P3` in Figure 2 are derived from this class. New problems can be added by deriving them from `ODE_NUMERIC`.

`ODE_SOLUTION` contains the last obtained a priori and tight enclosures of the solution and the value of t where the tight enclosure is computed. `ODE_SOLUTION` contains also a pointer to a file that stores information from the preceding steps (e.g., enclosures of the solution and stepsizes).

Method Classes. Class `ODE_SOLVER` is a general description of a method that "solves" an `ODE_NUMERIC` problem. The `ODE_SOLVER` class declares the pure virtual function `Integrate`, whose definition is not provided in it. As a result, instances of `ODE_SOLVER` cannot be created. This class also contains class `METHOD_CONTROL`, which includes different flags (encapsulated in `FLAGS`) and statistics collected during the integration (encapsulated in `STATISTICS`).

Class `VODE_SOLVER` implements a validated solver by defining the function `Integrate`. We have divided this solver into four abstract methods: for selecting an order, selecting a stepsize, and computing initial and tight enclosures of the solution. These methods are realized by the abstract classes `ORDER_CONTROL`, `STEP_CONTROL`, `INIT_ENCL`, and `TIGHT_ENCL`, respectively. Their purpose is to provide general interfaces to particular methods. A new method can be added by deriving it from these abstract classes. `Integrate` performs an integration by calling objects that are instances of classes derived from `ORDER_CONTROL`, `STEP_CONTROL`, `INIT_ENCL`, and `TIGHT_ENCL`. We have used polymorphism in implementing this function: it calls virtual functions

[4] How to specify and interpret error tolerances will be discussed in the documentation of VNODE.

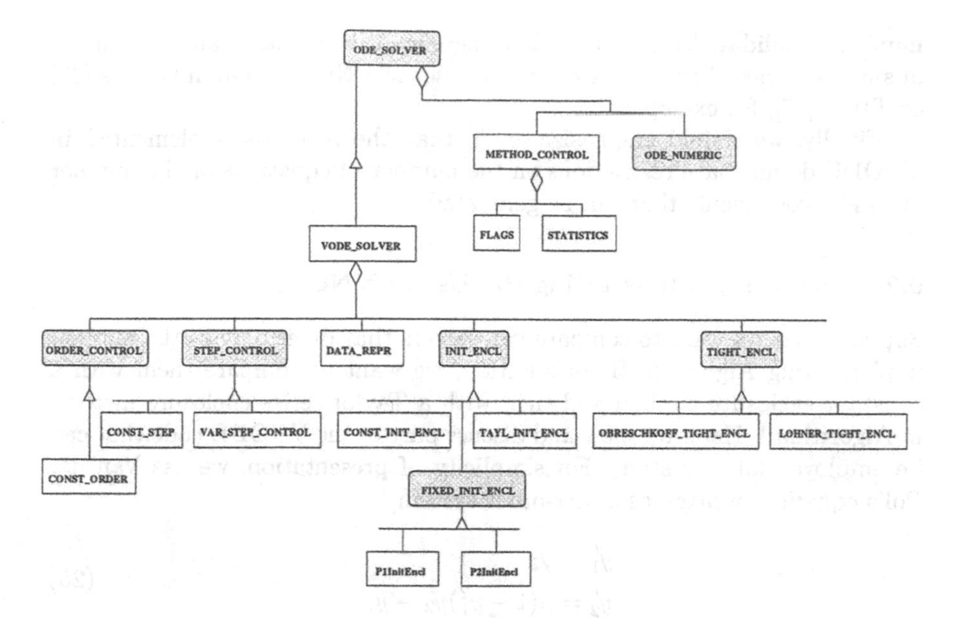

Fig. 3. Method classes.

that are declared, but not defined, in these abstract classes. During execution, depending on the type (class) of the object, a particular function will be called (see Section 9.2 for an example).

Currently, VNODE does not implement variable-order methods, and the class `ORDER_CONTROL` has only one derived class, `CONST_ORDER`, whose role is to return a constant order. For selecting a stepsize, `CONST_STEP` returns a constant stepsize on each step, and `VAR_STEP_CONTROL` implements a stepsize selection scheme based on controlling the width of the remainder term [34].

There are two methods for validating existence and uniqueness of the solution in VNODE: a constant enclosure method (`CONST_INIT_ENCL`) and a Taylor series method (`TAYL_INIT_ENCL`). The purpose of `FIXED_INIT_ENCL` is to compute a priori enclosures of the solution from the formula for the true solution, if the problem has a closed form solution. This class turns out to be helpful when we want to isolate the influence of Algorithm I, because it often reduces the input stepsize.

There are two methods for computing a tight enclosure of the solution: an interval Hermite-Obreschkoff method (`OBRESCHKOFF_TIGHT_ENCL`) and the method by Lohner (`LOHNER_TIGHT_ENCL`). They both use Lohner's QR-factorization technique for reducing the wrapping effect. The `VODE_SOLVER` class also has a pointer to `DATA_REPR`, which is responsible for generating and storing Taylor coefficients and their Jacobians.

As we briefly discussed in Section 6, interval methods for IVPs for ODEs are not as developed as point methods for IVPs for ODEs. As a result, the number of validated ODE methods is significantly smaller than the number of standard ones. Thus, we have fewer classes in VNODE than in Godess [38] or TIDE [17], for example.

Finally, we should emphasize again that the methods implemented in VNODE do not have restrictions on the number of equations or the number of Taylor coefficients that can be generated.

9.2 An Example Illustrating the Use of VNODE

Suppose that we want to compare two solvers that differ only in the method implementing Algorithm II. In addition, we want to compare them with a constant enclosure method and then with a Taylor series enclosure method in Algorithm I. Here, we show and discuss part of the VNODE code that can be employed for this study. For simplicity of presentation, we use Van der Pol's equation, written as a first-order system,

$$\begin{aligned} y_1' &= y_2 \\ y_2' &= \mu(1 - y_1^2)y_2 - y_1. \end{aligned} \tag{25}$$

where $\mu \in \mathbb{R}$. In a traditional ODE solver, we provide a function for computing the right side. In a validated solver, we also have to provide functions for generating Taylor coefficients and their Jacobians. Since we use an AD package for generating such coefficients, we have to specify a function for computing the right side of (25) for this package. We write the template function

```
template <class YTYPE> void VDPtemplate(YTYPE *yp, const YTYPE *y)
{
  yp[0] =  y[1];
  yp[1] =  MU*(1-sqr(y[0]))*y[1] - y[0];
}
```

which is used by FADBAD/TADIFF and IADOL-C to store the computation graph, and by VNODE to create a function for computing the right side. Then we derive a class `VDP` from `ODE_NUMERIC`. Since the details about the declaration of `VDP` are not essential to understand our example, we omit this declaration.

Figure 4 shows a typical use of VNODE classes. First, we create an `ODE_NUMERIC` object,[5] `VDP`, and load the initial condition, the interval of integration, and tolerance by calling the function `LoadProblemParam` (Part A). For testing, it is convenient to have a function that supplies different sets of data depending on the parameter to this function.

Then, we create methods and return pointers to them (Part B), as described below. `ITS` and `IHO` are pointers to objects for computing enclosures

[5] `Ptr` stands for pointer in Figure 4.

```
// ...
// A. Create the ODE problem.
PtrODENumeric ODE = new VDP;
ODE->LoadProblemParam(1);

// B. Create the methods.
int K, P, Q;
K = 11;          // order
P = Q = (K-1)/2;

PtrTightEncl ITS = new LOHNER_TIGHT_ENCL(K);
PtrTightEncl IHO = new OBRESCHKOFF_TIGHT_ENCL(P,Q);

PtrInitEncl  InitEncl  =
               new CONST_INIT_ENCL(ODE->Size, new VDPTaylGenVAR);
PtrStepCtrl  StepCtrl  = new VAR_STEP_CONTROL(ODE->Size);
PtrOrderCtrl OrderCtrl = new CONST_ORDER(K);
PtrDataRepr  DataRepr  =
               new TAYLOR_EXPANSION<VDPTaylGenODE,VDPTaylGenVAR>;

// Part C. Create the solvers and integrate.
PtrVODESolver SolverITS = new
   VODE_SOLVER(ODE, DataRepr, OrderCtrl, StepCtrl, InitEncl, ITS);

PtrVODESolver SolverIHO = new
   VODE_SOLVER(ODE, DataRepr, OrderCtrl, StepCtrl, InitEncl, IHO);

SolverITS->Integrate();
SolverIHO->Integrate();

// Part D. Replace the method implementing Algorithm I and integrate.
InitEncl = new TAYL_INIT_ENCL(ODE->Size, new VDPTaylGenODE,
    new VDPTaylGenVAR);

SolverITS->SetInitEncl(InitEncl);
SolverIHO->SetInitEncl(InitEncl);

SolverITS->Integrate();
SolverIHO->Integrate();
// ...
```

Fig. 4. The test code.

using Lohner's and the IHO methods, respectively. `InitEncl` is a pointer to an object for validating existence and uniqueness of the solution with the constant enclosure method; `StepControl` refers to an object that implements a variable stepsize control; and `OrderControl` points to an object that provides a constant value for the order.

The purpose of class `TAYLOR_EXPANSION` is to generate and store Taylor coefficients and their Jacobians. It is a template class, for which instances are created by specifying a class for generating Taylor coefficients and a class for generating Jacobians of Taylor coefficients. Here, we create such an instance with parameters `VDPTaylGenODE` and `VDPTaylGenVar`, which are classes[6] for generating Taylor coefficients and their Jacobians for (25).

In part C, we create two solvers, `SolverITS` and `SolverIHO` and integrate the problem by calling the `Integrate` function on these solvers.[7] Note that they differ only in the method for computing a tight enclosure of the solution. Thus, we can isolate and compare the two methods implementing Algorithm II.

Now, in part D, we want to replace the constant enclosure method for validating existence and uniqueness of the solution with a Taylor series method and repeat the same integrations. We create an instance of `TAYL_INIT_ENCL` by

```
InitEncl = new TAYL_INIT_ENCL(ODE->Size, new VDPTaylGenODE,
                                         new VDPTaylGenVAR);
```

set it by calling the `SetInitEncl` function, and integrate.

We explain how class `INIT_ENCL` works; the same idea is used in the other abstract classes. `INIT_ENCL` is an abstract class since it contains the pure virtual function

```
virtual void Validate( ... ) = 0;
```

(for simplicity, we leave out the parameters). Each of the derived classes of `INIT_ENCL` must declare a function with the same name and parameters and specify the body of the function. In `Integrate`, there is a call to `Validate`. During execution, depending on the object set, the appropriate `Validate` function will be called. We use polymorphism here: the function that is called is determined by the type of object during the execution of the program. In our example, the method for validating existence and uniqueness is replaced, but the integrator function is not changed. A user who wants to implement a new Algorithm I has to define a derived class of `INIT_ENCL` and an associated `Validate` function.

[6] We do not describe these classes here.

[7] We omit the details about extracting data after an integration.

10 Notes from Our Experience

Here, we present some general notes and then discuss exception handling issues and wrapping effect in generating interval Taylor coefficients.

The most difficult part in designing VNODE was to choose a structure that would allow us to implement existing schemes efficiently while providing the flexibility to allow us to add other methods later without requiring an extensive modification of existing classes. This was complicated by the lack of a broad range of validated methods for ODEs to test our proposed structure. Although our design is appropriate for the few promising existing validated methods and for the extensions of standard methods to validated methods that we currently envision, we cannot, of course, guarantee that the VNODE structure will accommodate all novel schemes that may be developed in the future. However, we have made several extensions to VNODE already and plan several more soon. These are all easy to fit within the existing structure of VNODE.

As noted earlier, VNODE does not support variable-order methods yet. There are now different classes for order and stepsize control, but it may be more convenient to merge them into one class. Future experience will show if we should keep them separate.

In earlier versions of VNODE, we created new problems by instantiating the `ODE_NUMERIC` class. Now, for a particular problem, we derive a class from it. Although the current version is "more object-oriented", creating problems by instantiating the `ODE_NUMERIC` class seems simpler than deriving problem classes.

We decided to implement classes for generating Taylor coefficients, so that when a new package is implemented, the user can easily substitute the new one for an old one. Given a function for computing the right-hand side of an ODE, a package for generating Taylor coefficients should provide (at least) the following operations: *generate* coefficients (and store them), *read* coefficients, *sum* coefficients, and *scale* coefficients. These four "basic" operations are sufficient for building a validated ODE solver based on the theory discussed in this chapter.

10.1 Exception Handling

We discuss two situations, in the context of PROFIL/BIAS, in which a flexible exception handling mechanism of the interval-arithmetic package would facilitate the development of validated ODE software and, we believe, interval software in general.

PROFIL/BIAS clears the exception flags of the processor when setting the rounding mode. With respect to division by an interval containing zero, this package can be compiled such that it interrupts (default mode) or continues the execution, if such a division occurs. (The user cannot switch between these two modes after PROFIL/BIAS has been compiled.) Although terminating program execution when division by zero occurs is a reasonable action in many applications, in the following situation, it is desirable to raise a flag and continue the execution.

When validating existence and uniqueness of the solution, we normally guess a priori bounds for the solution. Then, we verify whether these bounds satisfy some condition (as in (4.3); see also [34]). It may happen that we have guessed bounds that are too wide and contain zero. If there is a division by such a bound, we want to recognize division by zero without interrupting the execution, so we can try with a priori bounds that do not contain zero.

PROFIL/BIAS continues the execution if there is a NaN or an infinity. In some circumstances, as in the example described below, we prefer to interrupt the computations immediately if a NaN occurs.

After computing tight bounds for the solution, we intersect the tight and a priori enclosures: the true solution must be contained in the intersection. The `Intersect` function from PROFIL/BIAS returns true or false, depending on whether two intervals intersect. If the implementation is correct, this function must return true when computing an intersection of tight and a priori bounds. Even when the algorithms are implemented correctly, it is often difficult to predict if a floating-point exception will occur, and it is possible that the tight or a priori bounds contain a NaN. Then this function returns false. Although determining what has happened may not be very difficult, this may be time consuming. If the computations stop immediately when a NaN occurs, it would be easier to locate where a NaN has occurred.

10.2 Wrapping Effect in Generating Taylor Coefficients

Generating point Taylor coefficients for the solution of an IVP for an ODE is not a difficult task, though not trivial to implement. In the interval case, however, there may be a wrapping effect that would lead to large overestimation in these coefficients.

Consider the constant coefficient problem

$$y' = By, \quad y(0) \in [y_0]. \tag{26}$$

In practice, the relation (12) is used for generating interval Taylor coefficients. As a result, the ith coefficient is computed by

$$[y]_i = \frac{1}{i} B[y]_{i-1} = \frac{1}{i!} B\big(B \cdots (B\,(B[y_0])) \cdots\big). \tag{27}$$

Therefore, the computation of the ith Taylor coefficient may involve i wrappings. In general, this implies that the computed enclosure of the kth Taylor coefficient, $f^{[k]}([\tilde{y}_j])$, on $[t_j, t_{j+1}]$ may be a significant overestimation of the range of $y^{(k)}(t)/k!$ on $[t_j, t_{j+1}]$. As a result, a variable stepsize method that controls the width of $h_j^k f^{[k]}([\tilde{y}_j])$ may impose a stepsize limitation much smaller than one would expect. In this example, it would be preferable to compute the coefficients directly by

$$[y]_i = \frac{1}{i!} B^i [y_0],$$

which involves at most one wrapping.

We do not know how to reduce the wrapping effect in generating interval Taylor coefficients. The TADIFF package implements the relation (12), thus generating coefficients for (26) as in (27). We have not verified whether IADOL-C produces interval Taylor coefficients for (26) as in (27), but we expect that this is the case.

Acknowledgments

This work was supported in part by the Natural Sciences and Engineering Research Council of Canada, the Information Technology Research Centre of Ontario, and Communications and Information Technology Ontario.

References

1. G. Alefeld and J. Herzberger. *Introduction to Interval Computations.* Academic Press, New York, 1983.
2. S. Balay, W. D. Gropp, L. C. McInnes, and B. F. Smith. Parallelism in object-oriented numerical software libraries. In E. Arge, A. M. Bruaset, and H. P. Langtangen, editors, *Modern Software Tools in Scientific Computing,* pages 163–202. Birkhäuser, Boston, 1997. See also the PETSc home page http://www.mcs.anl.gov/petsc/.
3. C. Bendsten and O. Stauning. FADBAD, a flexible C++ package for automatic differentiation using the forward and backward methods. Technical Report 1996-x5-94, Department of Mathematical Modelling, Technical University of Denmark, DK-2800, Lyngby, Denmark, August 1996. FADBAD is available at http://www.imm.dtu.dk/fadbad.html.
4. C. Bendsten and O. Stauning. TADIFF, a flexible C++ package for automatic differentiation using Taylor series. Technical Report 1997-x5-94, Department of Mathematical Modelling, Technical University of Denmark, DK-2800, Lyngby, Denmark, April 1997. TADIFF is available at http://www.imm.dtu.dk/fadbad.html.
5. M. Berz. COSY INFINITY version 8 reference manual. Technical Report MSUCL–1088, National Superconducting Cyclotron Lab., Michigan State University, East Lansing, Mich., 1997. COSY INFINITY is available at http://www.beamtheory.nscl.msu.edu/cosy/.
6. M. Berz and K. Makino. Verified integration of ODEs and flows using differential algebraic methods on high-order Taylor models. *Reliable Computing,* 4:361–369, 1998.
7. G. Booch. *Object-Oriented Analysis and Design.* The Benjamin/Cummings Publishing Company Inc., Rational, Santa Clara, California, 2nd edition, 1994.

8. J. O. Coplien. *Advanced C++ Programming Styles and Idioms.* Addison-Wesley, AT & T Bell Laboratories, 1992.
9. G. F. Corliss and L. B. Rall. Adaptive, self-validating quadrature. *SIAM J. Sci. Stat. Comput.*, 8(5):831–847, 1987.
10. G. F. Corliss and R. Rihm. Validating an a priori enclosure using high-order Taylor series. In G. Alefeld and A. Frommer, editors, *Scientific Computing, Computer Arithmetic, and Validated Numerics*, pages 228–238. Akademie Verlag, Berlin, 1996.
11. V. K. Decyk, C. D. Norton, and B. K. Szymanski. Expressing object-oriented concepts in Fortran 90. *ACM Fortran Forum*, 16(1):13–18, April 1997.
12. P. Eijgenraam. *The Solution of Initial Value Problems Using Interval Arithmetic.* Mathematical Centre Tracts No. 144. Stichting Mathematisch Centrum, Amsterdam, 1981.
13. M. A. Ellis and B. Stroustrup. *The Annotated C++ Reference Manual.* Addison–Wesley, 1990.
14. A. Griewank, D. Juedes, and J. Utke. ADOL-C, a package for the automatic differentiation of algorithms written in C/C++. *ACM Trans. Math. Softw.*, 22(2):131–167, June 1996.
15. E. R. Hansen. *Global Optimization Using Interval Analysis.* Marcel Dekker, New York, 1992.
16. A. Hohmann. An implementation of extrapolation codes in C++. Technical Report TR 93-08, Konrad-Zuse-Zentrum für Informationstechnik, Berlin, Germany, 1993. Available at http://www.zib.de/bib/pub/pw/index.en.html.
17. A. Hohmann. TIDE, 1993. Available at http://www.rt.e-technik.tu-darmstadt.de/~mali/DOC/tide/welcome.html.
18. R. V. Iwaarden. IADOL-C, personal communications, 1997. IADOL-C is available through the author. E-mail: vaniwaar@metsci.com.
19. D. Juedes. A taxonomy of automatic differentiation tools. In A. Griewank and G. F. Corliss, editors, *Automatic Differentiation of Algorithms: Theory, Implementation, and Application*, pages 315–329. SIAM, Philadelphia, Penn., 1991.
20. R. B. Kearfott. Fortran 90 environment for research and prototyping of enclosure algorithms for constrained and unconstrained nonlinear equations. *ACM Trans. Math. Softw.*, 21(1):63–78, March 1995.
21. R. B. Kearfott. Interval computations: Introduction, uses, and resources. *Euromath Bulletin*, 2(1):95–112, 1996.
22. R. B. Kearfott. INTERVAL_ARITHMETIC: A Fortran 90 module for an interval data type. *ACM Trans. Math. Softw.*, 22(4):385–392, 1996.
23. R. B. Kearfott, M. Dawande, K. Du, and C. Hu. Algorithm 737: INTLIB: A portable Fortran 77 interval standard function library. *ACM Trans. Math. Softw.*, 20(4):447–459, December 1995.
24. R. Klatte, U. Kulisch, M. Neaga, D. Ratz, and C. Ullrich. *Pascal–XSC: Language Reference with Examples.* Springer-Verlag, Berlin, 1992.
25. O. Knüppel. PROFIL/BIAS – a fast interval library. *Computing*, 53(3–4):277–287, 1994. PROFIL/BIAS is available at http://www.ti3.tu-harburg.de/Software/PROFILEnglisch.html.
26. F. Korn and C. Ullrich. Verified solution of linear systems based on common software libraries. *Interval Computations*, 3:116–132, 1993.
27. F. Krückeberg. Ordinary differential equations. In E. Hansen, editor, *Topics in Interval Analysis*, pages 91–97. Clarendon Press, Oxford, 1969.

28. H. P. Langtangen. *Computational Partial Differential Equations – Numerical Methods and Diffpack Programming.* Springer-Verlag, 1999.
29. R. J. Lohner. Enclosing the solutions of ordinary initial and boundary value problems. In E. W. Kaucher, U. W. Kulisch, and C. Ullrich, editors, *Computer Arithmetic: Scientific Computation and Programming Languages*, pages 255–286. Wiley-Teubner Series in Computer Science, Stuttgart, 1987.
30. R. J. Lohner. *Einschließung der Lösung gewöhnlicher Anfangs- und Randwertaufgaben und Anwendungen.* PhD thesis, Universität Karlsruhe, 1988. AWA is available at ftp://iamk4515.mathematik.uni-karlsruhe.de/pub/awa/.
31. R. J. Lohner. Step size and order control in the verified solution of IVP with ODE's, 1995. SciCADE'95 International Conference on Scientific Computation and Differential Equations, Stanford, California, March 28 – April 1, 1995.
32. M. Metcalf and J. Reid. *Fortran 90 Explained.* Oxford University Press, Oxford, England, 1990.
33. R. E. Moore. *Interval Analysis.* Prentice-Hall, Englewood Cliffs, N.J., 1966.
34. N. S. Nedialkov. *Computing Rigorous Bounds on the Solution of an Initial Value Problem for an Ordinary Differential Equation.* PhD thesis, Department of Computer Science, University of Toronto, Toronto, Canada, M5S 3G4, February 1999. Available at http://www.cs.toronto.edu/NA/reports.html.
35. N. S. Nedialkov and K. R. Jackson. An interval Hermite-Obreschkoff method for computing rigorous bounds on the solution of an initial value problem for an ordinary differential equation. *Reliable Computing*, Submitted, October 1998. Available at http://www.cs.toronto.edu/NA/reports.html.
36. N. S. Nedialkov, K. R. Jackson, and G. F. Corliss. Validated solutions of initial value problems for ordinary differential equations. *Appl. Math. Comp.*, To appear, 1999. Available at http://www.cs.toronto.edu/NA/reports.html.
37. A. Neumaier. *Interval Methods for Systems of Equations.* Cambridge University Press, Cambridge, 1990.
38. H. Olsson. Object-oriented solvers for initial value problems. In E. Arge, A. M. Bruaset, and H. P. Langtangen, editors, *Modern Software Tools for Scientific Computing.* Birkhäuser, 1997. Godess is available at http://www.-cs.lth.se/home/Hans_Olsson/Godess/.
39. L. B. Rall. *Automatic Differentiation: Techniques and Applications*, volume 120 of *Lecture Notes in Computer Science.* Springer Verlag, Berlin, 1981.
40. R. Rihm. On a class of enclosure methods for initial value problems. *Computing*, 53:369–377, 1994.
41. A. D. Robinson. C++ gets faster for scientific computing. *Computers in Physics*, 10(5):458–462, Sep/Oct 1996.
42. J. Rumbaugh, M. Blaha, W. Premerlani, F. Eddy, and W. Lorensen. *Object-Oriented Modeling and Design.* Prentice Hall, New York, 1991.
43. M. J. Schulte, V. Zelov, A. Akkas, and J. C. Burley. The interval-enhanced GNU Fortran compiler. *Reliable Computing*, Submitted, October 1998.
44. B. Speelpenning. *Compiling Fast Partial Derivatives of Functions Given by Algorithms.* PhD thesis, Department of Computer Science, University of Illinois at Urbana-Champaign, Urbana-Champaign, Ill., January 1980.
45. H. Spreuer and E. Adams. On the existence and the verified determination of homoclinic and heteroclinic orbits of the origin for the Lorenz system. *Computing Suppl.*, 9:233–246, 1993.

46. O. Stauning. *Automatic Validation of Numerical Solutions*. PhD thesis, Technical University of Denmark, Lyngby, Denmark, October 1997. This thesis and ADIODES are available at http://www.imm.dtu.dk/fadbad.html#download.
47. S. J. Sullivan and B. G. Zorn. Numerical analysis using nonprocedural paradigms. *ACM Trans. Math. Softw.*, 21(3):267–298, Sept 1995.
48. T. Veldhuizen. Expression templates. *C++ Report*, 7(5):26–31, June 1995.
49. T. Veldhuizen. Scientific computing: C++ versus Fortran. *Dr. Dobb's Journal*, 34, November 1997.
50. T. Veldhuizen and M. E. Jernigan. Will C++ be faster than Fortran? In *Proceedings of the 1st International Scientific Computing in Object-Oriented Parallel Environments (ISCOPE'97)*, Lecture Notes in Computer Science. Springer-Verlag, 1997.
51. G. Wanner. On the integration of stiff differential equations. In *Proceedings of the Colloquium on Numerical Analysis*, volume 37 of *Internat. Ser. Numer. Math.*, pages 209–226, Basel, 1977. Birkhäuser.
52. W. Yang and G. F. Corliss. Bibliography of computational differentiation. In M. Berz, C. H. Bischof, G. F. Corliss, and A. Griewank, editors, *Computational Differentiation: Techniques, Applications, and Tools*, pages 393–418. SIAM, Philadelphia, Penn., 1996.

The Evolution and Testing of a Medium Sized Numerical Package

David Barnes and Tim Hopkins

Computing Laboratory, University of Kent, UK
Email: {T.R.Hopkins,D.J.Barnes}@ukc.ac.uk

Abstract. We investigate the evolution of a medium sized software package, LAPACK, through its public releases over the last six years and establish a correlation, at a subprogram level, between a simply computable software metric value and the number of coding errors detected in the released routines. We also quantify the code changes made between issues of the package and attempt to categorize the reasons for these changes.

We then consider the testing strategy used with LAPACK. Currently this consists of a large number of mainly self-checking driver programs along with sets of configuration files. These suites of test codes run a very large number of test cases and consume significant amounts of cpu time. We attempt to quantify how successful this testing strategy is from the viewpoint of the coverage of the executable statements within the routines being tested.

1 Introduction

Much time, effort and money is now spent in the maintenance and upgrading of software; this includes making changes to existing code in order to correct errors as well as adding new code to extend functionality. Some sources suggest that as much as 80% of all money spent on software goes on post-release maintenance [11]. When any type of change is made, programmers need to have enough understanding of the code to be confident that any changes they make do not have a detrimental effect on the overall performance of the software. All too often a change that fixes one particular problem causes the code to act incorrectly in other circumstances.

We are interested in investigating whether it is possible, via the use of some quantitative measurements, to determine which parts of a software package are the most difficult to understand and thus, probably, the most likely to contain undetected errors and the most likely to cause further problems if subjected to code changes and updates.

Prior work in this area ([12] and [14]) has shown that, for a number of rather simple codes, software metrics can be used successfully to identify problem routines in a library of Fortran subroutines. This paper extends this work by applying it to a much larger body of code which has gone through a number of revisions both to extend functionality and to correct errors. This allows us to identify where it has been necessary to make changes to the code and why, and to correlate the occurrence of errors in the software with the values of a particular software metric.

Our hope is that it will be possible to identify, prior to its release, whether a package contains subprograms which are likely to provide future maintenance problems. This would allow authors to rethink (and possibly reimplement) parts of a package in order to simplify logic and structure and, hence, improve maintainability and understandability. Our ultimate goal is to be able to improve the quality and reliability of released software by automatically identifying future problem subprograms.

LAPACK [1] was an NSF supported, collaborative project to provide a modern, comprehensive library of numerical linear algebra software. It contains software for the solution of linear systems; standard, non-symmetric and generalized eigenproblems; linear least squares problems and singular value decomposition. The package serves as an efficient update to the Eispack ([19] and [9]) and Linpack [8] libraries that were developed in the 1970's. The complete package is available from netlib (see http://www.netlib.org/bib/mirror.html for information as to where your nearest server is). Mostly written in standard Fortran 77 [2], LAPACK uses the BLAS Level 2 [7] and Level 3 [6] routines as building blocks. LAPACK both extends the functionality and improves the accuracy of its two predecessors. The use of block algorithms helps to provide good performance from LAPACK routines on modern workstations as well as supercomputers and the project has spawned a number of additional projects which have produced, for example, a distributed memory [3] and a Fortran 90 [23] version of at least subsets of the complete library.

An integral part of the complete software package is the very extensive test suite which includes a number of test problems scaled at the extremes of the arithmetic range of the target platform. Such a test suite proved invaluable in the porting exercise which involved the package being implemented on a large number of compiler/hardware combinations prior to its release.

In Section 2, following a short introduction to LAPACK, we provide a detailed analysis of the size of the package and the extent of the source changes made between successive versions. We also categorize all the changes made to the executable statements and obtain a count of the number of routines that have had failures fixed. We then report on a strong connection between the size of a relatively simple software code metric and a substantial fraction of the routines in which failures have been corrected.

In Section 4 we look quantitatively at how well the testing material supplied with the package exercises the LAPACK code and suggest how the use of a software tool may improve this testing process. Finally we present our conclusions.

2 The LAPACK Library Source Code

The LAPACK routines consist of both user callable and support procedures; in what follows we do not differentiate between these. The source directory of the original release, 1.0, consisted of 930 files (only two files, dlamch and slamch contain more than one subprogram; the six routines in each of these files being used to compute machine parameters for the available double and single precision arithmetics). Table 1 shows how the number of files has increased with successive releases of the package along with the release dates of each version.

Version	No. of files	Release Date
1.0	930	29 February 1992
1.0a	932	30 June 1992
1.0b	932	31 October 1992
1.1	1002	31 March 1993
2.0	1080	30 September 1994

Table 1. Number of library source files for each released version.

A number of straightforward metrics exist for sizing software, for example, the number of lines in the source files. This is somewhat crude and we present in Table 2 a more detailed view of the size of the package. The column headed 'executable' shows the number of executable statements in the entire package whilst that headed 'non-exec' gives the number of declarative and other non-executable statements. The third column gives the total number of code statements being the sum of the previous two columns. The final two columns provide the total number of comment lines in the code and the total number of blank lines and blank comment lines.

The large number of non-executable statements is partially due to the use of the NAGWare 77 declarization standardizer [17] which generates separate declaration blocks for subroutine arguments, local variables, parameter values, functions, etc. This is actually no bad thing as it aids the maintenance of the code by allowing a reader to immediately identify the type and scope of each identifier. A ratio of executable to non-executable statements of 1.8 is, however, on the low side, as this implies that there are relatively

Version	Executable	Non-exec	Lines	Comments	Blank
1.0	59684	35010	94694	143249	52581
1.0a	59954	35104	95058	143439	52679
1.0b	59897	34876	94773	142185	52243
1.1	67473	38312	105785	156077	57516
2.0	76134	41986	118120	169813	62618

Table 2. Statement counts by type.

small amounts of code packaged amidst large quantities of declarations which generally makes code difficult to read and assimilate.

Comments form an important part of the documentation of any software and this is especially the case for LAPACK where the description of the arguments to all procedures (both user callable and internal) is detailed enough to allow the use of the routine without the need for a separate printed manual. This accounts for the high level of commenting, around 1.5 non-blank comments per executable line. There is also heavy use of blank comment lines (or totally blank lines); such lines act as an aid to readability within both the textual information and the source code.

A more detailed view of code size may be obtained by considering the operators and operands that make up the source. These are defined abstractly in Halstead [10] and there appears to be no general agreement as to which language tokens are considered operators and which operands for any particular programming language. The values given in Table 3 were generated using the nag_metrics tool [17] which defines an operator to be

- a primitive operator, for example, $*$, $+$, $.EQ.$, $.AND.$ etc,
- a statement which counts as an operator, for example, `ASSIGN`, `IF`, `ELSE IF`, `GOTO`, `PRINT`, `READ` and `WRITE`,
- a pair of parentheses, an end-of-statement, or a comma

and operands as

- constants,
- name of variables and constants,
- strings (all strings are considered distinct from each other).

Thus the declarative part of any program is ignored by this metric as it is not considered to add to the complexity of the code.

Version	Operators	Operands	Total	% Increase
1.0	370784	325907	696691	
1.0a	371605	326524	698129	0.02
1.0b	370928	326089	697017	−0.02
1.1	415626	364816	780442	11.9
2.0	468487	411122	879609	12.7

Table 3. Total number of operators and operands.

At each new version of the package complete routines were added and deleted and changes were made to routines that were common to both the new and previous versions. Table 4 show the distribution of affected routines. Changes to program units common to successive versions have been categorized depending on whether only comments, only non-comments, or both comments and non-comments were changed. This shows that although there were textual changes to 1690 routines over the four revisions 1144 of these involved changes to comment lines only (this accounts for 64% of all the changed routines).

Version	Added	Deleted	Total	Changed c/only	s/only	Both
1.0 →1.0a	2	0	147	75	34	38
1.0a→1.0b	2	2	339	236	33	70
1.0b→1.1	72	2	570	554	0	16
1.1 →2.0	84	6	634	279	128	227

Table 4. Routine changes at each version.

We analyzed the routines that had been changed between releases by running the two versions through the Unix file comparison tool *diff* and processing the output to count the number of changed comment and non-comment statements.

Diff classifies changes in three ways, lines in the new version that did not appear in the old (App), lines in the old version that did not appear in the new (Del) and blocks of lines that have changed between the two (Changes). Table 5 gives the totals for comment and non-comment statements according to the categories for all the changed routines at each version. It should be noted here that some changes to statements are due to changes in statement labels which occur when a label is either inserted or deleted and the code is passed through the NAGWare Fortran 77 source code formatter, nag_polish [17]. In a few cases diff exaggerates the number of changes due to synchronization problems which may occur if a line happens to be repeated or sections of code are moved. No attempt was made to compensate for this; indeed it may be argued that a move of a section of code should be treated as both a delete and an add. Resynchronization problems appeared to be relatively few and far between and, it was felt, they were unlikely to perturb the final results by more than a few percent. In all cases comment changes reflecting the new version number and release date have been ignored.

Finally Table 6 gives a breakdown of the non-comment and comment lines added and deleted via complete routines between releases.

Although the release notes made available with each new revision gave some details of which routines had had bug fixes applied to them this information was far from complete. It is not safe to assume that all non-comment code changes are necessarily bug fixes. In order to determine the nature of the changes to routines at each release, a visual inspection of each altered single precision complex and double precision real routine was conducted us-

Version	Comments			Non-comments		
			Changes			Changes
	Del	App	(old/new)	Del	App	(old/new)
1.0 →1.0a	38	134	550/484	69	312	775/808
1.0a→1.0b	258	605	2183/1573	384	593	1325/1722
1.0b→1.1	227	70	9428/8354	34	36	724/752
1.1 →2.0	541	527	4393/4688	449	730	2363/2274

Table 5. Interversion statement changes from diff.

ing a graphical file difference tool [4]. As a result we have categorized the code changes between all at each revision as being one of

i: enhanced `INFO` checks or diagnostics and `INFO` bug fix (usually an extended check on the problem size parameter `N`),
pr: further uses of machine parameters (for example, the use of `DLAMCH('Precision')` in `DLATBS` at version 1.0a to derive platform dependent values),
c: cosmetic changes (for example, the use of `DLASET` in place of `DLAZRO` in the routine `DPTEQR` at version 2.0),
en: enhancement (for example, the addition of equilibration functionality to the routine `CGBSVX` at version 1.0b),
ef: efficiency change (for example, the quick return from the routine `CGEQPF` in the case when `M` or `N` is zero at version 1.1),
re: removal of redundant code (for example, a `CABS2` calculation in `CGEEQU` was not required and was removed at version 1.0a),
mb: minor bug (typically a few lines of changed code; for example, the changes made to `DLAGS2` at version 2.0 to add an additional variable and to modify a conditional expression to use it),
Mb: major bug (a relatively large code change; for example, the changes made to the routine `DSTEQR` at version 2.0).

Such a classification provides a much firmer base from which to investigate a possible correlation between complexity and coding errors.

Version	Added		Deleted	
	Com	Non-com	Com	Non-com
1.0 →1.0a	160	86	0	0
1.0a→1.0b	99	65	160	86
1.0b→1.1	14351	10196	160	86
1.1 →2.0	13770	12248	311	129

Table 6. Added and deleted routines by comment and non-comment lines.

Routines in the LAPACK library are in one of four precisions; single or double precision, real or complex. While the single and double precision versions of a routine can generally be automatically generated from one another using a tool like nag_apt [17], the real and complex parts of the package are often algorithmically quite different. For this reason we only consider the single precision complex and the double precision real routines in the remaining sections of this paper. These routines may be identified by their name starting with either a C (single precision complex) or a D (double precision real).

3 The Path Count Metric

Table 7 provides a summary of the number of single precision complex and double precision real routines that fall into each category for successive releases of the package. We were interested in discovering whether there was any relationship between those routines needing bug fixes and any software complexity metric values. One metric, a version of Nejmeh's path count metric [16], is calculated by the QA Fortran tool [18]. This metric is an estimate of the upper bound on the number of possible static paths through a program unit (based solely on syntax). Note that some paths so defined may not be executable but such impossible routes cannot usually be determined by simple visual inspection. The path count complexity of a function is defined as the product of the path complexities of the individual constructions. Thus, for example, the path complexity of a simple if-then-else statement is 2 while three consecutive if-then-else statements would have an associated value of $2^3 = 8$. Three nested if statements would have a path count of 4. This metric provides a useful measure of the effort required to test the code stringently as well as giving an indication of code comprehensibility and maintainability. A reduction in the path count caused by restructuring code would imply the elimination of paths through the code which were originally either impossible to execute or irrelevant to the computation. This would be likely to reduce the time spent in the testing phase of the software development. An example of the use of this metric to identify problem code within a small library of relatively small Fortran routines may be found in [12] and [13].

The metric value returned by the package has a maximum value of 5×10^9 although, realistically, a program unit can be classified as too complex to test fully when this value exceeds 10^5. Table 8 gives the number of routines in which faults were correlated against the value of the path metric. (We differentiate here between routines that were introduced before version 2 and those that were added at version 2 and cannot, therefore, have had source changes applied to them.) This data clearly shows a high correlation between routines identified as complex by the path count metric and those having had bug fixes applied to them. 41% of all the bugs occurred in routines with a path count metric in excess of 10^5 and these routines constitute just 16% of the total number of subprograms making up the library. The chance of a bug

	1.0a C, D	1.0b C, D	1.1 C, D	2.0 C, D
mb	9, 9	12, 9	1, 3	16,14
Mb	3, 3	1, 1	0, 0	1, 2
re	2, 0	1, 1	0, 0	0, 0
i	3, 5	8,10	1, 1	9, 9
pr	3, 4	1, 0	0, 0	3, 4
en	0, 0	5, 5	0, 0	1, 2
ef	0, 0	3, 1	1, 1	1, 1
c	1, 0	0, 1	0, 1	105,15

Table 7. Number of routines affected by each category of source code change.

occurring in these routines would appear to be around 6 times more likely than in routines with a path count of less than 10^5. It should also be noted that a large number of the routines added at version 2.0 have extremely large path counts and, from our current analysis, we would expect a high percentage of these to require patches to be applied in forthcoming releases.

	$\lfloor \log_{10}(\text{Path Counts}) \rfloor$					%
	> 8	7	6	5	≤ 5	> 5
v2.0	5	1	0	4	26	28
<2.0	10	6	10	8	285	16
Faults	8	3	7	3	30	41
%routines	80	50	70	37	11	–

Table 8. Occurence of Faults against Path Count Metric.

4 Testing

The test suite forms an integral part of the LAPACK software package. The code was designed to be *transportable*; all new LAPACK code was to be portable while efficiency of the package as a whole was to be platform dependent. This was achieved by coding in standard Fortran 77 [2] for portability of the higher level routines and using platform specific versions of the BLAS to obtain high efficiency. Thus, by using the BLAS Levels 2 and 3 definition codes ([7], [6]) the entire package may be made portable at the price of suboptimal execution speed.

Two test suites are provided with the released package; one for checking the installation and the other for producing platform dependent timings. All the tests are self checking in that the only output that a user has to check is in the form of summaries of the number of tests applied to each user routine along with the number of successes and counts and details of any failures.

Matrix data is either generated randomly or specially constructed (see [5] for more details). In both cases the testing software uses a metric to decide whether each computed solution is 'correct'. Test cases are also generated to exercise the routines on data at the extremes of the floating point arithmetic ranges. The number, and to a minor extent, the range of tests applied, may be controlled by the user at a data file level. However this generally precludes the user from forcing execution through specific sections of the code which usually requires specially constructed data. To all intents and purposes the test strategy must be categorized as white box (or glass box) testing, where test cases are selected by considering their effect on the code rather than just testing their adherence to the specification as is the case with black box testing.

We were interested in determining quantitatively how effective this strategy was in exercising the code. The minimal requirement of a test suite should be to ensure that all the executable statements in the package are executed at least once. Note that this is very different from path coverage as defined earlier.

We used the NAGWare 77 source code instrumenter and execution analysis tools nag_profile and nag_history [17] to determine cumulatively the number of times each basic block was executed. (A basic block is a straight line section of code, i.e., it does not contain any transfer of control statements.) This allowed us to determine how many basic blocks were not exercised by the test suite. We ran the instrumented code on both the installation and timing test suites.

	Routines	Basic Blocks	% Executed
C install	258	12019	89.56
C timing	124	5324	72.84
D install	263	12852	89.20
D timing	143	6615	72.41

Table 9. Basic block execution for installation and timing test suites.

Table 9 compares the number and percentage of basic blocks executed using both test suites.

Further analysis showed that of the 1388 unexecuted blocks, 214 (15.4%) were concerned with the checking of input arguments. The worst case was the routine, DGEGS, used for computing the general Schur factorization which has 48% of its basic blocks untested.

We looked in detail at the routine DGBBRD. This routine consisted of 111 basic blocks containing 124 executable statements. The installation test omitted to cover 13 basic blocks of which 11 were concerned with checking the consistency of the input arguments. The final two were in the main body of the code.

The code operates on a banded, rectangular ($M \times N$), matrix where the user provides the number of sub- and super-diagonals, KL and KU respectively. The configuration file used to provide data for this routine specified M, N and K, the total bandwidth. Code within the test routine then splits K into KL and KU. Although a total of 1500 calls are made to DGBBRD the test data failed to generate the special case $M = N = 2, KU = 0$ and $KL > 0$. The problem was that with M and N both greater than zero the code to set KL and KU could not generate $KU = 0$ and $KL > 0$. Providing data for a separate test is straightforward and the two previously untested blocks executed successfully.

As well as using the profiling information from nag_profile to identify the basic blocks of code that were not being executed by the test data, we can also use it to check that individual tests (or batches of tests) actually contribute to the overall statement coverage. Ideally we would like to minimize the number of test cases being run to obtain maximum coverage. We note here that it may not be possible to exercise 100% of the basic blocks, for example, there may be defensive code that exits if convergence is not attained although no numerical examples are known which trip this condition.

Using DGBBRD again as our example, we obtained individual basic block coverage profiles for each of the 20 M and N pairs (each pair generates a number of calls to the routine for different KL and KU values). It was found that two of these 20 tests covered 83 of the 98 executed blocks and four could be chosen to cover all 98. The effect of using just this minimal number of tests was to reduce the execution time for testing this routine by a factor of four without any loss of code coverage. A further reduction in the execution time could be made by reducing the number of (KL, KU) pairs chosen for a given (M, N) pair.

The above analysis could be applied to all the test drivers in order to reduce the total number of tests being executed whilst maximizing the code coverage.

5 Conclusion

We have analyzed the source code changes made between successive versions of the LAPACK software library and we have presented strong evidence that the path count software metric is a good indicator of routines that are likely to require post release maintenance. We note that many of the newly introduced routines have metric values indicative of problem code.

By using a profiling tool we have been able to measure how well the installation testing software, provided with the package, executes the LAPACK sources. The coverage is extremely good although we believe that it could be improved further. In addition we have shown that, from the point of view of code coverage, many of the tests do not contribute as they fail to exercise any blocks of code that are not already executed by other tests. We are certain that by analyzing the output from this tool it would be possible both to reduce the number of tests necessary to obtain the attained code coverage and to improve the coverage by pinpointing untested sections of code.

One of the most worrying trends is that there appears to be a definite trend towards very complex routines being added to the library. Of the new routines introduced at version 2.0 almost 30% had path counts in excess of 10^5; this is almost double the percentage of routines introduced prior to that version. Whilst it may be argued that new routines are solving more complex problems it is possible to structure these codes so that the components are far simpler from a software complexity viewpoint and are thus much easier to test thoroughly. It is highly likely that over half of the routines with a path count of 10^5 or more will require bug fixes in the near future.

It is difficult to compare the quality of LAPACK with other libraries of numerical software since, as far as we know, no other public domain numerical package has either a complete source code change history or a complete set of the sources of all the relevant releases available. That apart we believe that LAPACK is a high quality software package with a bug fix being applied on average for approximately every 600 executable statements.

A similar complexity analysis to the one performed in this paper could be applied to any Fortran 77 software; extracting the required metric values is very simple via QA Fortran, and the NagWare 77 profiling tool allows a relatively painless analysis of the statement coverage. Far more difficult is the automatic generation of test data; there appear to be no tools available, at least not in the public domain, that would help in this area. Generating test drivers and data to ensure a high percentage of statement coverage is thus both difficult and time consuming.

The path count metric detailed in Section 3 could be used in a similar way for both C and Fortran 90 and indeed for almost all imperative languages. For object-oriented languages a different approach would be necessary and there is not as yet any general consensus on which metrics are most appropriate. A general discussion of object-oriented software metrics may be found in Lorenz and Kidd [15].

It is our intention to extend the work presented in this paper to investigate the detection of errors in released software. We propose using the path count metric to identify possible problem routines and to subject these routines to extensive white box testing in an attempt to exercise as many paths through the code as possible. Current work ([21], [20] and [22]) provides some hope of generating data to exercise specified paths through code and this would certainly open up new possibilities for fault detection.

References

1. E. Anderson, Z. Bai, C. Bischof, J. Demmel, J. J. Dongarra, J. Du Croz, A. Greenbaum, S. Hammarling, A. McKenney, S. Ostrouchov, and D. Sorensen. *LAPACK: users' guide*. SIAM, Philadelphia, second edition, 1995.
2. ANSI. *Programming Language Fortran X3.9-1978*. American National Standards Institute, New York, 1979.
3. L.S. Blackford, J. Choi, A. Cleary, E. D'Azevedo, J. Demmel, I. Dhillon, J. Dongarra, S. Hammarling, G. Henry, A. Petitet, K. Stanley, D. Walker, and R.C. Whaley. *ScaLAPACK Users' Guide*. SIAM, Philadelphia, 1997.
4. David Barnes, Mark Russell, and Mark Wheadon. Developing and adapting UNIX tools for workstations. In *Autumn 1988 Conference Proceedings*, pages 321–333. European UNIX systems User Group, October 1988.
5. J. Demmel and A. McKenney. A test matrix generation suite. Technical Report MCS-P69-0389, Argonne National Laboratories, Illinois, March 1989.
6. J. J. Dongarra, J. Du Croz, I. S. Duff, and S. Hammarling. Algorithm 679: A set of level 3 basic linear algebra subprograms. *ACM Trans. Math. Softw.*, 16(1):18–28, March 1990.
7. J. J. Dongarra, J. Du Croz, S. Hammarling, and R. J. Hanson. Algorithm 656: An extended set of basic linear algebra subprograms: Model implementation and test programs. *ACM Trans. Math. Softw.*, 14(1):18–32, March 1988.
8. J. J. Dongarra, C. B. Moler, J. R. Bunch, and G. W. Stewart. *LINPACK: Users' Guide*. SIAM, Philadelphia, 1979.
9. B. S. Garbow, J. M. Boyle, J. J. Dongarra, and C. B. Moler. *Matrix Eigensystem Routines – EISPACK Guide Extension*, volume 51 of *Lecture notes in computer science*. Springer-Verlag, New York, 1977.
10. M.H. Halstead. *Elements of Software Science*. Operating and Programming Systems Series. Elsevier, New York, 1977.
11. Les Hatton. Does OO sync with how we think? *IEEE Software*, pages 46–54, May/June 1998.
12. T.R. Hopkins. Restructuring software: A case study. *Software — Practice and Experience*, 26(8):967–982, July 1996.
13. T.R. Hopkins. Is the quality of numerical subroutine code improving? In E. Arge, A.M. Bruaset, and H.P. Langtangen, editors, *Modern Software Tools for Scientific Computing*, pages 311–324. Birkhäuser Verlag, Basel, 1997.
14. T.R. Hopkins. Restructuring the BLAS Level-1 routine for computing the modified Givens transformation. *ACM SIGNUM*, 32(4):2–14, October 1997.
15. M. Lorenz and J. Kidd. *Object-Oriented Software Metrics*. Object-Oriented Series. Prentice Hall, Englewood Cliffs, New Jersey, 1994.
16. B. A. Nejmeh. NPATH: A measure of execution path complexity and its applications. *Commun. ACM*, 31(2):188–200, 1988.
17. Numerical Algorithms Group Ltd., Oxford, UK. *NAGWare f77 Tools*, second edition, September 1992.
18. Programming Research Ltd, Hersham, Surrey. *QA Fortran 6.0*, 1992.
19. B. T. Smith, J. M. Boyle, J. J. Dongarra, B. S. Garbow, Y. Ikebe, V. C. Klema, and C. B. Moler. *Matrix Eigensystem Routines – EISPACK Guide*, volume 6 of *Lecture notes in computer science*. Springer-Verlag, New York, second edition, 1976.

20. Nigel Tracey, John Clark, and Keith Mander. Automated program flaw finding using simulated annealing. In *Software Engineering Notes, Proceedings of the International Symposium on Software Testing and Analysis*, volume 23, pages 73–81. ACM/SIGSOFT, March 1998.
21. Nigel Tracey, John Clark, and Keith Mander. The way forward for unifying dynamic test-case generation: The optimisation-based approach. In *International Workshop on Dependable Computing and Its Applications (DCIA)*, pages 169–180. IFIP, January 1998.
22. Nigel Tracey, John Clark, Keith Mander, and John McDermid. An automated framework for structural test-data generation. In *Proceedings of the International Conference on Automated Software Engineering*. IEEE, October 1998.
23. J. Wasniewski and J.J. Dongarra. High performance linear algebra package – LAPACK90. Technical Report CS-98-384, University of Tennessee, Knoxville, April 1998.

A Availability of Tools

The main software tools mentioned in the paper are available as follows:

QA Fortran: Programming Research Limited, 1/11 Molesey Road, Hersham, Surrey, KT12 4RH, UK.
(http://www.prqa.co.uk/qafort.htm)
NagWare 77: NAG Ltd, Wilkinson House, Jordan Hill Road, Oxford, OX2 8DR, UK.
(http://www.nag.co.uk/nagware/NANB.html)
Perl: http://www.perl.com/pace/pub.
Vdiff: A graphical file comparator, Kent Software Tools.
(http://www.cs.ukc.ac.uk/development/kst/).

Other analysis of the code reported in the paper was performed using bespoke Perl scripts.

An Object-Oriented Approach to the Finite Element Modeling and Design of Material Processes

Nicholas Zabaras

Sibley School of Mechanical and Aerospace Engineering,
Cornell University, USA
Email: zabaras@cornell.edu

Abstract. An object-oriented approach to the analysis and design of metal forming and directional solidification processes is presented. A number of specific ideas are introduced on how a deformation or a solidification process can be partitioned into subproblems that can be independently developed and tested in the form of class hierarchies. Class development based on mathematical and/or physical arguments is emphasized. General ideas are provided to demonstrate that an OOP approach to the FEM modeling and design of material processes can lead to efficient simulators that are easily maintainable and expandable. A metal forming example is presented to demonstrate the ability of the deformation simulator to handle the analysis of industrial metal forming processes. Also, a design directional solidification example is presented to demonstrate the substantial benefits of using OOP techniques for the design of processes with a desired solidification morphology.

1 Introduction

Various commercial FEM software codes have been developed for the implementation of a variety of material processes and in particular forming and solidification processes (e.g. [1], [13]). These codes have been developed with the non-expert user in mind and generally provide only limited flexibility for the implementation of particular desired algorithms or processes. The user, however, is allowed to implement a particular material model, boundary conditions, geometry (including the finite element grid), etc. as required by the problem at hand.

In this communication, a brief presentation is given of an object-oriented programming (OOP) approach to modeling forming and solidification processes. The discussion is restricted to the development of OOP formulations that allow the user, using a well defined programming structure, to implement his/her own algorithms and/or a particular process.

The fundamentals of object-oriented programming for scientific applications are reviewed in [6]. Various applications of OOP to engineering problems can be found in [21] and [7]. Diffpack is an example of a C++ based object-oriented environment developed to address the modeling of continuum systems governed by partial differential equations (PDEs) using both finite differences and finite elements [16]. Basic finite element operations (e.g. finite element definition, integration rules, assembly of the stiffness matrix and load vectors, application of essential and natural boundary conditions, definition of continuum fields over finite element grids, etc.) have been introduced in Diffpack using class hierarchies. This provides the flexibility of solving a boundary value problem by only having to declare the particulars of the problem that may include the coefficients of the PDE, and the essential and/or natural boundary conditions. Such information is introduced by using virtual functions in hierarchies derived from base classes currently developed in Diffpack.

The major advantage of using an OOP environment such as Diffpack is that the user can utilize as much or as little of the provided class structure as it is appropriate in a particular application. Such flexibility may look against the objective of general purpose simulators developed with procedural languages, however, it provides an enormous flexibility to address various algorithms and processes that do not necessarily fit within the program structure already developed. Such flexibility does not exist with most of the current simulators based on procedural languages.

As it is impossible in one paper to present even the basic aspects of forming and solidification processes, this paper is restricted to a few mathematical and physical aspects of forming and solidification processes for which the essentials of an OOP structure are presented. However, references to the particular models implemented in the presented simulators will be given for further consultation and details.

Finite element simulation of a material process follows different approaches depending on the mathematical and physical structure of the process. For example, solidification processes are usually modeled as a coupled system of PDEs each of them linked to a particular physical mechanism (heat or solute diffusion, fluid (melt) flow, etc.) [26]. On the other hand, forming processes are considered as a set of coupled problems each of them represented with PDEs but also with variational inequalities (as for example is the case of modeling the contact/frictional constraints) [24] and [17].

The features of OOP that are explored here are inheritance and polymorphism. They lead to code expandability, maintainability and reusability. In the present work, we have used the widely used Diffpack library of classes for analysis of systems governed by partial differential equations, so as to concentrate only on the implementation of the particular physical mechanisms in the form of classes [16]. Depending on the type of process examined, a common approach to OOP implementation is by partitioning the original problem in various subproblems based on physical and/or mathematical arguments. Examples of both of these cases will be shown later in this paper. For each subproblem defined, appropriate class hierarchies are introduced to allow the user to implement process/problem specific functionality via virtual functions (functors). The base classes define the general structure of the problem, whereas the derived classes are implementing functionality specific to each process. This, for example, is important while modeling forming processes in order to account for the significant differences in the nature of boundary conditions between various forming processes (e.g. rolling, forging, extrusion, etc.). In addition, this structure can for example account for the similar nature of the deformation at a particular material point in the Lagrangian analysis of all forming processes.

Each subproblem can be modelled and tested independently of the other and many algorithms can be used for the coupling of all subproblems. It is here demonstrated via specific examples that an OOP approach to material processing is ideal for (1) development and testing of the various subproblems (2) coupling of the various subproblems (3) generating a generic structure that can be used as prototype for implementation of other processes, material models, and/or algorithms and (4) for a reliable memory management.

The basic structure of this paper is as follows. Forming processes are introduced in Section 2. A brief descriptive definition of a typical forming problem is presented in Section 2.1 using a Lagrangian framework. The main algorithm is summarized in Section 2.2. The developed object-oriented class structure based on the physics of each subproblem is reviewed in Section 2.3. An example of a plane strain closed-die forging process is summarized in Section 2.4.

A directional solidification simulator is presented in Section 3. A brief problem definition is given in Section 3.1 and the main algorithm is discussed in Section 3.2. The OOP implementation of FEM algorithms is presented in Section 3.3. As a demonstration of a design simulator, Section 4 presents the definition, analysis and OOP implementation of a design binary alloy solidification problem for a desired stable growth. Such problems are here formulated as functional optimization problems. The design algorithm considered is an implementation of the adjoint method in the L_2 space. Appropriate continuum sensitivity and adjoint problems are defined as required for the implementation of functional optimization techniques (e.g. the conjugate gradient method) [45] and [43]. It is shown that an OOP approach to such design problems provides an ideal framework in which the similarities in the mathematical structure of the various subproblems (direct, sensitivity and adjoint fluid flow, heat and mass transfer, etc., problems) are taken into account. As a particular case of the design methodology, an example problem is presented in Section 4.4 for the design of the boundary heat fluxes in the directional solidification of a pure material that lead to desired interface heat fluxes and growth velocity. Finally, the paper concludes with a discussion on the efficiency of OOP techniques for the implementation of FEM formulations of material processes.

2 Lagrangian Analysis of Forming Processes

A research object-oriented simulator has been developed for the analysis of large deformations of isotropic, hyperelastic, viscoplastic microvoided materials including damage as well as thermal coupling. A generic OOP structure has been introduced to account for various constitutive models, various integration algorithms of constitutive models, various contact algorithms, various thermo-mechanical coupling algorithms, process-dependent conditions (including die shape, boundary conditions), etc. No attempt is made here to review all details, but the generic mathematical structure of some of the above mentioned subproblems is identified in order to motivate the selected object-oriented environment for their implementation. In Section 2.1 a brief definition of the deformation problem is given followed in Section 2.2 by a presentation of the main subproblems to be examined. Section 2.3 presents some of the key elements of the OOP approach to the modeling of these type of processes. More details on the precise problem definition, FEM algorithm and OOP structure can be found in [46]–[48].

2.1 Problem Description

In a typical deformation problem, one seeks the history-dependent solution (in terms of incremental displacements, stresses, strains, state variables, etc.) that satisfies the equilibrium equations, the kinematic equations (strain/displacement relations), the constitutive equations (elastic and inelastic including the evolution of the material state) and the contact constraint equations. A Lagrangian framework of analysis is considered here in order to allow ourselves to evaluate residual stresses. Consider the smooth deformation mapping defined as $\boldsymbol{x} = \phi(\boldsymbol{X}, t) : \boldsymbol{B}_o \rightarrow \mathcal{R}^3$ which maps particles $\boldsymbol{X}$ in the reference configuration $\boldsymbol{B}_o$ to their positions $\boldsymbol{x}$ in the deformed configuration $\boldsymbol{B} \subset \mathcal{R}^3$ at time t. The deformation gradient $\boldsymbol{F}$ with respect to the reference configuration $\boldsymbol{B}_o$ is given by,

$$\boldsymbol{F}(\boldsymbol{X}, t) = \frac{\partial \phi(\boldsymbol{X}, t)}{\partial \boldsymbol{X}}, \qquad \det \boldsymbol{F} > 0. \tag{1}$$

The deformation gradient is decomposed into thermal, plastic, and elastic parts as follows:

$$\boldsymbol{F} = \boldsymbol{F}^e\, \boldsymbol{F}^p\, \boldsymbol{F}^\theta, \qquad \det \boldsymbol{F}^p > 0,\ \det \boldsymbol{F}^e > 0,\ \det \boldsymbol{F}^\theta > 0, \tag{2}$$

where $\boldsymbol{F}^e$ is the elastic deformation gradient, $\boldsymbol{F}^p$, the plastic deformation gradient and $\boldsymbol{F}^\theta$ is the thermal part of the deformation gradient. A graphical representation of equation (2) is given in Figure 1. Assuming isotropic thermal expansion, the evolution of the intermediate thermally expanded unstressed configuration is given as follows:

$$\dot{\boldsymbol{F}}^\theta\, {\boldsymbol{F}^\theta}^{-1} = \beta\, \dot{\theta} \boldsymbol{I}, \tag{3}$$

where β is the thermal expansion coefficient, $\dot{\theta}$ the temperature rate and $\boldsymbol{I}$ the second-order identity tensor.

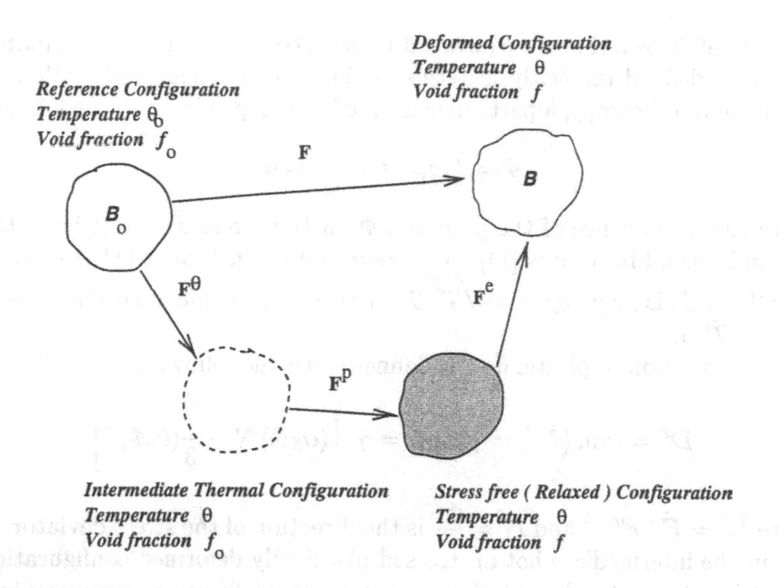

Fig. 1. A kinematic framework for the constitutive modeling of thermo-inelastic deformations coupled with damage. The intermediate thermally expanded hot unstressed configuration and the intermediate hot plastically deformed relaxed (unstressed) configuration are shown.

From the polar decomposition of the elastic deformation gradient, we can write the following:

$$\boldsymbol{F}^e = \boldsymbol{R}^e\,\boldsymbol{U}^e. \tag{4}$$

In the constitutive equations to be defined here, logarithmic stretches are used as strain measures. The elastic strain denoted by $\bar{\boldsymbol{E}}^e$ is defined in the intermediate hot unstressed configuration as

$$\bar{\boldsymbol{E}}^e = \ln \boldsymbol{U}^e. \tag{5}$$

The corresponding conjugate stress measure $\bar{\boldsymbol{T}}$ used in this model is the pullback of the Kirchhoff stress with respect to $\boldsymbol{R}^e$,

$$\bar{\boldsymbol{T}} = \det(\boldsymbol{U}^e)\;\boldsymbol{R}^{eT}\;\boldsymbol{T}\;\boldsymbol{R}^e, \tag{6}$$

where $\boldsymbol{T}$ represents the Cauchy stress [41].

Damage in the form of void growth is considered in the present analysis. The body is assumed to be a continuum despite the presence of microvoids. The internal state variables are taken to be $(s,\ f)$, where s represents the scalar resistance to plastic flow offered by the matrix material and f is the void volume fraction in the deformed configuration.

It is well known that the equivalent tensile stress σ_m of the matrix material should be defined implicitly in terms of the Cauchy stress and f. With the assumption of isotropy, a particular form of this dependency is the following:

$$\Phi = \Phi(\sigma_m, f, p, \mathcal{S}) = 0 \tag{7}$$

where the dependence of the potential Φ on the stress $\bar{\boldsymbol{T}}$ is restricted to its first and second invariants [14], [40]. Here, the norm $\mathcal{S}$ of the stress deviator $\bar{\boldsymbol{T}}' = \bar{\boldsymbol{T}} + p\,\boldsymbol{I}$, is given by $\mathcal{S} = \sqrt{\bar{\boldsymbol{T}}' \cdot \bar{\boldsymbol{T}}'}$, where p is the mean normal pressure, $p = -\text{tr}\, \bar{\boldsymbol{T}}/3$.

The evolution of plastic flow is defined with the following normality rule:

$$\bar{\boldsymbol{D}}^p = \text{sym}\left(\bar{\boldsymbol{L}}^p\right) = \dot{\gamma}\, \partial_{\bar{\boldsymbol{T}}} \Phi = \dot{\gamma} \left[(\partial_{\mathcal{S}} \Phi) \boldsymbol{N} - \frac{1}{3} (\partial_p \Phi) \boldsymbol{I} \right] \tag{8}$$

where $\bar{\boldsymbol{L}}^p = \dot{\boldsymbol{F}}^p\, \boldsymbol{F}^{p-1}$ and $\boldsymbol{N} = \frac{\bar{\boldsymbol{T}}'}{\mathcal{S}}$ is the direction of the stress deviator. The spin in the intermediate hot unstressed plastically deformed configuration is immaterial for isotropic materials and is assumed to vanish resulting in the following equations [41]:

$$\begin{aligned} \bar{\boldsymbol{D}}^p &= \bar{\boldsymbol{L}}^p, \\ \bar{\boldsymbol{W}}^p &= 0. \end{aligned} \tag{9}$$

The parameter $\dot{\gamma}$ in equation (8) is determined using the work equivalence relation $\bar{\boldsymbol{T}}' \cdot \bar{\boldsymbol{D}}^p = (1 - f)\sigma_m \dot{\bar{\epsilon}}_m^p$.

The evolution of the equivalent tensile plastic strain in the matrix $\bar{\epsilon}_m^p$ is specified via uniaxial experiments as:

$$\dot{\bar{\epsilon}}_m^p = f(\sigma_m, s, \theta), \tag{10}$$

and the evolution of the isotropic scalar resistance s is also obtained from experiments and has the form,

$$\dot{s} = g(\sigma_m, s, \theta) = h(\sigma_m, s, \theta)\dot{\bar{\epsilon}}_m^p - \dot{r}(s, \theta), \tag{11}$$

where $\dot{r}(s, \theta)$ is the static recovery function. For an example of such a model see [9]. The evolution equation for the void fraction is essentially a statement concerning conservation of mass and takes the form:

$$\dot{f} = (1 - f)\text{tr}(\bar{\boldsymbol{D}}^p). \tag{12}$$

The hyperelastic constitutive model used here is as follows [41]:

$$\bar{\boldsymbol{T}} = \leq \left[\bar{\boldsymbol{E}}^e\right], \tag{13}$$

where the isotropic elastic moduli $\mathcal{L}^e$ are generally functions of f and θ [10].

Finally, in the absence of external heat sources, the balance of energy in the current configuration takes the following form:

$$\rho c \dot{\theta} = \mathcal{W}_{\text{mech}} + \nabla \cdot k \ \nabla \theta, \tag{14}$$

where c is the specific heat capacity per unit mass (generally a function of f and θ), ρ the density in the current configuration and k the conductivity. In general the thermophysical properties are considered to be functions of f and θ. The mechanical dissipation $\mathcal{W}_{\text{mech}}$ is usually specified in terms of the plastic power by the following empirical law,

$$\mathcal{W}_{\text{mech}} = \omega \ \bar{\boldsymbol{T}} \cdot \bar{\boldsymbol{D}}^p, \tag{15}$$

where the constant dissipation factor ω represents the fraction of the plastic work that is dissipated as heat.

2.2 FEM Algorithm

The overall algorithm is presented in Figure 2 (assuming for simplicity isothermal conditions). The incremental deformation problem consists of a Newton-Raphson iteration for the calculation of the incremental displacements. Such a step requires full linearization of the principle of virtual work (including contributions from contact and from the assumed strain model used to model the near-incompressibility conditions, [46]) as well as a calculation of the consistent linearized material moduli.

In the *constitutive integration incremental subproblem*, the variables $(\boldsymbol{T}, s, f, \boldsymbol{F}^e)$ are determined at the end of the time step (time t_{n+1}), given the body configuration $\boldsymbol{B}_{n+1}$ and the temperature field at time t_{n+1}. The body configuration $\boldsymbol{B}_n$, the temperature field at time t_n and the variables $(\boldsymbol{T}, s, f, \boldsymbol{F}^e)$ at the beginning of the time step (time t_n) are known. Thus, the solution is advanced within the incremental solution scheme by integrating the flow rule (equation (8)), the evolution equation for the state variable s (equation (11)) and the evolution equation for the void volume fraction f (equation (12)) over a finite time step $\Delta t = t_{n+1} - t_n$ [24], [41], [46]. In the present simulator [46]–[48], a full Newton-Raphson scheme has been developed for the calculation of the incremental displacements. Radial-return algorithms have been implemented for both rate-dependent and rate-independent models of dense [41], [46] and micro-voided materials [34], [3]–[52].

In the *contact/friction subproblem*, given the configuration $\boldsymbol{B}_{n+1}$, the die location and shape at time t_{n+1}, as well as estimates of the contact tractions (e.g. from the previous time step or a previous Newton-Raphson iteration), one updates the regions of contact as well as the applied contact tractions at t_{n+1}. The contact subproblem is coupled with the *kinematic incremental problem*, where given $(\boldsymbol{T}_{n+1}, s_{n+1}, f_{n+1}, \bar{\boldsymbol{F}}^p_{n+1})$, the configuration $\boldsymbol{B}_n$ and the applied boundary conditions at t_{n+1} (including the contact tractions), one calculates (updates) the configuration B_{n+1}. Various algorithms have been developed to implement contact and friction The augmented Lagrangian formulation of Simo and Laursen was found appropriate for fully implicit Lagrangian formulations [46].

For thermomechanical processes including hot forming, a coupling of a thermal analysis with the deformation algorithm of Figure 2 must be introduced. Various implicit, explicit and semi-implicit algorithms can be found in [34] and [19]–[33].

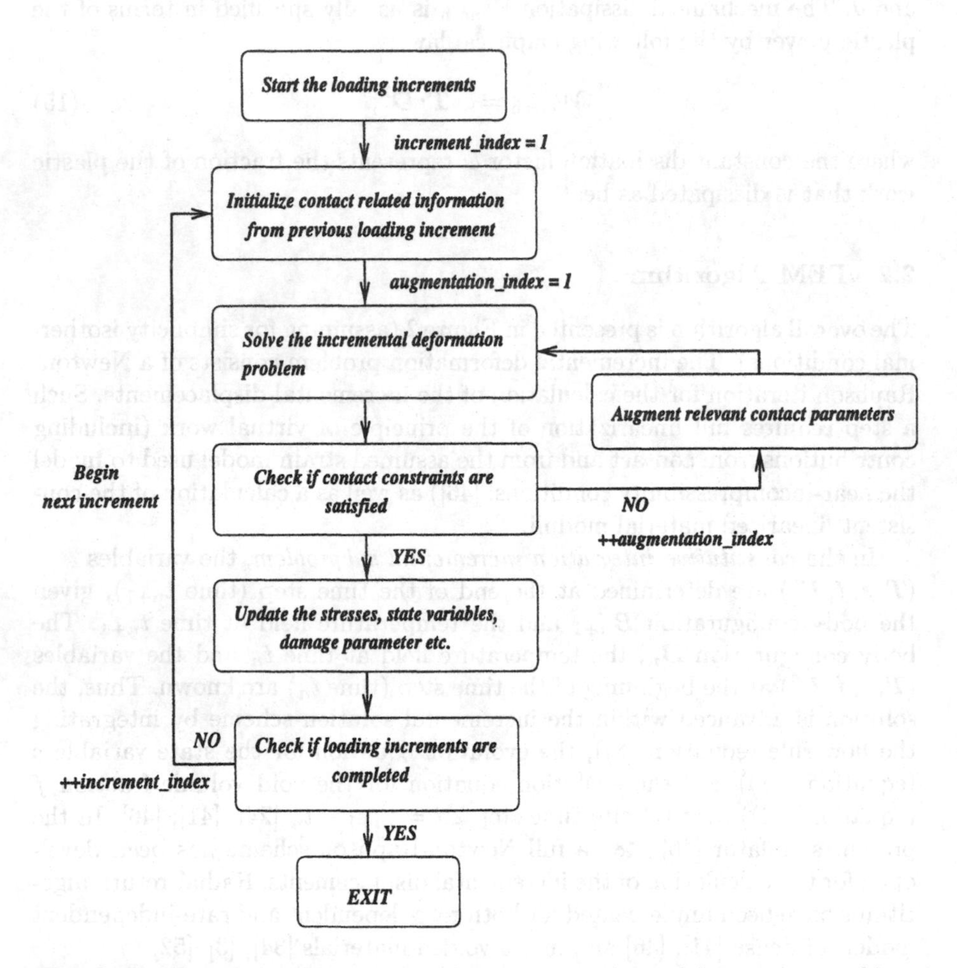

Fig. 2. The overall algorithm for the solution of an isothermal deformation problem. The contact algorithm is based on an augmented Lagrangian implicit formulation [17].

2.3 An OOP Approach to the FEM Analysis of Forming Processes

Figure 3 presents the main object-oriented simulator LargeDef introduced to model large deformations of inelastic bodies. Most of the members of this class are developed based on physical and algorithmic arguments. For example, the material constitutive model, the integration of the constitutive model, the die geometry, the contact/friction model and its integration, the description of the local kinematics at a material point, etc. are implemented in the form of classes. Such a class structure allows us to easily implement a number of coupling algorithms for the various subproblems without paying special attention to the formulation and algorithmic details of each of them.

The classes ConstitutiveBaseUDC and ConstitutiveIntegrationUDC were introduced to describe the general functionality of inelastic constitutive models and of integration schemes for such models, respectively. The generic structure of the class ConstitutiveBaseUDC is shown in Figure 4. Its member classes ElasticUDC, PStrainUDC, StateEvolveUDC and DamageUDC are introduced to define the elastic properties (equation (13)), the evolution of the tensile plastic strain (equation (10)), the evolution of state (equations (11) and (12)) and the functional form of the damage potential (equation (7)), respectively. As shown in Figure 4, the actual material properties are defined in classes derived from the above base classes.

Radial-return like algorithms have currently been implemented based on a generic structure provided in the class ConstitutiveIntegrationUDC. In derived classes, the constitutive integration algorithms are provided for isotropic rate-dependent and rate-independent models with and without temperature/damage effects [46]–[48].

The base class Die provides the parametrization of the die surface, whereas the class Contact defines the algorithm for the integration of the contact problem [46]. Dimension-specific and process-specific dies are defined in hierarchies derived from the base classes Die and Contact.

Various aspects of the non-linear kinematics are accounted for in class Deformation. This class defines the relation between the deformation gradient and the incremental displacements given the reference grid and it also provides the functionality for the calculation of the deformation gradient given the reference and current (deformed) grids.

A matrix class MatDef(real) has been introduced to represent tensors (e.g. the deformation gradient at a Gauss point). This matrix class allows various common tensor operations (e.g. the RU decomposition for the deformation gradient, tensor algebra operations, etc.) [46]. The representation of the various tensor variables (stress tensor, rotation and stretch tensors, plastic deformation gradient, etc.) that appear in the linearized form of the principle of virtual work as MatDef objects facilitates the calculation of the tangent stiffness and the formation of a linear system of equations for the incremental displacements.

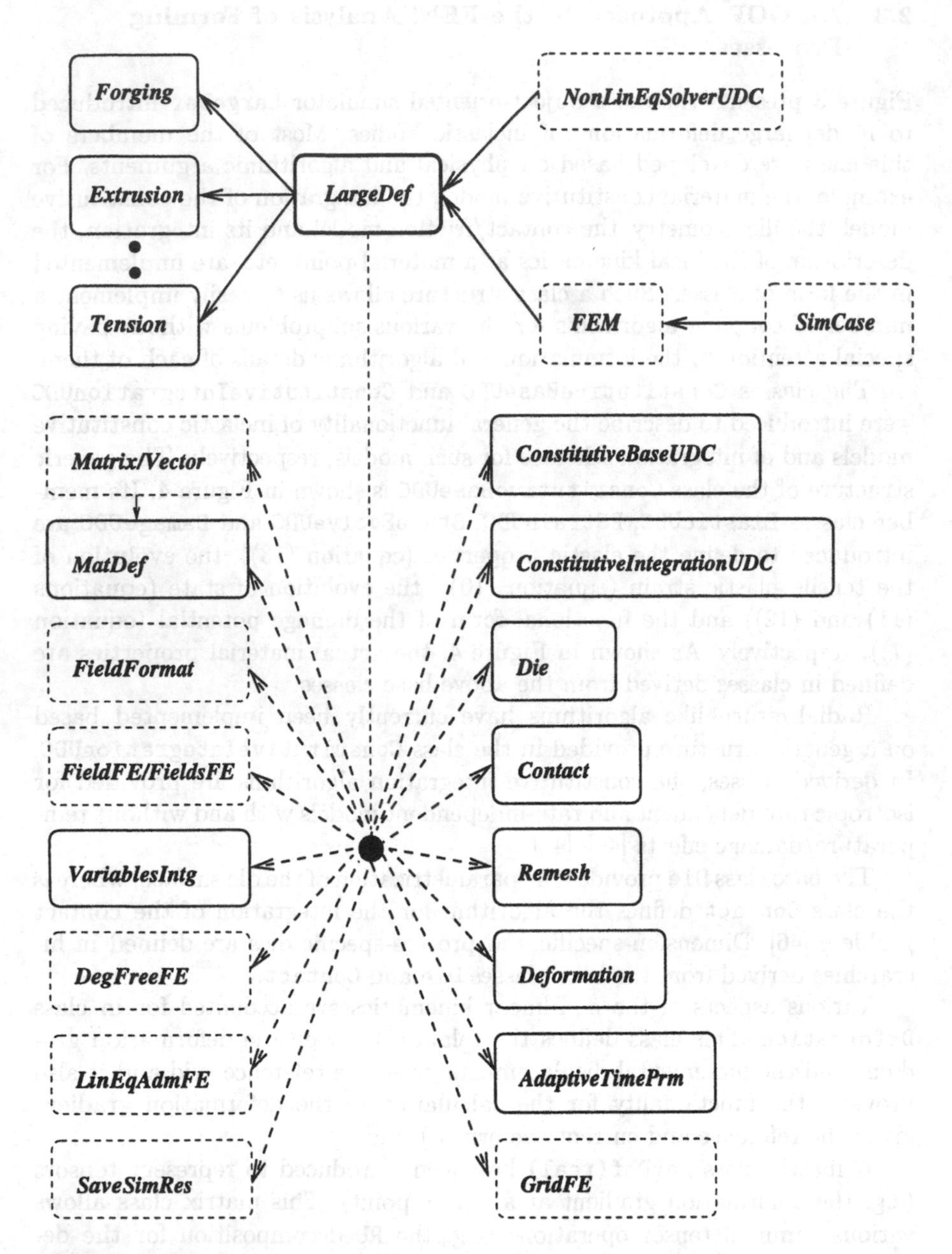

Fig. 3. A schematic of the finite deformation class LargeDef. In this and the following class schematics, the solid arrow lines indicate an is-a relation, whereas the dashed arrow lines represent a has-a relation. Classes within solid line boxes are particular to this simulator, whereas classes within dotted line boxes are part of the Diffpack libraries.

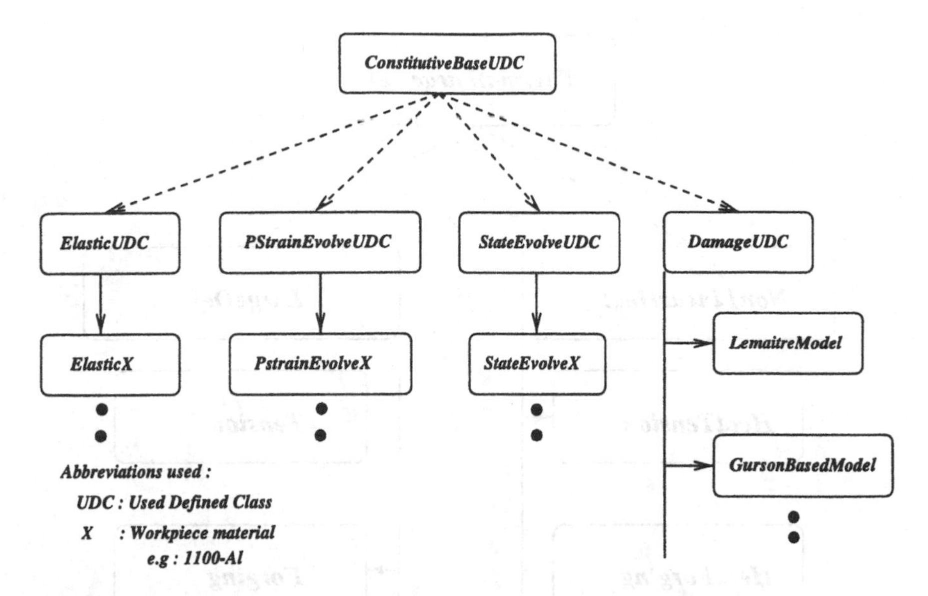

Fig. 4. A schematic of the base class `ConstitutiveBaseUDC` and its members `ElasticUDC`, `PStrainUDC`, `StateEvolveUDC` and `DamageUDC`. A `String X` is introduced to identify each constitutive model, with `X` usually taken to be the name of the particular model. For example, the classes `ElasticX`, `PStrainX`, `StateEvolveX` define the constitutive model `X`. The `String X` is used to bind pointers to the base classes with objects of the derived classes where the specific functionality of each constitutive model is defined. Note that a particular constitutive model with given thermoelastic properties and evolution laws for the tensile plastic strain and state variables can be combined with various type of damage models (e.g. a Gurson based model or the Lemaitre model).

The calculation of the tangent stiffness is performed with the virtual function `integrands()` inherited from the basic Diffpack class `FEM`. Diffpack provides the functionality for essential boundary conditions and here the problem-dependent information is provided in derived classes (e.g. `Forging`, etc.).

To model coupled thermomechanical processes, a simulator with the name `ThermalLargeDef` has been developed with `LargeDef` and `NonLinearHeat` as members. The present heat transfer analysis implemented in `NonLinearHeat` provides the integration of equation (14). The thermal/deformation coupling arises mainly from the heat source given by equation (15), the temperature dependence of the mechanical response, and the change in geometry due to deformation (which may lead to a change in the nature of the applied thermal boundary conditions). As we discussed earlier, process-dependent forming conditions are implemented in classes derived from `LargeDef` (e.g. `Tension`, `Forging`, `Extrusion`, `Rolling`, etc.). Similarly, thermal process-dependent

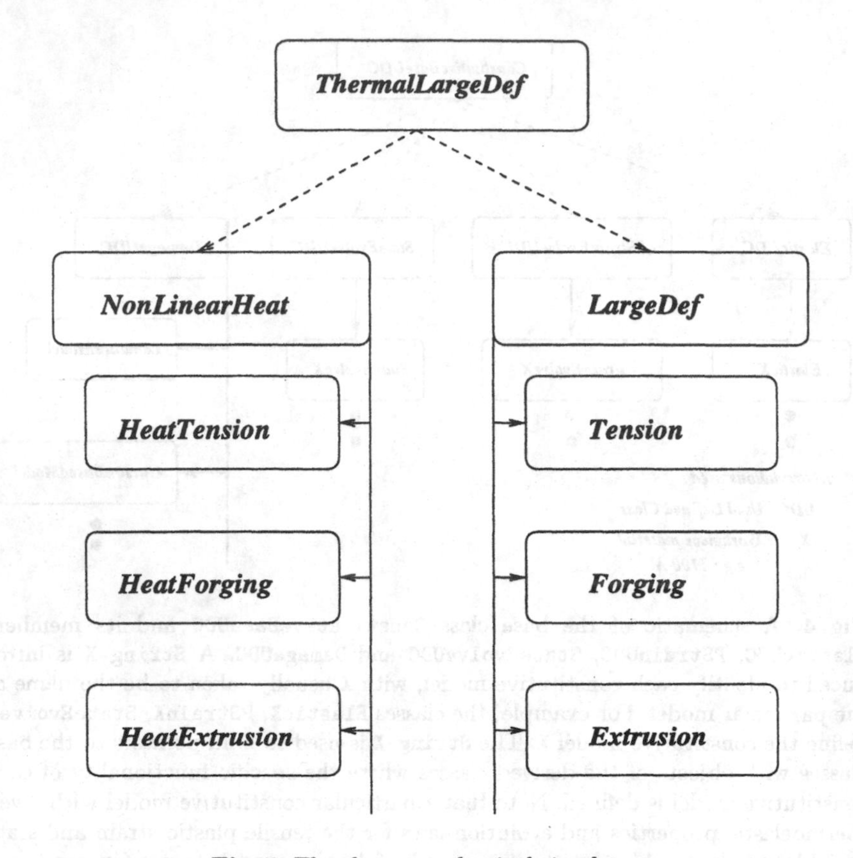

Fig. 5. The thermomechanical simulator.

conditions are implemented in classes derived from `NonLinearHeat`. These hierarchies are shown in Figure 5.

The representation of continuous variables as `FieldsFE` objects is essential in the development of the present deformation simulator [16]. Such fields provide the nodal field values in a given finite element mesh but also allow the use of various postprocessing operations available in Diffpack as well as easy data transfers after remeshing or when restarting a previously interrupted run.

The use of *smart pointers* is an essential tool for memory binding. The class `Handle` is used widely in Diffpack to provide intelligent pointers with reference counting [16]. This class allows the automatic deletion of an object only when all pointers that are bound to this particular object have been deleted. This is important since in large scale simulators, it is customary to have more than one pointer bound to a given object. As an example, the pointers to the objects representing the plastic work rate at each Gauss point in the deformation and heat transfer simulators point to the same point in memory. Thus duplication of memory is avoided and changing the plastic work in the deformation simulator implies that the same changes will be in effect in the thermal analysis. This pointer binding process is applied extensively in the developed simulator. For more details, [46] provides a number of specific examples.

Finally, as in most Lagrangian large-deformation simulators, **LargeDef** provides functionality for remeshing (class **Remesh**), for transferring data (in a **FieldFE** format) between different finite element grids (class **VariablesIntg**), for automatic time stepping (class **AdaptiveTimeStep**) and other (see Figure 3). The interested reader should consult [46]–[48] for further details.

2.4 An Analysis of a Plane Strain Closed-Die Forging Process

An example problem is presented here to briefly demonstrate the ability of the developed OOP simulator in modeling forming problems of industrial importance. For the detailed implementation of various forming processes including CPU requirements of the present simulator see [46]–[48].

The influence of damage on macroscopic deformation characteristics is shown in this axisymmetric isothermal closed die forging example of an 1100-Al cylindrical workpiece. The workpiece is assumed to be initially voided with $f_o = 0.05$. An effective forging process requires that these existing voids be eliminated if the forged product is to achieve reasonable mechanical property levels.

Unlike conventional fully-dense materials, the plastic deformation of a voided material undergoes volume change. As such, the macroscopic deformation response of voided materials poses practical challenges in the design of *flashless* closed die forging processes.

The current process is analyzed for 2 preforms. The cylindrical preforms are chosen such that the initial volume equals that of the die cavity. Preform A has an initial radius of 6.31 mm and an initial height of 10.0 mm. Preform B has an initial radius of 7.0 mm and an initial height of 8.12 mm. A realistic friction coefficient of 0.1 is assumed at the die-workpiece interface and a nominal strain rate of 0.01 is applied during the forging process. Due to the symmetry of the problem, only one-fourth of the geometry is modelled.

The deformed mesh at the final stage is shown together with the initial mesh for preform A and preform B in Figures 6 and 7, respectively. It is evident from Figures 6 and 7 that there is a significant overall decrease in the volume of the workpiece as a result of void closure and the workpiece does not fill up the die cavity exactly. An accurate estimate of this volume change, which is process-dependent, is necessary for the design of *flashless* closed die forging processes.

The force required to forge the product for both the preforms is shown in Figure 8. The void fraction distribution in the final product for preform A and preform B are shown in Figures 9 and 10, respectively. It is evident from Figure 8 that the required work in obtaining the final product is significantly smaller for preform B than for preform A. However, the void fraction

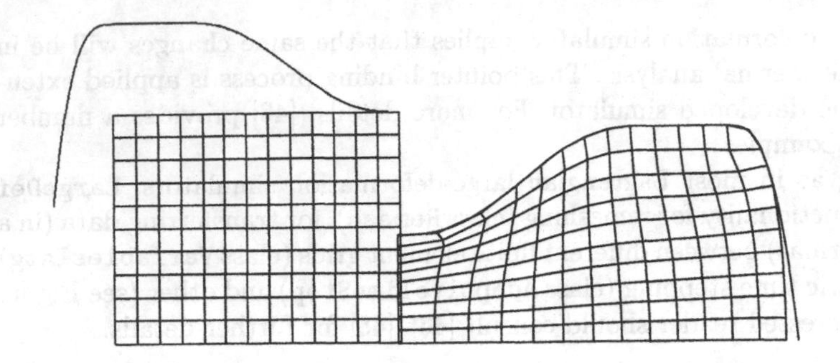

Fig. 6. Initial mesh and final mesh for the axisymmetric closed die forging of a cylindrical billet for preform A.

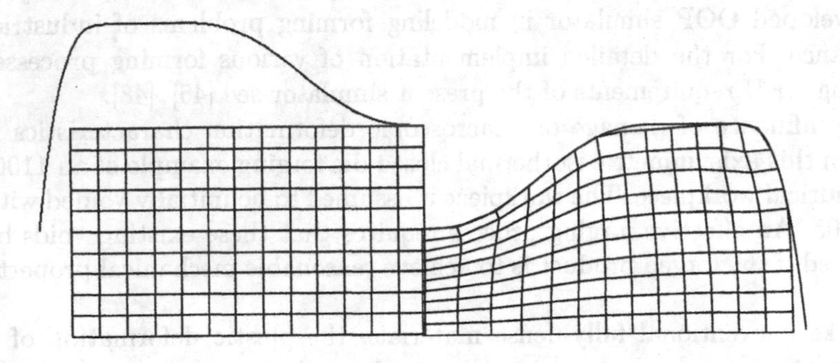

Fig. 7. Initial mesh and final mesh for the axisymmetric closed die forging of a cylindrical billet for preform B.

distribution in the forged product indicates that for preform A, the final void fraction has decreased significantly in all the regions of the workpiece. This was not the case with preform B with some regions in the outer radii of the forged product showing an increase in void fraction of about 6%.

The pressure contours at the final stage of deformation for preform A and preform B are shown in Figures 11 and 12, respectively and are quite different from each other. The results clearly indicate the important role that the preform geometry plays in determining the quality of the forged product. As a result of the change in the initial geometry, the workpiece was subjected to a different contact history and the damage distribution in the final product was significantly different.

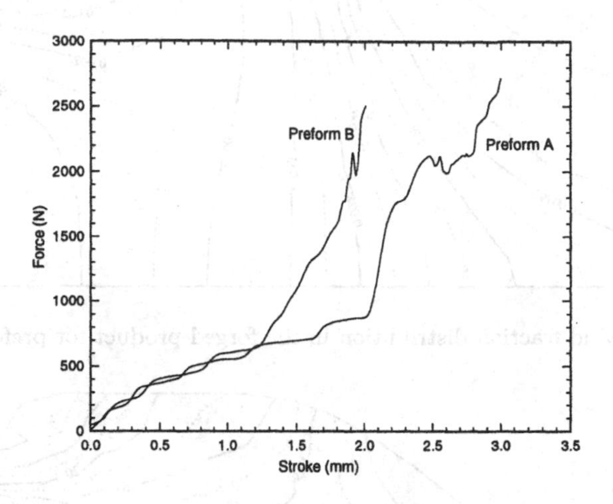

Fig. 8. Force required to forge preforms A and B.

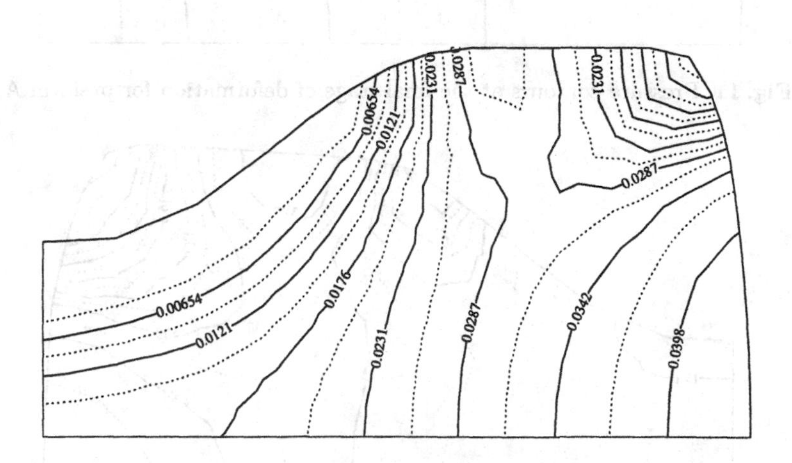

Fig. 9. Void fraction distribution in the forged product for preform A.

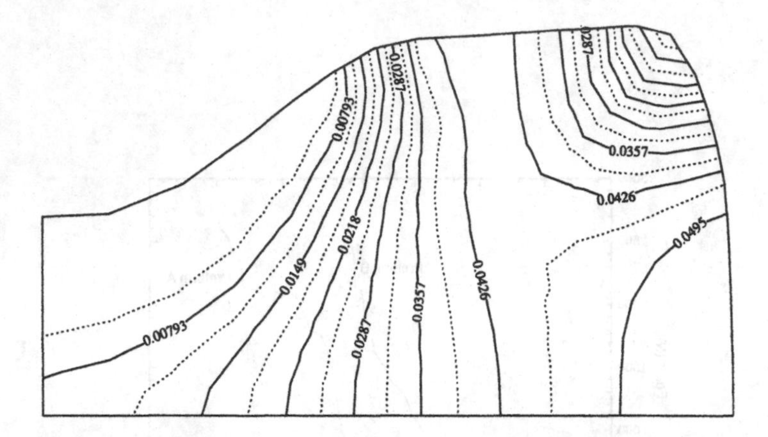

Fig. 10. Void fraction distribution in the forged product for preform B.

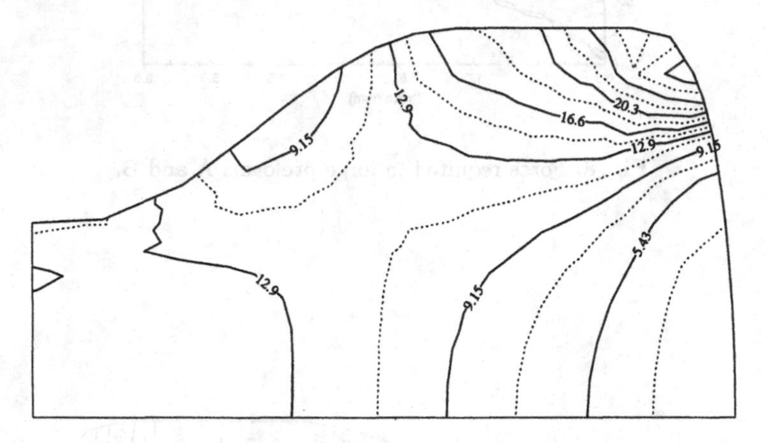

Fig. 11. Pressure contours at the final stage of deformation for preform A.

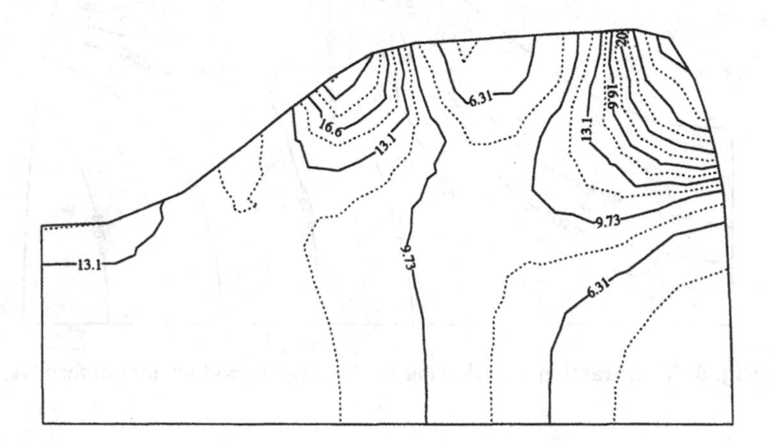

Fig. 12. Pressure contours at the final stage of deformation for preform B.

3 Analysis and Design of Directional Solidification Processes

Some *direct* binary alloy directional solidification problems are introduced here as well as a *design* solidification problem. In particular, Section 3.1 presents the direct problem definition. Section 3.2 briefly discusses a front tracking FEM modeling of direct solidification problems. In Section 3.3, the common structure of the various subproblems of the direct problem (e.g. of the heat transfer and melt flow subproblems) is identified. An appropriate class hierarchy is then introduced for the representation of the subproblems.

Such hierarchies are shown in Section 4 to be very efficient tools for the analysis of design solidification problems in which the boundary cooling/heating is calculated such that desired growth conditions at the freezing front are achieved. In particular, Section 4.1 introduces the definition of a design solidification problem, Section 4.2 presents an adjoint method formulation using an infinite dimensional optimization scheme and Section 4.3 presents the OOP structure of the developed design simulator. Section 4.4 presents the solution of a design problem in which the boundary heat fluxes are calculated such that a desired freezing front velocity and interface heat fluxes are obtained in the solidification of a pure material in the presence of a magnetic field.

3.1 Problem Description

A binary alloy melt is considered at a uniform concentration $\hat{c}_i$ and a uniform temperature $\hat{T}_i$. At time $t = 0^+$, a cooling heat flux is applied at the side Γ_{os} of the mold wall to bring the boundary temperature to the freezing temperature corresponding to concentration $\hat{c}_i$ (see Figure 13). Solidification starts and proceeds from Γ_{os} to Γ_{ol}. It takes place in the presence of a strong externally applied magnetic field $\mathbf{B_o}$ which is at an angle γ to the direction of solidification.

The melt flow is initially driven by thermal and solutal gradients which in turn induce density gradients. The flow in the presence of a magnetic field also gives rise to a Laplace force term in the Navier-Stokes equation. It is here assumed that the induced magnetic field is negligible with respect to the imposed constant magnetic field $\mathbf{B_o}$ and that the electric charge density, ρ_e, is small.

To simplify the presentation, only the governing equations in the melt are presented here. Let L be a characteristic length of the domain, ρ the density, k the conductivity, α the thermal diffusivity, D the solute diffusivity, σ the electrical conductivity and ν the kinematic viscosity. All the thermophysical properties are assumed to be constant. The characteristic scale for time is taken as L^2/α and for velocity as α/L. The dimensionless temperature θ is taken as $\theta = (\hat{T} - T_o)/\Delta T$ where $\hat{T}$, T_o and ΔT are the temperature, reference temperature and reference temperature drop, respectively. Likewise, the dimensionless concentration field c is defined as $(\hat{c}-c_o)/\Delta c$ where $\hat{c}$, c_o and Δc are the concentration, reference concentration and reference concentration drop, respectively. The dimensionless form m of the slop of the liquidus line, is taken as $\hat{m}\Delta c/\Delta T$, whereas the dimensionless form of the melting point of the pure material T_m is defined as $\theta_m = (\hat{T}_m - T_o + \hat{m}c_o)/\Delta T$. The characteristic scale for the electric potential is taken as αB_o. The symbol ϕ is here used to denote the nondimensional electric potential. Other dimensionless parameters used later in this paper include the partition coefficient κ and the parameter $\delta = c_o/\Delta c$ [27].

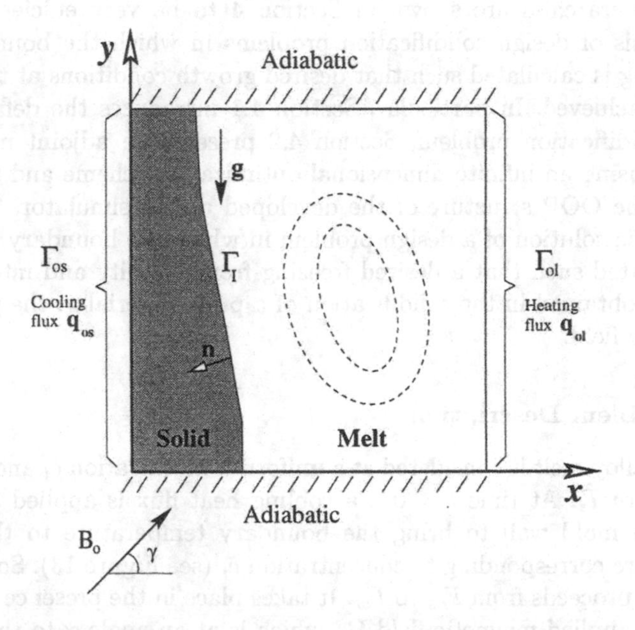

Fig. 13. Schematic of the direct and design alloy solidification problem. The solid region is denoted as Ω_s and the liquid region as Ω_l. These regions share the interface Γ_I. The region Ω_l has a boundary Γ_l which consists of Γ_I (the solid-liquid interface), Γ_{ol} and the remaining boundary Γ_{hl} (the top and bottom mold walls). Similarly Ω_s has boundary Γ_s, which includes Γ_I, Γ_{os} and the top and bottom mold boundary Γ_{hs}.

The key dimensionless quantities are the Prandtl number ($\mathrm{Pr} = \nu/\alpha$), the Lewis number ($\mathrm{Le} = \alpha/D$), the Stefan number ($\mathrm{Ste} = (c\Delta T_o)/L_H$, with L_H the latent heat), the thermal Rayleigh number ($\mathrm{Ra_T} = g\beta_T \Delta T L^3/\nu\alpha$, with β_T the thermal expansion coefficient), the solutal Rayleigh number ($\mathrm{Ra_c} = g\beta_c \Delta c L^3/\nu\alpha$, with β_c the solutal expansion coefficient), and the Hartmann number ($\mathrm{Ha} = \left[\left(\frac{\sigma}{\rho\nu}\right)^{1/2} B_o L\right]$) (see [27] for further review of these definitions). The governing PDEs are summarized in a dimensionless form in Figure 14 together with typical initial conditions for the field variables. In addition to these equations, one must consider the solute balance at the freezing front, the Stefan condition, the equilibrium condition at the front relating the temperature and concentration (via the liquidus line of the phase diagram) and appropriate boundary conditions for all field variables.

The system of PDEs in the melt has to be solved simultaneously with the heat conduction problem in the solid phase to calculate the main field variables $(\theta, c, \boldsymbol{v}, \phi)$ as well as the location of the freezing front. More details on the definition of directional alloy solidification problems can be found in [26] and [27]. Issues related to the physical feasibility of such processes with an a-priori assumed stable growth (i.e. with a sharp interface) are discussed in [42] and [27].

For reasons to be explained later, the heat fluxes q_{os} and q_{ol} are considered as parameters to this direct problem. In particular, Figure 14 shows the governing PDEs in the melt considering the heat flux $q_{ol}(\boldsymbol{x}, t)$, $(\boldsymbol{x}, t) \in \Gamma_{ol} \times [0, t_{max}]$ as a (functional) parameter.

$$\frac{\partial \theta}{\partial t} + \boldsymbol{v} \cdot \nabla \theta = \nabla^2 \theta \tag{16}$$

$$\frac{\partial \boldsymbol{v}}{\partial t} + (\nabla \boldsymbol{v})\, \boldsymbol{v} = -\nabla p + \mathrm{Pr} \nabla^2 \boldsymbol{v} - \mathrm{Ra_T} \mathrm{Pr} \theta \mathbf{e_g} + \mathrm{Ra_c} \mathrm{Pr} c \mathbf{e_g} + \mathrm{Ha}^2 \mathrm{Pr} \left[-\nabla \phi + \boldsymbol{v} \times \mathbf{e_B} \right] \times \mathbf{e_B} \tag{17}$$

$$\nabla \cdot \boldsymbol{v} = 0 \tag{18}$$

$$\frac{\partial c}{\partial t} + \boldsymbol{v} \cdot \nabla c = Le^{-1} \nabla^2 c \tag{19}$$

$$\nabla^2 \phi = \nabla \cdot (\boldsymbol{v} \times \mathbf{e_B}) \tag{20}$$

$$\theta(\boldsymbol{x}, 0) = \theta_i, \quad c(\boldsymbol{x}, 0) = c_i, \quad \boldsymbol{v}(\boldsymbol{x}, 0) = \mathbf{0}, \quad \boldsymbol{x} \in \Omega_l(0) \tag{21}$$

Fig. 14. The governing PDEs and initial conditions for the direct problem in the melt in terms of $\theta(\boldsymbol{x}, t; q_o)$, $\boldsymbol{v}(\boldsymbol{x}, t; q_o)$, $c(\boldsymbol{x}, t; q_o)$ and $\phi(\boldsymbol{x}, t; q_o)$, with $(\boldsymbol{x}, t) \in \Omega_l(t) \times [0, t_{max}]$. Also, $\mathbf{e_g}$ and $\mathbf{e_B}$ denote the unit vectors in the direction of gravity and magnetic field, respectively.

3.2 FEM Algorithm

From the problem statement given in Figure 14, it becomes clear that the overall direct solidification problem can be considered as the coupling of four main PDE based subproblems (the flow, heat diffusion/convection, mass diffusion/convection and the magnetic field potential problem). In addition, the motion of the freezing interface must be considered in the analysis.

Let us first recognize that the three transport problems have the same convection/diffusion form. Any algorithm addressing such convection/diffusion problems can be applied for the solution of these three problems. A Petrov-Galerkin analysis with the consistent penalty method was applied in [26] to produce a spatially discretized weak form for such problems. Other techniques can be applied as well [37]–[39].

Most FEM techniques for advection/diffusion problems finally lead to a system of discretized equations of the following form:

$$[\mathbf{M}]\{\dot{\alpha}\} + [\mathbf{C}]\{\alpha\} = \{F\} \tag{22}$$

where $\{\alpha\}$ is the vector of the unknown degrees of freedom (nodal velocity components). Various time integration schemes can be applied for the solution of the above ODE. A predictor-corrector scheme was used in [26] following the work of [8]. Other time stepping techniques can for example be found in [37].

The solution of the electric potential subproblem is not discussed in any detail here as a simple Galerkin formulation can be applied once the velocity field $\boldsymbol{v}$ is calculated.

The various time sequential algorithms for each subproblem (the flow, heat and solute diffusion/convection and electromagnetics subproblems) must be coupled within each time step. The calculation of the freezing front motion is an integrable part of the overall algorithm [26]. In this work, a deforming FEM technique is used to track the freezing interface. In such formulations, one must account for the apparent convection resulting from the mesh motion. More on the selected mesh motion for directional solidification problems can be found in [26].

3.3 An OOP Approach to the FEM Analysis of Solidification processes

In the previous section, three main convection/diffusion subproblems were identified that in a spatially discretized form fit in the category of the algebraic system given in the model equation (22). A general abstract base class `ConvectionDiffusion` is defined to manage the solution process for the model scalar-convection diffusion equation. The general structure of the class `ConvectionDiffusion` is shown in Figure 15. Further information for object-oriented programming of convection/diffusion problems can be found in [28]–[32].

The above development of an abstract base class that solves a general model equation helps in *avoiding repetitive codes.* As such all three of the above diffusion/convection subproblems can be described by the `ConvectionDiffusion` class introduced here. The main differences between the heat transfer, solute transport and melt flow mechanisms are in the coefficients of the corresponding PDEs as well as in the initial and boundary conditions. This information particular to each PDE can be introduced using the virtual functions `integrands()`, `setIC()` and `essBC`, respectively [16]. For example, the virtual function `FEM:integrands()` is used to define the weak Petrov-Galerkin form of each PDE and perform the stiffness and load calculations at a particular Gauss point. These details are obvious and will not be further discussed here.

Experimentation with a different integration scheme for (22) will only require the development of a new derived class from `ConvectionDiffusion` where the particulars of the new algorithm are implemented. The binding of a pointer to an object of the base class `ConvectionDiffusion` to the particular algorithm is performed at run time. The particular predictor/corrector scheme used in [26] is here implemented with a derived class, called `PredictorCorrector`.

To further emphasize the importance of recognizing the common structure of the three convection/diffusion subproblems of Figure 14, we will next present the definition, solution algorithm and OOP implementation of an inverse design directional solidification problem.

4 On the Optimum Design of Directional Solidification Processes

A typical inverse design problem is one governed by a system of PDEs with insufficient or no boundary conditions in part of the boundary but overspecified conditions in another part of the boundary or within the domain. These problems are generally ill-posed and a quasi-solution is being sought in some optimization sense [47]. The type of approach employed here for the solution of such inverse problems is the so called *iterative regularization technique* [2], [23]. In simple terms, the problem is formulated as a functional optimization problem over the whole time and space domain [47]. For example, in an inverse heat transfer problem, the overspecified data in the part Γ_I of the boundary Γ can be the temperature θ_I and heat flux q_I, whereas the heat flux q_o (and temperature) are unknown in another part Γ_o of the boundary. The cost function in this example is defined as the error $||\theta(\boldsymbol{x}, t; q_o) - \theta_I||^2_{L_2(\Gamma_I \times [0, t_{max}])}$ between the computed (via a well-posed direct analysis) solution $\theta(\boldsymbol{x}, t; q_o)$ using q_o and q_I as boundary conditions on Γ_o and Γ_I, respectively, and the given temperature θ_I on Γ_I. The gradient of such cost functionals is calculated with a *continuum adjoint problem*, whereas the step size in each optimization step requires the solution of a *continuum sensitivity problem*. In addition to

```
class ConvectionDiffusion: public ... {
protected:
  Handle(GridFE)         grid;              // finite element grid
  Handle(DegFreeFE)      dof;               // mapping of nodal values
                                            // <->linear system
  Handle(TimePrm)        tip;               // time loop parameters
  Handle(FieldsFE)       alpha;             // FE field, the primary unknown
  Handle(FieldsFE)       alpha_prev;        // solution in the previous time step
  Handle(FieldsFE)       alpha_dot;         // rate field
  Handle(FieldsFE)       alpha_dot_prev;    // rate field in the previous time step
  real                   gamma;             // time integration parameter
  real                   err_tolerance;     // error tolerance for the
                                            // nonlinear solver
  ...
  Vec(real)              linear_solution;   // solution of the linear subsystem
  Handle(LinEqAdmFE)     lineq;             // linear system and solution
  ...
  virtual void timeLoop();
  virtual void solveAtThisTimeLevel();
  virtual void fillEssBC();
  virtual void integrands(ElmMatVec& elmat, FiniteElement& fe);
  virtual void calcElmMatVec(int elm_no, ElmMatVec& elmat,
                             FiniteElement& fe);
  virtual void integrands4side(int side, int boind, ElmMatVec& elmat,
                             FiniteElement& fe);
  virtual void setIC() = 0;                 // initial conditions
  virtual real essBC(int nno,               // essential boundary conditions
                  const Ptv(real)& x) = 0;
public:
  ...
  virtual void define(MenuSystem& menu, int level = MAIN);
  virtual void scan(MenuSystem& menu);
  virtual void adm(MenuSystem& menu);
  virtual void solveProblem();
  virtual void storeResults() = 0;
  virtual void updateDataStructures();
};
```

Fig. 15. The main public and protected members of the class `ConvectionDiffusion` introduced for carrying out program administration and the various numerical steps in the algorithms.

these problems, the solution of the well-posed direct problem is required as well.

4.1 Thermal Design of Binary Alloy Solidification to Achieve a Stable Desired Growth

The design solidification problem of interest here can be stated as follows:

> *Find the boundary heating/cooling fluxes on the left and right vertical walls of Figure 13 such that a stable desired interface growth is achieved.*

Similar design problems can be stated in which the path in the state space $G - \boldsymbol{v}_f$ is defined [15], [35] (here G denotes the interface flux and $\boldsymbol{v}_f$ the growth velocity). Since the growth velocity and the interface heat fluxes control the obtained solidification microstructures, such problems can be used to design solidification processes that lead to desired properties in the cast product [15], [35], [42], [27].

The overspecified conditions in the present inverse design problems are defined at the freezing front, whereas no thermal conditions are available on part of the outside boundary (boundaries Γ_{ol} and Γ_{os} (see Figure 13)). The two overspecified conditions on the interface include the growth velocity $\boldsymbol{v}_f$ and the stability condition. The last condition has been interpreted in [27] as a relation between the thermal gradient and the solute gradient at the interface. In particular, the following form of the stability condition is considered [27]:

$$\frac{\partial \theta}{\partial n}(\boldsymbol{x}, t; q_{ol}) = m\frac{\partial c}{\partial n}(\boldsymbol{x}, t; q_{ol}) + \epsilon(\boldsymbol{x}, t), \qquad (\boldsymbol{x}, t) \in \Gamma_I(t) \times [0, t_{max}]. \quad (23)$$

Note that the vector $\boldsymbol{n}$ used above to define the normal derivatives at the interface points towards the solid region (see Figure 13).

There are various possibilities for the selection of the parameter $\epsilon \leq 0$ [42] and [27]. Equation (23) is the dimensionless form of the well known constitutional undercooling stability condition [15] and the parameter ϵ can here be interpreted as the level of 'over-stability' desired in the solidification system.

In addition to the stability condition and $\boldsymbol{v}_f$, the solute balance and the temperature/concentration equilibrium relation are also provided at the freezing front. The overall design problem can now be considered with the solution of two inverse problems, one in the solid phase and another in the liquid melt.

The optimization scheme for the inverse melt flow problem is briefly presented next. The unknown heat flux q_{ol} on the boundary Γ_{ol} is assumed to be an $L_2(\Gamma_{0l} \times [0, t_{max}])$ function. Using part of the overspecified data on the freezing front and for a guess value of the unknown heat flux q_{ol}, one can solve a direct problem with the PDEs and initial conditions of Figure 14 and the boundary/interface conditions given in Figure 16.

In particular, the conditions considered at the freezing front as part of this direct problem are the desired front velocity $\boldsymbol{v}_f$, the stability condition given by equation (23) or its equivalent form (27) and the solute balance condition (equation (29)). The solution of this direct problem provides the values of the temperature and concentration fields at the freezing front. The discrepancy of these values from the ones corresponding to thermodynamic equilibrium is the driving force for the optimization algorithm.

$$\boldsymbol{v}(\boldsymbol{x}, t; q_{ol}) = 0, \quad \boldsymbol{x} \in \Gamma_l(t) \tag{24}$$

$$\frac{\partial \theta}{\partial n}(\boldsymbol{x}, t; q_{ol}) = 0, \quad \boldsymbol{x} \in \Gamma_{hl}(t) \tag{25}$$

$$\frac{\partial \theta}{\partial n}(\boldsymbol{x}, t; q_{ol}) = q_{ol}(\boldsymbol{x}, t), \quad \boldsymbol{x} \in \Gamma_{ol}(t) \tag{26}$$

$$\frac{\partial \theta}{\partial n}(\boldsymbol{x}, t; q_{ol}) = \epsilon(\boldsymbol{x}, t) + mLe(\kappa - 1)\boldsymbol{v}_f \cdot \boldsymbol{n}(c(\boldsymbol{x}, t; q_{ol}) + \delta), \ \boldsymbol{x} \in \Gamma_I(t) \tag{27}$$

$$\frac{\partial c}{\partial n}(\boldsymbol{x}, t; q_{ol}) = 0, \quad \boldsymbol{x} \in \Gamma_{hl}(t) \cup \Gamma_{ol}(t) \tag{28}$$

$$\frac{\partial c}{\partial n}(\boldsymbol{x}, t; q_{ol}) = Le(\kappa - 1)\boldsymbol{v}_f \cdot \boldsymbol{n}(c(\boldsymbol{x}, t; q_{ol}) + \delta), \ \boldsymbol{x} \in \Gamma_I(t) \tag{29}$$

$$\frac{\partial \phi}{\partial n}(\boldsymbol{x}, t; q_{ol}) = 0, \quad \boldsymbol{x} \in \Gamma_l(t) \tag{30}$$

Fig. 16. The boundary conditions given here together with the equations of Figure 14 define $\theta(\boldsymbol{x}, t; q_{ol})$, $\boldsymbol{v}(\boldsymbol{x}, t; q_{ol})$, $c(\boldsymbol{x}, t; q_{ol})$ and $\phi(\boldsymbol{x}, t; q_{ol})$ for $t \in [0, t_{max}]$.

For an arbitrary q_{ol}, the following cost functional is defined:

$$J(q_{ol}) = \frac{1}{2} ||\theta(\boldsymbol{x}, t; q_{ol}) - (\theta_m + mc(\boldsymbol{x}, t; q_{ol}))||^2_{L_2(\Gamma_I \times [0, t_{max}])} \tag{31}$$

to indicate the discrepancy of the calculated temperature (using the direct problem defined in Figure 14 and Figure 16) from the concentration-dependent liquidus temperature at the interface. Finally, the *control problem* of interest is defined as follows:

Find $\bar{q}_{ol} \in L_2(\Gamma_{0l} \times [0, t_{max}])$ such that

$$J(\bar{q}_{ol}) \leq J(q_{ol}), \forall q_{ol} \in L_2(\Gamma_{0l} \times [0, t_{max}]). \tag{32}$$

In this paper, our objective is to construct a minimizing sequence $q^k_{ol}(\boldsymbol{x}, t) \in L_2(\Gamma_{0l} \times [0, t_{max}])$, $k = 1, 2, \ldots$ that converges to at least a local minimum of $J(q_{ol})$.

4.2 The Adjoint Method

The main difficulty with the above optimization problem is the calculation of the gradient $J'(q_{ol})(\boldsymbol{x}, t)$ in the space $L_2(\Gamma_{0l} \times [0, t_{max}])$. Taking the directional derivative of the cost functional with respect to q_{ol} in the direction of the variation Δq_{ol} yields the following:

$$D_{\Delta q_{ol}} J(q_{ol}) = (J'(q_{ol}), \Delta q_{ol})_{\mathcal{U}} = (\Theta - mC, R)_{L_2(\Gamma_I \times [0, t_{max}])} \tag{33}$$

where the residual temperature field R is defined as $R(\boldsymbol{x}, t; q_{ol}) \equiv \theta(\boldsymbol{x}, t; q_{ol}) - (\theta_m + mc(\boldsymbol{x}, t; q_{ol}))$. The sensitivity fields corresponding to the magnetic potential, temperature, melt velocity and concentration fields are here denoted as $\Phi(\boldsymbol{x}, t; q_{ol}, \Delta q_{ol})$, $\Theta(\boldsymbol{x}, t; q_{ol}, \Delta q_{ol})$, $\boldsymbol{V}(\boldsymbol{x}, t; q_{ol}, \Delta q_{ol})$ and $C(\boldsymbol{x}, t; q_{ol}, \Delta q_{ol})$, respectively. These sensitivity fields are defined as directional derivatives of the direct fields calculated at q_{ol} in the direction Δq_{ol}. Due to the form of equation (33), it is easier from now on to work with the variable $R(\boldsymbol{x}, t; q_{ol})$ instead of the temperature field θ. The Figures 14 and 16 can be modified appropriately by expressing all equations in terms of $R, \boldsymbol{v}, c$ and ϕ. The sensitivity field corresponding to R is here denoted as $\Re(\boldsymbol{x}, t; q_{ol}, \Delta q_{ol})$. The governing PDEs for the sensitivity problem are given in Figure 17.

It is clear from equation (33) that the calculation of the gradient of the cost functional requires the evaluation of the adjoint field $\psi(\boldsymbol{x}, t; q_{ol})$ corresponding to the sensitivity field $\Re(\boldsymbol{x}, t; q_{ol}, \Delta q_{ol})$. In the process of evaluating ψ, one must also obtain the corresponding adjoint fields for melt velocity, electric field and concentration here denoted as $\boldsymbol{\phi}(\boldsymbol{x}, t; q_{ol})$, $\eta(\boldsymbol{x}, t; q_{ol})$ and $\rho(\boldsymbol{x}, t; q_{ol})$, respectively. The governing PDEs for the adjoint problem are shown in Figure 18 (for the derivation of these equations see [27]). In addition to the dimensionless parameters introduced earlier, $\bar{\mathrm{Ra}}_c$ and χ are here defined as $\bar{\mathrm{Ra}}_c = \mathrm{Ra_T}\left[\mathrm{m} - \frac{\mathrm{Ra_c}}{\mathrm{Ra_T}}\right]$ and $\chi = \mathrm{m}\frac{\mathrm{Ra_T}}{\mathrm{Ra_c}}(\mathrm{Le}^{-1} - 1)$, respectively [27].

It can be shown that the desired gradient $J'(q_{ol})$ of the cost functional $J(q_{ol})$ is given as follows [27]:

$$J'(q_{ol}) = \psi(\boldsymbol{x}, t; q_{ol}), \quad (\boldsymbol{x}, t) \in \Gamma_{ol} \times [0, t_{max}]. \tag{44}$$

$$\frac{\partial \Re}{\partial t} + \boldsymbol{v} \cdot \nabla \Re + \boldsymbol{V} \cdot \nabla R = \nabla^2 \Re + \mathrm{m}(1 - \mathrm{Le}^{-1}) \nabla^2 C \tag{34}$$

$$\frac{\partial \boldsymbol{V}}{\partial t} + (\nabla \boldsymbol{v})\, \boldsymbol{V} + (\nabla \boldsymbol{V})\, \boldsymbol{v} = -\nabla \Pi + \mathrm{Pr} \nabla^2 \boldsymbol{V} - \mathrm{Ra_T Pr} \Re \boldsymbol{e_g}$$
$$-\mathrm{Ra_T Pr}\left(\mathrm{m} - \frac{\mathrm{Ra_c}}{\mathrm{Ra_T}}\right) C \boldsymbol{e_g} + \mathrm{Ha^2 Pr}\left[-\nabla \Phi + \boldsymbol{V} \times \mathbf{e_B}\right] \times \mathbf{e_B} \tag{35}$$

$$\nabla \cdot \boldsymbol{V} = 0 \tag{36}$$

$$\frac{\partial C}{\partial t} + \boldsymbol{v} \cdot \nabla C + \boldsymbol{V} \cdot \nabla c = \mathrm{Le}^{-1} \nabla^2 C \tag{37}$$

$$\nabla^2 \Phi = \nabla \cdot (\boldsymbol{V} \times \mathbf{e_B}) \tag{38}$$

Fig. 17. PDEs governing the sensitivity problem that defines $\Re(\boldsymbol{x}, t; q_{ol}, \Delta q_{ol})$, $\boldsymbol{V}(\boldsymbol{x}, t; q_{ol}, \Delta q_{ol})$, $C(\boldsymbol{x}, t; q_{ol}, \Delta q_{ol})$ and $\Phi(\boldsymbol{x}, t; q_{ol}, \Delta q_{ol})$, with $(\boldsymbol{x}, t) \in \Omega_l(t) \times [0, t_{max}]$.

$$\frac{\partial \psi}{\partial t} + \boldsymbol{v} \cdot \nabla \psi = -\nabla^2 \psi + \boldsymbol{\phi} \cdot \boldsymbol{e_g} \tag{39}$$

$$\frac{\partial \boldsymbol{\phi}}{\partial t} + (\nabla \boldsymbol{\phi})\, \boldsymbol{v} - (\nabla \boldsymbol{v})^T \boldsymbol{\phi} = \nabla \pi - Pr \nabla^2 \boldsymbol{\phi} - Ha^2 Pr[\mathbf{e_B}(\boldsymbol{\phi} \cdot \mathbf{e_B}) - \boldsymbol{\phi}]$$
$$+ PrRa_T \psi \nabla R + Pr\bar{R}a_c \rho \nabla c + Ha^2 Pr[\nabla \eta \times \mathbf{e_B}] \tag{40}$$

$$\nabla \cdot \boldsymbol{\phi} = 0 \tag{41}$$

$$\nabla^2 \eta = \nabla \cdot (\boldsymbol{\phi} \times \mathbf{e_B}) \tag{42}$$

$$\frac{\partial \rho}{\partial t} + \boldsymbol{v} \cdot \nabla \rho(\boldsymbol{x}, t; q_{ol}) = -Le^{-1} \nabla^2 \rho + \boldsymbol{\phi} \cdot \boldsymbol{e_g} + \chi \nabla^2 \psi \tag{43}$$

Fig. 18. PDEs governing the adjoint problem that defines $\psi(\boldsymbol{x}, t; q_{ol})$, $\boldsymbol{\phi}(\boldsymbol{x}, t; q_{ol})$, $\eta(\boldsymbol{x}, t; q_{ol})$ and $\rho(\boldsymbol{x}, t; q_{ol})$, with $(\boldsymbol{x}, t) \in \Omega_l(t) \times [0, t_{max}]$.

A typical adjoint formulation for an inverse design problem is effectively an infinitedimensional optimization problem that can be approached with a variety of techniques such as the conjugate gradient method [18]. The solution procedure for the particular inverse solidification design problem is shown in Figure 19.

In the following section, an elegant technique is presented for the OOP implementation of the algorithm of Figure 19 as applied to the above design problem. The ideas here are general and applicable to a variety of optimum continuum design problems.

4.3 An OOP Approach to the Implementation of the Adjoint Method

The first task in the OOP design of the simulator is to recognize the similarities of the various subproblems. The governing PDEs for the direct, adjoint and sensitivity flow, heat and solute transport problems are of similar nature and fall under the general category of scalar or vector transient convection-diffusion equations. One possible design for the considered direct, adjoint and sensitivity problems (defined by the collection of 9 diffusion/convection like PDEs plus three simpler steady equations for the electromagnetics) would have been to declare 9 different classes, each one managing its own fields, grids, linear system solvers and nonlinear solution process. Such a strategy would not have taken into account the similarity in the problems involved and would thus involve repeated code. In addition, changes in the solution scheme for the integration of the PDEs as well as changes in the optimization algorithm could not be applied without a significant programming effort. Also, testing of the various algorithms for the adjoint and sensitivity subproblems would not have been easy following such an approach.

We here re-introduce the master class `ConvectionDiffusion` (see Figure 15) from which the classes corresponding to each subproblem of the direct, sensitivity and adjoint problems are derived. Next stage in the design of the simulator is to recognize specific differences that distinguish one PDE from another. For example what distinguishes the flow equation from the heat equation? The immediate task is to incorporate those differences in some form into the simulator. The fundamental principle of object-oriented programming requires that these changes should be incorporated without affecting the already existing code and should be flexible enough to allow further changes. In the present simulator the heat, flow and solute solvers of the direct, adjoint and sensitivity problems have been built on these lines. Nine classes, one for each PDE, are declared but they are all derived (or inherited) from the base class `ConvectionDiffusion` and hence contain only the *differences* between the various solvers such as the weak form of each PDE and the corresponding initial/boundary conditions. This construction helps in sharing common features such as the solution process, storing of fields etc. Moreover, this method provides more robustness as the base-class

Step A: Start with an initial guess q_{0l}^0 for the unknown function. Set $k = 0$.
Step B: Solve the direct problem with the current estimate of q_{0l} and using part of the overspecified data.
Step C: Solve the adjoint problem.
Step D: Calculate the search direction $\mathbf{p}_k$ in this optimization step.
Step E: Solve the sensitivity problem.
Step F: Calculate the step length α_k in the search direction.
Step G: Update the heat flux $q_{0l}^{k+1} = q_{0l}^k + \alpha_k \mathbf{p}_k$.
Step H: Check if the cost functional is reduced below the desired tolerance. If not, proceed to another optimization step (i.e. return to step B). If yes, STOP.

Fig. 19. A typical gradient algorithm for the solution of an inverse optimum design problem.

can be tested independently on simpler problems. This construction basically demonstrates the effectiveness of using the concept of *inheritance*.

With the above tasks completed, one needs to address the important process of combining the individual simulators to form the direct, adjoint and sensitivity subproblems. Three subproblem classes are declared (`Direct`, `Adjoint` and `Sensitivity`) to manage the time stepping and whole time domain solution of the three subproblems (see Figure 20). Each of these classes has pointers to the corresponding PDE classes and provides the required time stepping functionality. As it is customary with OOP implementation of complex systems, the development of the classes for the direct, sensitivity and adjoint problems was performed and tested in a number of steps. For example, a class hierarchy `Direct`, `DirectAlloy` and `DirectMagnetoAlloy` was implemented to model the direct analysis of natural convection of pure materials, double diffusive convection in alloys, and magneto-convection in alloys, respectively. Such class hierarchies are not shown in Figure 20.

Some details of the class `DirectAlloy` are given in Figure 21. In addition to the functionality and data of the class `Direct`, the concentration field is added as a protected member of the class `DirectAlloy`. To allow access of an object of the class `DirectAlloy` to the protected members of the class `Direct`, pointers to the flow and heat solvers are also declared as members of the class `DirectAlloy`. These pointers are appropriately bound to the flow and heat objects of the class `Direct`.

An appropriate binding between the various smart pointers (handles) is an essential element of an efficient OOP implementation of such problems. For example, in the class `Direct` (that has two protected data members pointing to objects of the classes `NavierStokes` and `NlHeat`), the velocity field v which is a member of the `NavierStokes` class needs to be passed on as input

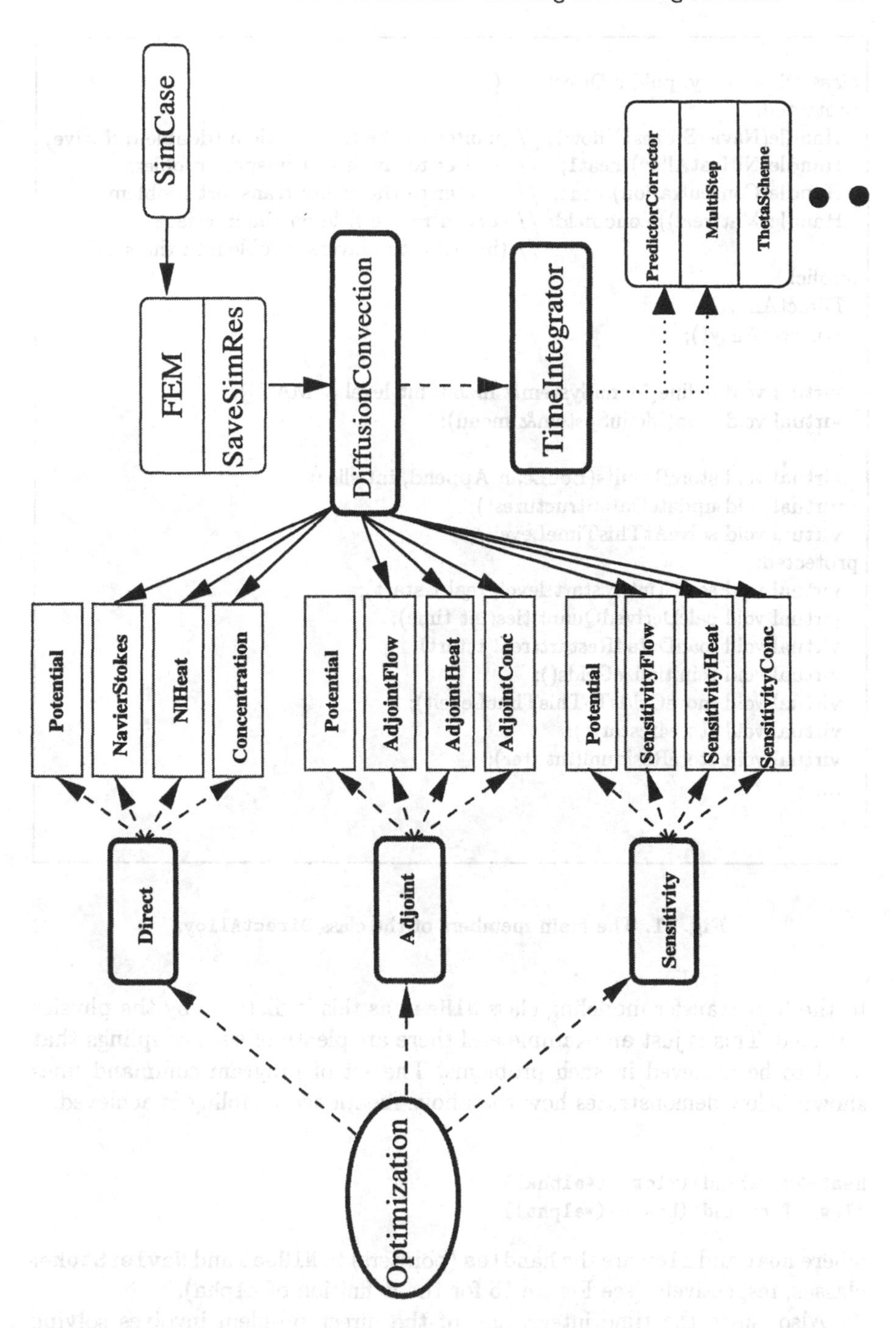

Fig. 20. A sketch of the inverse design simulator (Optimization), the base classes and some layers of the internal objects. A solid line indicates class derivation (is-a relationship), whereas dashed lines represent a pointer/handle (has-a relationship).

```
class DirectAlloy: public Direct, ... {
protected:
  Handle(NavierStokes1) flow1; // pointer to the flow problem (double-diffusive)
  Handle(NlHeatAlloy) heat1;  // pointer to the heat transport problem
  Handle(Concentration) conc; // pointer to the solute transport problem
  Handle(Mat(real)) conc_field; // concentration field on the interface
                                // (input to the inverse problem in the solid)
public:
  DirectAlloy();
  ~DirectAlloy();
  ...
  virtual void define(MenuSystem& menu, int level = MAIN);
  virtual void scan(MenuSystem& menu);
  ...
  virtual void storeResults(BooLean Append, int slice);
  virtual void updateDataStructures();
  virtual void solveAtThisTimeLevel();
protected:
  virtual void setIC(int restart_level, real t_start);
  virtual void calcDerivedQuantities(int time);
  virtual void loadData4Restart(real tstart);
  virtual void reinitializeGrids();
  virtual void moveGridsToThisTimeLevel();
  virtual void store4restart();
  virtual void storeResidual(int iter);
  ...
};
```

Fig. 21. The main members of the class `DirectAlloy`.

to the heat transfer modeling class `NlHeat` as this is dictated by the physics involved. This is just an example and there are plenty of such couplings that need to be achieved in such problems. The set of program command lines shown below demonstrates how the above mentioned coupling is achieved:

```
heat->v.rebind (flow->(*alpha));
flow->T.rebind (heat->(*alpha));
```

where `heat` and `flow` are the `handles` (pointers) to `NlHeat` and `NavierStokes` classes, respectively (see Figure 15 for the definition of `alpha`).

Also, since the time integration of the direct problem involves solving the heat, flow and solute diffusion subproblems simultaneously and the same being the case for the adjoint and sensitivity problems, the 'time integration objects' of the pointers to the subclasses `NlHeat`, `NavierStokes` and

`Concentration` need to be bound so that there exist only one 'time integration object'. An example of this process (in the class `Direct`) is shown in the command lines below:

```
heat->tip.rebind (*tip);
flow->tip.rebind (*tip);
```

heat→ tip. rebind (tip());
flow→ tip. rebind (tip());

where here `tip` refers to the object that manages the time integration, e.g. the initialization of the time loop, the time stepping process, etc. (see Figure 15).

The next important step after the declaration of the subproblem classes is the design of the main simulator. This task involves combining the individual solution routines to the overall optimization scheme. An abstract base class `Optimization` is defined for this purpose that has pointers to objects of the `Direct`, `Adjoint` and `Sensitivity` classes (Figure 20). The general structure of this class is shown in Figure 22. The class `Optimization` and all its subproblem classes are defined as abstract base classes. This is quite an important and elegant feature of an object-oriented implementation.

To allow the developed simulator to accommodate various thermophysical problems, initial conditions, boundary conditions, etc., the actual direct problem details are implemented as virtual functions (`functors` in the terminology of Diffpack) in derived classes from `NlHeat`, `NavierStokes` and `Concentration`. For example, if we declare a pure function `neumann` in the base class `NlHeat` and pass on the specific neumann boundary condition in a problem-dependent class, say `NlHeatDerv`, then the base class pointer could still access the problem-dependent function during runtime. This technique has been used extensively in the present simulator and has been shown to provide clarity and robustness to the developed code. More details on the OOP implementation of the adjoint method for complex transport systems are provided in [25].

4.4 An Example of the Design of Directional Solidification of Pure Materials with Desired Growth and Freezing Interface Heat Fluxes

The example to be addressed in this section consists of the inverse design of a directional solidification system through thermal boundary heat flux control in the presence of a vertical externally applied constant magnetic field. A square mold with aspect ratio of 1.0 is considered. It is initially occupied by liquid Aluminum whose properties are given in [26]. The ambient temperature was taken to be 25° C. The melt is initially superheated by an amount of $\Delta T_o = 200^o$ C. A mixed temperature/flux boundary condition was applied to one wall and all other walls were insulated as shown in Figure 23. The dimensionless variables governing the direct problem are listed in Table 1.

```
class Optimization: public ... {
protected:
  Handle(Direct)          direct;               // Pointer to direct problem
  Handle(Adjoint)         adjoint;              // Pointer to adjoint problem
  Handle(Sensitivity)     sensitivity;          // Pointer to sensitivity problem
  Handle(Mat(real))       heatFlux;             // Primary unknown: design flux
  Mat(real)               heatFlux_prev;        // Known flux at previous iter
  Handle(Mat(real))       gradient_Qo;          // ∇J(q_o^k)
  Mat(real)               gradient_Qo_prev;     // ∇J(q_o^{k-1})
  Handle(Mat(real))       directionVec;         // p^k
  Mat(real)               directionVec_prev;    // p^{k-1}
  Handle(Mat(real))       residual;             // θ - θ_m
  Handle(Mat(real))       sens_T_gamma_I;       // Θ on Γ_I
  ...
  int                     method;               // CGM update formula
public:
  ...
  virtual void define(MenuSystem& menu, int level = MAIN);
  virtual void scan(MenuSystem& menu);
  virtual void adm(MenuSystem& menu);
  virtual void driver();
  virtual void storeResults(int k);
  virtual void updateDataStructures();
protected:
  ...
  virtual void conjugateGradient();
  virtual real costFunctional();
  virtual void calculateDirectionVec(int iteration);
  virtual real innerProduct(Mat(real)& f, Mat(real)& g, int n);
  ...
};
```

Fig. 22. The main members of the class `Optimization`.

The penalty number to enforce incompressibility was chosen to be 10^8. A finite element mesh of 18×18 4-noded quadrilateral elements is used in the melt domain while a compatible grid of 15×18 is used in the solid domain.

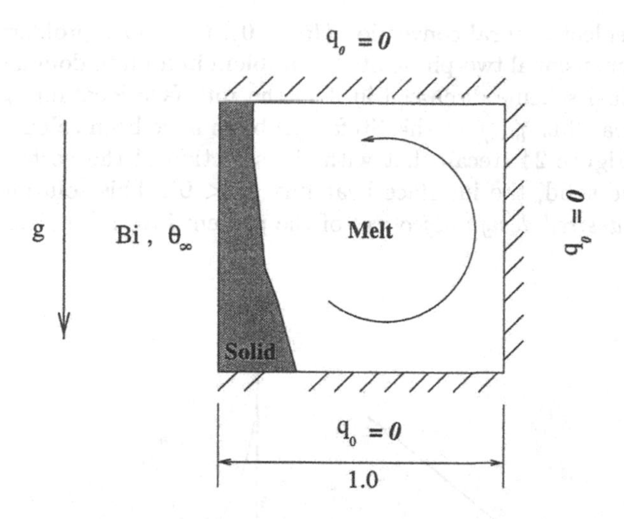

Fig. 23. Unidirectional solidification of Aluminum in a square enclosure. A constant vertical magnetic field is applied.

Note that as part of the moving FEM formulation, the FEM grids in the solid and liquid domains must share the same nodes at the solid/liquid interface. A small initial solid region (1 percent solid) was assumed at the start of the direct simulation. A uniform temperature gradient was assumed in the initial solid region. A complete simulation of the direct problem for various strengths of the magnetic field has been performed. Such simulations are similar to the ones examined in [26] without the presence of magnetic fields. The above direct problem constitutes what is referred to here as the *initial design*. A *reference design problem* or a "target state" is defined next.

In the above described direct convection based solidification problem (see Figure 23), the onset of natural convection results in a curved solid-liquid interface ($\frac{\partial s}{\partial y} \neq 0$), as well as in a vertical non-uniformity of the interfacial heat flux ($\frac{\partial q_I}{\partial y} \neq 0$). The interface location is here denoted as $s(y,t)$ and q_I is the heat flux on the liquid side of the freezing front.

	Symbol	Value
Prandtl number	Pr	0.0149
Rayleigh number	Ra	2×10^4
Stefan number	Ste	0.5317
Heat conductivity ratio	R_k	1.0
Specific heat ratio	R_C	1.0
Biot number	Bi	3.3

Table 1. Dimensionless variables governing the solidification of pure Aluminum.

If we neglect natural convection ($Ra = 0$), the above problem is reduced to a one-dimensional two-phase Stefan problem in a finite domain [11]. Using the numerical scheme discussed in [26], the interface location $s_r(t)$ and the interface heat flux $q_r(t)$ for this Stefan problem have been calculated and are shown in Figure 24 (recall that with the selection of the vector $\boldsymbol{n}$ pointing towards the solid, the interface heat flux $q_r < 0$). This solution is used to define the *desired design objectives* of the present inverse analysis.

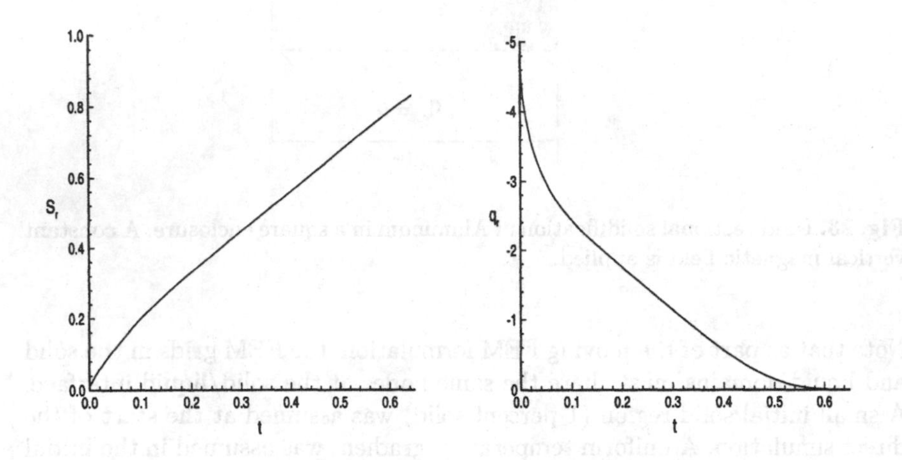

Fig. 24. The desired interface location and heat flux calculated from the solution of a Stefan problem (Ra = 0).

The design problem of interest can now be stated as follows:

Find the thermal condition on the left wall $x = 0$ and the right wall $x = 1$ such that with magneto-convection in the melt, a vertical interface $v_f = v_r(t)$ and a vertically uniform interfacial heat flux $q_I = q_r(t)$ are achieved.[a]

[a] This design problem is a particular case of the inverse design problem presented in Section 4.1 for stable desired growth of binary alloys. Here, $c = 0$ and ϵ in equation (23) is the desired interface heat flux. With $\epsilon \leq 0$, columnar growth is always stable in pure materials.

Zabaras and Yang [42] investigated earlier this problem without the presence of a magnetic field. They obtained an optimal heat flux with quite a complex spatial variation. Based on earlier work in the literature on electro-magnetic damping of melt flow, using the above defined inverse design problem one can investigate the effects that the magnetic field can have in the computed optimal heat flux solution. As such, various strengths of magnetic fields are considered in this analysis. Similar studies can also be conducted by varying the orientation of the magnetic field with respect to the direction of solidification.

The above design solidification problem is converted to two sub-design problems, one in the melt domain and another in the solid domain. Figures 25a, b give the complete definition of the two inverse design subproblems. Only details of the inverse design problem in the melt domain are considered here.

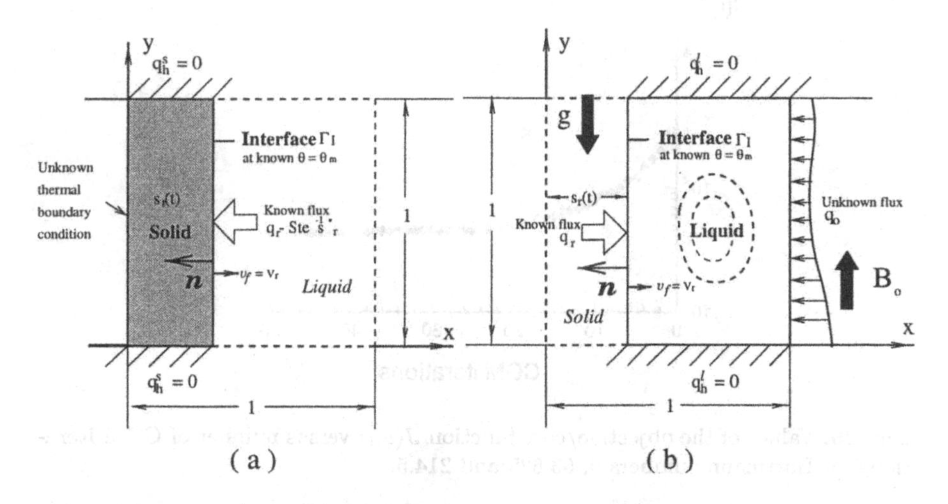

Fig. 25. Inverse design problems in the solid and melt domains for an Aluminum solidification system. With the particular selection of the vector $\boldsymbol{n}$, the Stefan condition (for $R_k = R_C = 1$) connecting the two inverse problems takes the form $\frac{\partial \theta_s}{\partial n} = \frac{\partial \theta_l}{\partial n} - \mathrm{Ste}^{-1}\dot{s}_r \equiv q_r - \mathrm{Ste}^{-1}\dot{s}_r$.

The inverse problem in the liquid domain solves for $q_o(y,t)$ at $x = 1$ with given freezing interface velocity $v_r(t)$ and heat flux $q_r(t)$ and the other mold walls being adiabatic. An initial guess $q_o^0(y,t) = 0$ is taken (this corresponds to the initial design introduced earlier). The time domain $[0, t_{max}]$ is taken with $t_{max} = 0.645$ which corresponds to solidification of about 85% of the initial melt.

Within each CGM iteration, the adjoint and sensitivity problems are solved with the same finite element algorithm as that used in the direct solidification problem. The spatial and temporal discretizations remain the same for all three problems. A deforming/moving front tracking FEM analysis is used for these type of solidification problems with a sharp solid/liquid freezing front [26]. The total number of time steps involved in the solution of each of the direct, adjoint and sensitivity problems is 825. The total computational time for each CGM iteration including the solutions of the three

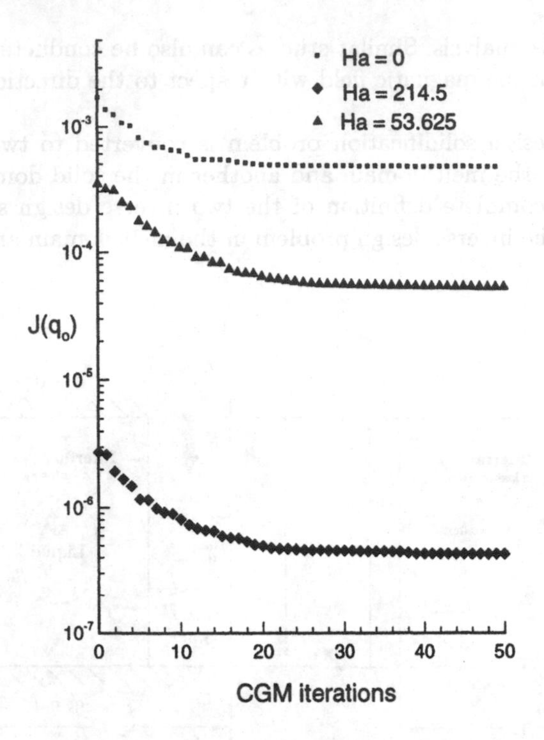

Fig. 26. Values of the objective/cost function $J(q_{ol})$ versus number of CGM iterations for Hartmann numbers 0, 53.625 and 214.5.

subproblems is about three hours CPU time on an IBM RS-6000 workstation. The convergence of the CGM method for the three cases with different magnetic field strengths is shown in Figure 26. The computations are stopped after 50 iterations as the cost functional reaches an asymptotic value.

The spatial and temporal variations of q_o^k at various iterations for the case of Hartmann number 53.625 are shown in Figure 27. The optimal solution ($k = 50$) exhibits the largest amount of heating and cooling at the corners of $x = 1$ and at the earlier stages of solidification. This is expected as the boundary heat flux q_{ol} should be adjusted early in time so that at later times it can influence the effects that convection has on the freezing interface.

From additional calculations with other strengths of the magnetic field, it is observed that the magnetic field strength decreases the order of magnitude of the cost function. This effect is due to damping of the flow caused by the applied magnetic field. The effect of magnetic damping reduces the role played by the thermal boundary fluxes in producing the desired interface data. Also, the optimal fluxes obtained with an externally applied magnetic field are shown to be smaller in magnitude and to vary in a more smoother manner compared to the optimal heat flux obtained in the case of no external magnetic field.

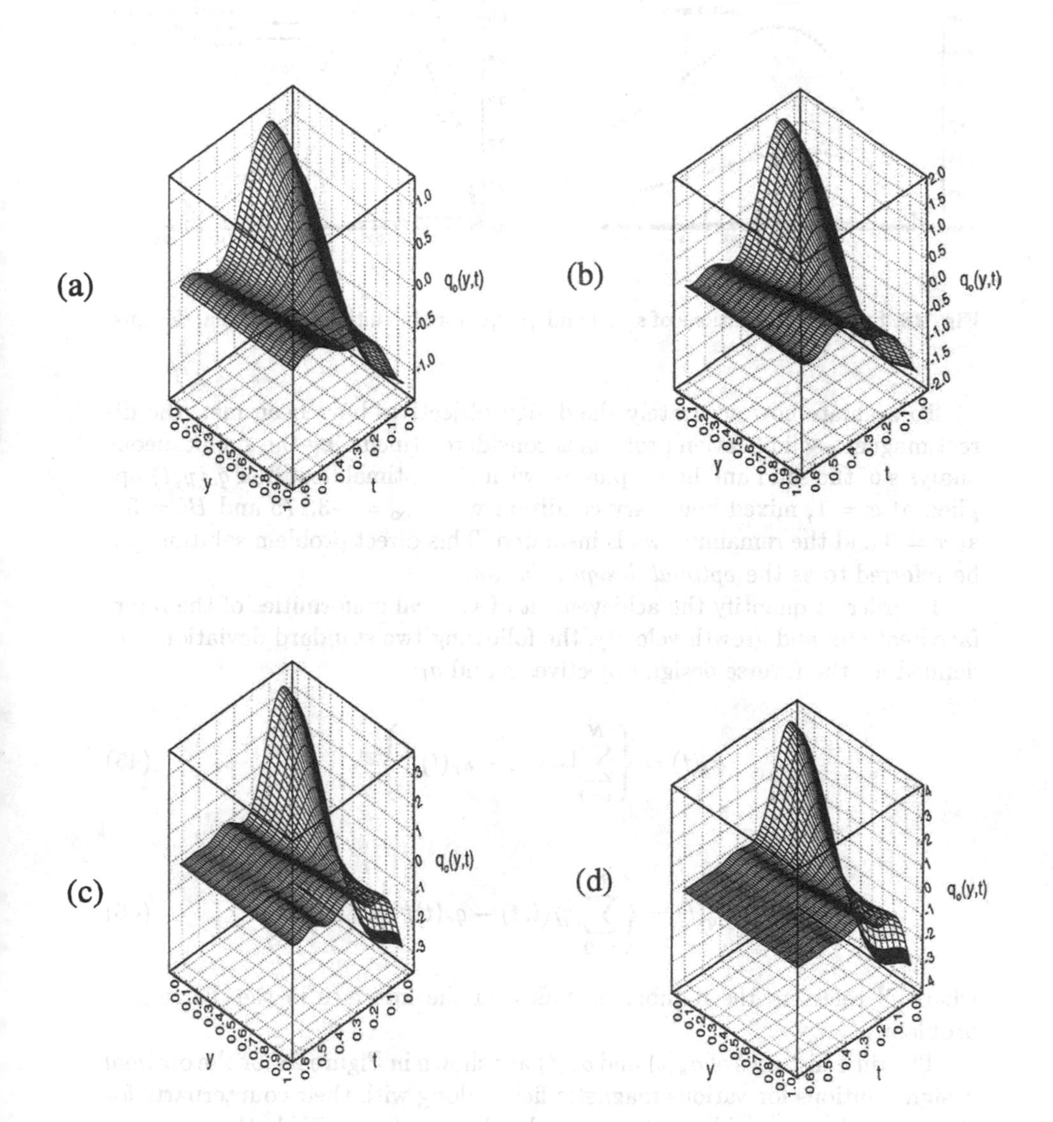

Fig. 27. Flux solution $q_o^k(y,t)$ at iterations (a) 5, (b) 10, (c) 20 and (d) 50 for Ha = 53.625.

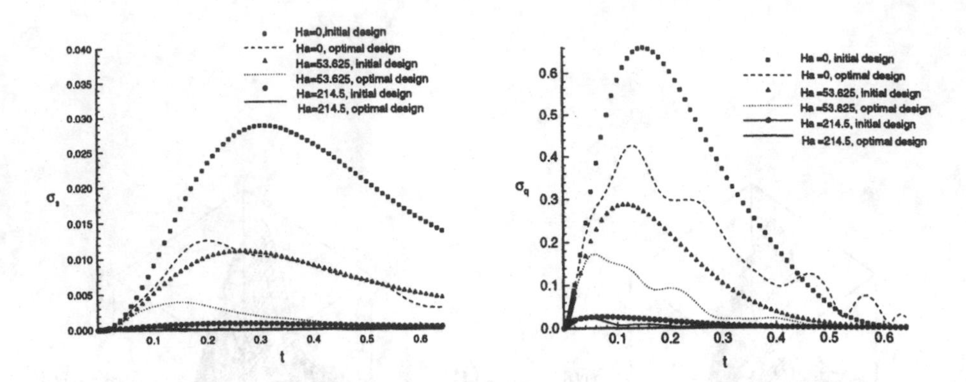

Fig. 28. Standard deviations of s(y,t) and q_I(y,t) for the initial and optimal designs.

To evaluate how accurately the design objectives have been met, the direct magneto-solidification problem is considered (including the simultaneous analysis of the solid and liquid phases) with the optimal heat flux $\bar{q}_o(y,t)$ applied at $x = 1$, mixed boundary condition with $\theta_\infty = -3.175$ and $Bi = 3.3$ at $x = 0$ and the remaining walls insulated. This direct problem solution will be referred to as the *optimal design solution.*

In order to quantify the achievement of vertical uniformities of the interface heat flux and growth velocity, the following two standard deviations are defined for the inverse design objectives s and q_I:

$$\sigma_s(t) = \left\{ \sum_{i=0}^{N} [s(i,t) - s_r(t)]^2 \right\}^{\frac{1}{2}} \tag{45}$$

$$\sigma_q(t) = \left\{ \sum_{i=0}^{N} [q_I(i,t) - q_r(t)]^2 \right\}^{\frac{1}{2}} \tag{46}$$

where N refers to the number of nodes on the interface in the discretized problem.

The time histories of $\sigma_s(t)$ and $\sigma_q(t)$ are shown in Figure 28 for the optimal design solutions for various magnetic fields along with their counterparts for the initial design problem. It can be clearly seen that with both an inverse design solution and an applied magnetic field the design solution gets closer to the reference problem and furthermore the solution improves with increasing magnetic field.

The freezing interface shapes obtained in the validation problems with various magnetic field strengths are plotted at a time which corresponds to mid-solidification, when the interface curvature is maximum for the initial design problem (see Figure 29). It can be concluded that application of a

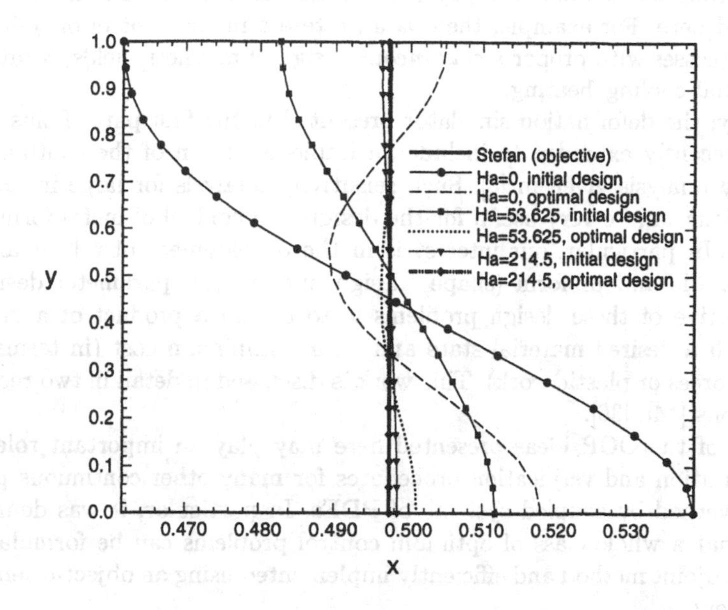

Fig. 29. Comparison of the interface locations at $\tau = 0.35$ for the various validation problems. These results demonstrate the effectiveness of both inverse thermal design and magnetic field in controlling the solidification system.

moderate magnetic field along with inverse thermal design almost achieves the required objective. The application of a high magnetic field, even though producing excellent agreement with the desired objective, may not be practically feasible.

5 Conclusions and Future Work

From the developments presented here, it is clear that significant advantages result from the object-oriented implementation of FEM algorithms for the analysis, design and control of material processes. The use of abstract classes and class hierarchies has been shown to reduce the required programming effort while it increases the readability, clarity and robustness of the developed codes. From the various preliminary simulations in both forming and solidification processing, it has been shown that the efficiency of the developed OOP simulators is comparable to that of codes developed using traditional programming approaches (including C or FORTRAN).

The presented solidification design simulator was recently used to design alloy solidification processes with desired stable growth and reduced segregation in the presence of magnetic fields [27]. The OOP structure of the present simulator can easily be extended to allow us to address other design solidification processes in which the physics is much more involved than the one considered here. For example, there is an interest in the control of solidification processes with proper *simultaneous* design of magnetic fields, rotation and thermal cooling/heating.

Finally, the deformation simulator presented in the first part of this paper was recently extended to include an implementation of the continuum sensitivity analysis given in [5]. Such sensitivity analysis for large inelastic deformations can be very useful for the design and control of metal forming processes. In particular, our interest is in the development of robust algorithms for die and preform (shape) design and process parameter design. The objective of these design problems is to obtain a product of a given shape with a desired material state and with a minimum cost (in terms of required forces or plastic work). This work is discussed in detail in two recent publications [44], [36].

Many of the OOP ideas presented here may play an important role in implementation and verification procedures for many other continuous processes governed by coupled systems of PDEs. In particular, it was demonstrated that a whole class of optimum control problems can be formulated with the adjoint method and efficiently implemented using an object-oriented environment.

Acknowledgments

The work presented here was funded by NSF grant DMII–9522613 to Cornell University. Additional support was provided by the Alcoa Laboratories (Dr. Paul Wang, program manager). The computational facilities provided by the Cornell Theory Center are acknowledged. The academic license for using the various libraries of Diffpack is appreciated. The contributions of R. Sampath in the development of the solidification simulator and of A. Srikanth and Y. Bao in the development of the deformation simulator are appreciated. Finally, the various constructive comments and suggestions of Prof. H. P. Langtangen of the University of Oslo and of three anonymous reviewers have contributed significantly to the improvement of the original manuscript.

References

1. ABAQUS. *Reference Manual.* Hibbit, Karlsson and Sorensen Inc., 100 Medway Street, Providence, RI, 02906–4402, 1989.
2. O. M. Alifanov. *Inverse Heat Transfer Problems.* Springer-Verlag, Berlin, 1994.
3. N. Aravas. On the numerical integration of a class of pressure-dependent plasticity models. *Int. J. Numer. Methods Engr.* 24 (1987) 1395–1416.
4. E. Arge, A. M. Bruaset and H. P. Langtangen (edts.). *Modern Software Tools for Scientific Computing.* Birkhäuser, Boston, 1997.

5. S. Badrinarayanan and N. Zabaras. A sensitivity analysis for the optimal design of metal forming processes. *Comp. Methods Appl. Mech. Engr.* 129 (1996) 319–348.
6. J. J. Barton and L. R. Nackman. *Scientific and Engineering C++*. Addison-Wesley, New York, 1994.
7. P. Bomme. *Intelligent Objects in Object-Oriented Engineering Environments.* Ph.D. Thesis, École Polytechnique Fédérale de Lausanne, Thèse No. 1763, 1998.
8. A. N. Brooks and T. J. R. Hughes. Streamline upwind/Petrov-Galerkin formulations for convection dominated flows with particular emphasis on the incompressible Navier-Stokes equations. *Comp. Methods Appl. Mech. Engr.* 32 (1982) 199-259.
9. S. B. Brown, K. H. Kim and L. Anand. An internal variable constitutive model for hot working of metals. *Int. J. Plasticity* 5 (1989) 95–130.
10. J. Budiansky. Thermal and thermoelastic properties of isotropic composites. *J. Composite Materials* 4 (1970) 286–295.
11. H. S. Carslaw and J. C. Jaeger. *Conduction of Heat in Solids.* Oxford University Press, 2nd Ed., New York, 1959.
12. M. Daehlen and A. Tveito (edts.). *Numerical Methods and Software Tools in Industrial Mathematics.* Birkäuser, Boston, 1997.
13. M. S. Engleman. *FIDAP*, v. 7.52. Fluid Dynamic International, Incorporated, 1996.
14. A. L. Gurson. Continuum theory of ductile rupture by void nucleation and growth: Part 1 - Yield criteria and flow rules for a porous ductile media. *J. Engr. Mat. Techn.* 99 (1977) 2–15.
15. W. Kurz and D. J. Fisher. *Fundamentals of Solidification.* Trans Tech Publications Ltd, Switzerland, 1989.
16. H. P. Langtangen. *Computational Partial Differential Equations – Numerical Methods and Diffpack Programming.* Springer-Verlag, New York, 1999.
17. T. A. Laursen and J. C. Simo. On the formulation and numerical treatment of finite deformation frictional contact problems. *Nonlinear Computational Mechanics - State of the Art*, (edts.) P. Wriggers and W. Wager. Springer Verlag, Berlin, (1991) 716–736.
18. D. G. Luenberger. *Optimization by vector space methods (Series in Decision and Control).* J. Wiley and Sons, New York, 1997.
19. A.M. Lush. *Computational procedures for finite element analysis of hot working.* Ph.D. Thesis, MIT, 1990.
20. A. M. Lush. Thermo-mechanically-coupled finite element analysis of hot-working using an implicit constitutive time integration scheme. *Numerical Methods in Industrial Forming Processes* (edts. J.-L. Chenot, R.D. Wood and O.C. Zienkiewicz) (1992) 281–286.
21. R. I. Mackie. Object oriented programming of the finite element method. *Int. J. Numer. Methods Engr.* 35 (1992) 425-436.
22. R. I. Mackie. Using objects to handle complexity in FE software. *Engineering with Computers* 13 (1997) 99-111.
23. G. I. Marchuck. *Adjoint Equations and Analysis of Complex Systems.* Kluwer Academic Publishers, Boston, 1995.
24. B. Moran, M. Ortiz and C. F. Shih. Formulation of implicit finite element methods for multiplicative finite deformation plasticity. *Int. J. Numer. Methods Engr.* 29 (1990) 483–514.

25. R. Sampath and N. Zabaras. An object-oriented implementation of adjoint techniques for the design of complex continuum systems. *Int. J. Numer. Methods Engr.*, submitted for publication.
26. R. Sampath and N. Zabaras. An object oriented implementation of a front tracking finite element method for directional solidification processes. *Int. J. Numer. Methods Engr.* 44(9) (1999) 1227–1265.
27. R. Sampath and N. Zabaras. Inverse thermal design and control of solidification processes in the presence of a strong magnetic field. *J. Comp. Physics*, submitted for publication.
28. R. Sampath and N. Zabaras. A diffpack implementation of the Brooks/Hughes Streamline-upwind/Petrov-Galerkin FEM formulation for the incompressible Navier-Stokes equations. Research Report MM-97-01, Sibley School of Mechanical and Aerospace Engineering, Cornell University, (URL: http://www.mae.cornell.edu/research/zabaras), 1997
29. R. Sampath and N. Zabaras. Finite element simulation of buoyancy driven flows using an object-oriented approach. Research Report MM-97-02, Sibley School of Mechanical and Aerospace Engineering, Cornell University (URL: http://www.mae.cornell.edu/research/zabaras), 1997.
30. R. Sampath and N. Zabaras. FEM simulation of double diffusive convection using an object-oriented approach. Research Report MM-97-03, Sibley School of Mechanical and Aerospace Engineering, Cornell University (URL: http://www.mae.cornell.edu/research/zabaras), 1997.
31. R. Sampath and N. Zabaras. FEM simulation of alloy solidification in the presence of magnetic fields. Research Report MM-97-04, Sibley School of Mechanical and Aerospace Engineering, Cornell University (URL: http://www.mae.cornell.edu/research/zabaras), 1997.
32. R. Sampath and N. Zabaras. Design and object-oriented implementation of a preconditioned-stabilized incompressible Navier-Stokes solver using equal-order interpolation velocity-pressure elements. Research Report MM-99-01, Sibley School of Mechanical and Aerospace Engineering, Cornell University (URL: http://www.mae.cornell.edu/research/zabaras), 1999.
33. J. C. Simo and C. Miehe. Associative coupled thermoplasticity at finite strains: Formulation, numerical analysis and implementation. *Comp. Methods Appl. Mech. Engr.* 98 (1992) 41–104.
34. A. Srikanth and N. Zabaras. A computational model for the finite element analysis of thermoplasticity coupled with ductile damage at finite strains. *Int. J. Num. Meth. Eng.*, in press.
35. N. Zabaras. Adjoint methods for inverse free convection problems with application to solidification processes. *Computational Methods for Optimal Design and Control* (edts. J. Borggaard, E. Cliff, S. Schreck and J. Burns). Birkäuser Series in Progress in Systems and Control Theory, Birkäuser (1998) 391–426.
36. A. Srikanth and N. Zabaras. Preform design and shape optimization in metal forming. *Comp. Methods Appl. Mech. Engr.*, submitted for publication.
37. T.E. Tezduyar. Stabilized finite element formulations for incompressible flow computations. *Advances in Applied Mechanics* 28 (1991) 1-44.
38. T. E. Tezduyar, M. Behr and J. Liou. A new strategy for finite element computations involving moving boundaries and interfaces - the deforming-spatial-domain/space-time procedure: I. The concept and the preliminary tests. *Comp. Meth. Appl. Mech. Engr.* 94 (1992) 339-351.

39. T. E. Tezduyar, M. Behr, S. Mittal and A. A. Johnson. Computation of unsteady incompressible flows with the finite element method - space-time formulations, iterative strategies and massively parallel implementations. *New Methods in Transient Analysis* (edts. P. Smolinski, W. K. Liu, G. Hulbert and K. Tamma, ASME, New York) AMD 143 (1992) 7-24.
40. V. Tvergaard and A. Needleman. Elastic-Viscoplastic analysis of ductile fracture. *Finite Inelastic Deformations - Theory and Applications*, (edts. D. Bedso and E. Stein). IUTAM symposium Hannover, Germany (1991) 3–14.
41. G. Weber and L. Anand. Finite deformation constitutive equations and a time integration procedure for isotropic, hyperelastic-viscoplastic solids. *Comp. Methods Appl. Mech. Engr.*, 79 (1990) 173–202.
42. G. Yang and N. Zabaras. The adjoint method for an inverse design problem in the directional solidification of binary alloys. *J. Comp. Phys.* 140(2) (1988) 432-452.
43. G. Yang and N. Zabaras. An adjoint method for the inverse design of solidification processes with natural convection. *Int. J. Num. Meth. Eng.* 42 (1998) 1121–1144.
44. N. Zabaras, Y. Bao, A. Srikanth and W. G. Frazier. A continuum sensitivity analysis for metal forming processes with application to die design problems. *Int. J. Numer. Methods Engr.*, submitted for publication.
45. N. Zabaras and T. Hung Nguyen. Control of the freezing interface morphology in solidification processes in the presence of natural convection. *Int. J. Num. Meth. Eng.* 38 (1995) 1555-1578.
46. N. Zabaras and A. Srikanth. An object oriented programming approach to the Lagrangian FEM analysis of large inelastic deformations and metal forming processes. *Int. J. Num. Meth. Eng.*, in press.
47. N. Zabaras and G. Yang. A functional optimization formulation and FEM implementation of an inverse natural convection problem. *Comp. Meth. Appl. Mech. Eng.* 144(3-4) (1997) 245-274.
48. N. Zabaras and A. Srikanth. Using objects to model finite deformation plasticity. *Engineering with Computers*, in press.
49. A. Zavaliangos, L. Anand, B. F. von Turkovich. Deformation processing. *Annals of the CIRP* 40 (1991) 267–271.
50. A. Zavaliangos and L. Anand. Thermal aspects of shear localization in microporous viscoplastic solids. *Int. J. Numer. Methods Engr.* 33 (1992) 595–634.
51. Z. L. Zhang. On the accuracy of numerical integration algorithms for Gurson-based pressure dependent elastoplastic constitutive models. *Comp. Methods Appl. Mech. Engr.* 121 (1995) 15-28.
52. Z. L. Zhang. Explicit consistent tangent moduli with a return mapping algorithm for pressure-dependent elastoplasticity models. *Comp. Methods Appl. Mech. Engr.* 121 (1995) 29-44.
53. T. Zimmermann, Y. Dubois-Pèlerin and P. Bomme. Object-oriented finite element programming: I Governing principles. *Comp. Methods Appl. Mech. Engr.* 98 (1992) 291-303.

Object-Oriented Field Recovery and Error Estimation in Finite Element Methods

Knut Morten Okstad and Trond Kvamsdal

SINTEF Applied Mathematics, Trondheim, Norway
Email: {Knut.M.Okstad, Trond.Kvamsdal}@math.sintef.no

Abstract. In this chapter we study an object-oriented implementation of procedures for field recovery and recovery-based error estimation. The field recovery is based on the superconvergent patch recovery technique by Zienkiewicz and Zhu. The core of the current implementation is problem independent, and is organized as a set of C++ classes based on the software library Diffpack. The use of the developed program module is demonstrated on an isotropic linear elasticity problem and on a stationary Navier–Stokes problem. For both example problems, analytical solutions are available. The *exact* error may therefore be computed in addition to the estimated error, enabling us to study the effectivity of the estimator. The computational efficiency of the object-oriented program module is assessed by comparing the time consumption with a similar program implemented in FORTRAN.

1 Introduction

Computer programs for numerical solution of Partial Differential Equations (PDEs) are traditionally coded in the FORTRAN programming language. The computational intensive parts of such programs, based either on the Finite Element (FE) method or some other numerical procedure, consist mainly of various vector and matrix operations. FORTRAN has therefore been regarded as the best computer language for implementing these programs, as it is particularly suitable for manipulation of large arrays. However, as the computers of today become more and more powerful and the compilers more efficient, the advantage of FORTRAN over alternative languages such as C and C++ is shrinking. Moreover, the developments within the FE methods have by no means halted, as new procedures with increasing complexities are being proposed continuously. This results in more complicated software, and therefore object-oriented programming has attracted an increasing popularity within the field of numerical computing during the 1990s. Languages supporting object-oriented programming have predefined mechanisms for modular design, re-use of code and for creation of flexible applications, etc.

Another topic within the field of FE computing that has attracted an increasing interest during the last decade is error estimation and adaptivity. However, although the topic has been under active research since the late 1970s, there are still only a few commercially available FE codes that have reliable methods for error estimation and adaptive refinement available. The increased use of object-oriented techniques to facilitate implementation of highly complex adaptive finite element algorithms may change this picture in a few years.

The present study is concerned with the use of object-oriented programming techniques in the development of a generic program module for *a posteriori* error estimation of FE computations. Our work is based on the object-oriented software tool Diffpack [3,4], which is a comprehensive C++ library for numerical solution of PDEs. It contains a large amount of classes for representation of and manipulation on basic numerical objects such as vectors and matrices with different storage schemes, and more abstract objects such as grids and fields. The developed program module features recovery-based error estimation based on the so-called Superconvergent Patch Recovery (SPR) method [14,15]. The program is demonstrated on a two-dimensional (2D) linear elasticity problem as well as on a 2D stationary fluid dynamics problem based on the Navier–Stokes equations.

2 Field Recovery by Means of SPR

2.1 Introduction

Often, we are more interested in the derivatives of the solution of a FE analysis rather than the solution itself. This is usually the case for elasticity problems, for instance, where stresses (or strains) may be of more interest than the displacements, which are the primary unknowns. Since the derivatives of the FE solution are discontinuous across element boundaries, there has been a need for recovery of continuous field derivatives since the early days of the FE method, for the purpose of visualization of such results.

Recovery of continuous field derivatives has traditionally been performed by means of direct nodal averaging or some global least-squares fit approach. Such methods are quite straight-forward to implement and yield sufficient results for visualization purposes. More recently, field recovery has become a popular means for estimating the error in the FE solution, and in such applications more accurate recovery schemes than the above mentioned methods are needed. This resulted in the introduction of the SPR-method by Zienkiewicz and Zhu [14,15], where a local least-squares fit is performed over small patches of elements instead of doing it globally.

The SPR-procedure utilizes the concept of *superconvergent* points. These are certain locations within each element where the FE solution has higher accuracy than elsewhere. The rate of convergence is also higher at these points than at other locations. For quadrilateral elements, the points with superconvergent properties for the solution derivatives coincide with the reduced Gauss quadrature points, also referred to as Barlow points. Bilinear elements have thus one such point, located at the centroid.

2.2 Governing Equations

Let $\boldsymbol{v}^h$ denote the piece-wise continuous field (computed by the analysis code), which we want to recover an improved version of, i.e. $\boldsymbol{v}^h \notin C^0(\Omega)$ but $\boldsymbol{v}^h|_{\Omega_e} \in C^0(\Omega_e)$, where Ω and Ω_e denote the whole domain and the element domain, respectively. The corresponding recovered field, $\boldsymbol{v}^*$, is expressed as a polynomial

$$\boldsymbol{v}^* = \mathbf{P}\,\mathbf{a} \tag{1}$$

where $\mathbf{P}$ is a matrix of polynomial terms and $\mathbf{a}$ is a vector of unknown coefficients which are to be determined.

The polynomial expansion (1) is usually chosen to be of the same complete order as the basis functions used to interpolate the primary variables of the FE problem. Using first order elements, $\mathbf{P}$ should therefore contain the bi-linear monomials, i.e.

$$\mathbf{P} = \begin{bmatrix} \mathbf{P}_1 & & \\ & \ddots & \\ & & \mathbf{P}_n \end{bmatrix}; \quad \mathbf{P}_i = [1, x, y, xy] \tag{2}$$

with n denoting the dimension of the vector $\boldsymbol{v}^*$.

Alternatively, the matrix $\mathbf{P}$ may be chosen such that the field $\boldsymbol{v}^*$ *a priori* satisfies locally the governing PDE of the FE problem. For instance, if the field to be recovered is $\boldsymbol{\sigma}^* = \{\sigma^*_{11}, \sigma^*_{22}, \sigma^*_{12}\}^{\mathrm{T}}$, the stress field of a 2D elasticity problem, the polynomial may be chosen such that the recovered stress field satisfies the equilibrium equation

$$\sigma_{ij,i} + f_i = 0. \tag{3}$$

The polynomial $\boldsymbol{\sigma}^* = \mathbf{P}\,\mathbf{a}$ is then said to be *statically admissible* [6,8] and the matrix $\mathbf{P}$ may be like

$$\mathbf{P} = \begin{bmatrix} 1 & 0 & 0 & x & 0 & y & 0 & xy & 0 & x^2 & y^2 & 0 \\ 0 & 1 & 0 & 0 & y & 0 & x & 0 & xy & y^2 & 0 & x^2 \\ 0 & 0 & 1 & -y & -x & 0 & 0 & -\frac{1}{2}y^2 & -\frac{1}{2}x^2 & -2xy & 0 & 0 \end{bmatrix}. \tag{4}$$

When the body forces f_i are non-zero there will be additional terms in the statically admissible field $\boldsymbol{\sigma}^*$ corresponding to the particular solution of Equation (3) (see [6] for details).

The polynomial (1) is defined over small patches of elements surrounding each nodal point (see Figure 1). The coefficients $\mathbf{a}$ are then determined from a least-squares fit of the field $\boldsymbol{v}^*$ to the values of $\boldsymbol{v}^h$ at the superconvergent points within the elements in the patch. This results in the following linear system of equations to be assembled and solved for each patch

$$\left[\sum_{i=1}^{n_{sp}} \mathbf{P}^{\mathrm{T}}(\boldsymbol{x}_i)\,\mathbf{P}(\boldsymbol{x}_i)\right] \mathbf{a} = \left\{\sum_{i=1}^{n_{sp}} \mathbf{P}^{\mathrm{T}}(\boldsymbol{x}_i)\,\boldsymbol{v}_i^h\right\} \tag{5}$$

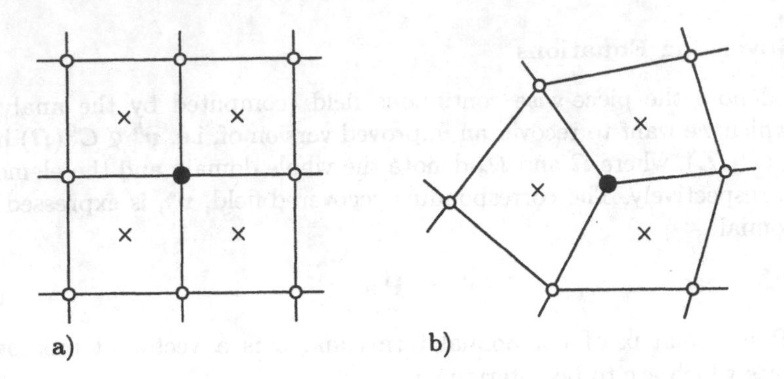

Fig. 1. Examples of nodal patches. (o) nodal points; (•) the patch node (node defining the nodal patch); (×) result sampling points (superconvergent points).

where x_i are the coordinates of the i'th superconvergent point, v_i^h is the computed value of the field v^h at that point, and n_{sp} denotes the total number of superconvergent points in the patch.

When field recovery is used for error estimation, we are normally interested in recovered values at element-interior points (integration points) rather than nodal values. Since a specific element belongs to more than one patch, the patch recovery does not provide a unique result value at such points. To construct a global recovered field, Blacker and Belytschko [2] proposed to conjoin the polynomial expansions (1), for all patches containing the actual element using the nodal shape functions as weighting functions. Adopting this approach, the recovered field within a bi-linear element e is evaluated through

$$v_e^*(x) = \sum_{a=1}^{4} N^a(x)\, v_a^*(x) \tag{6}$$

where $N^a(x)$ denotes the bi-linear shape function associated with element node a and $v_a^*(x)$ is a local recovered field of the form (1). If the element node a is an interior node, $v_a^*(x)$ is evaluated on the patch of elements surrounding that node. Nodes lying on the exterior boundary, however, are rarely connected to enough elements to form a valid patch[1]. For such nodes $v_a^*(x)$ is instead evaluated on the patch (or patches) associated with the other node(s) that are connected to node a through an interior element boundary (see Figure 2). If more than one patch is connected in this manner to a boundary node (Figure 2c)), the corresponding values for v_a^* computed on each patch are averaged.

[1] With first order elements containing one superconvergent point each, there must be at least three elements in the patch in order to be able to recover a linear field over the patch of elements.

2.3 Representation of Fields in Diffpack

Any numerical quantity that may be regarded as a function of space (and time) is a field. The field may be constant; it may vary according to some known function; or it may be unknown. In the latter case, the field is usually represented by an array of numerical values that are to be determined through some numerical procedure, such as the FE method. Each such value is associated with a certain spatial point and a predefined interpolation scheme is then used to determine the field values at intermediate points.

The various types of fields are organized in Diffpack in two class hierarchies; one for scalar fields and one for vector fields. We could alternatively consider a vector field simply as a vector of individual scalar fields and then avoid the vector field class hierarchy, but since many of the computations performed during the field evaluation would be similar for each field component it is more efficient to have separate classes for each vector field type. The vector field hierarchy is illustrated in Figure 3. The scalar field hierarchy has a similar structure.

The class `Fields` is a base class for general vector fields. It contains no data itself except for a `String` containing the name of the field, but defines a set of virtual member functions for evaluating the field:

```
class Fields
{
protected:
  String fieldname;

public:
  Fields (const char* name = NULL);
 ~Fields ();

  virtual int getNoValues () const;
```

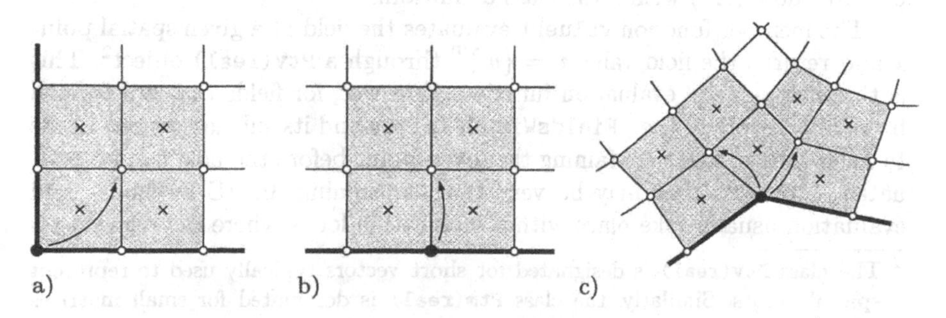

Fig. 2. Nodal patches associated with a node (•) on the exterior boundary. a) Node connected to exterior element boundaries only. b) Node connected to one interior element boundary. c) Node connected to two interior element boundaries.

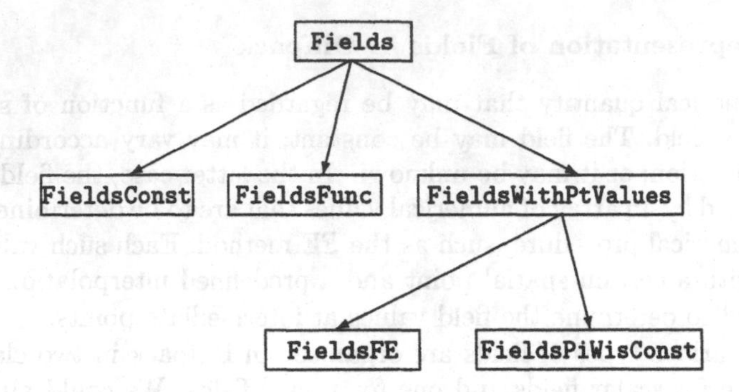

Fig. 3. The Fields class hierarchy. The solid arrows indicate class inheritance ("is a"-relationship).

```
  virtual void valuePt (Ptv(real)& v, const Ptv(real)& x);
  virtual void valueFEM (Ptv(real)& v, FiniteElement& fe);
  virtual void valueNode (Ptv(real)& v, const int node);

  virtual void gradientPt (Ptm(real)& g, const Ptv(real)& x);
  virtual void gradientFEM (Ptm(real)& g, FiniteElement& fe);
  virtual void gradientNode (Ptm(real)& g, const int node);

  virtual real divergencePt (const Ptv(real)& x);
  virtual real divergenceFEM (FiniteElement& fe);
  virtual real divergenceNode (const int node);

  ...
};
```

We here only show the most important members of the class. When a class contains additional data members and/or member functions, this is indicated by three dots (...) within the class definition.

The member function `valuePt` evaluates the field at a given spatial point `x` and returns the field value $\boldsymbol{v} = \{v_i\}^{\mathrm{T}}$ through a `Ptv(real)` object[2]. This is the most general evaluation function. However, for fields that are defined by pointwise values (i.e., `FieldsWithPtValues` and its subclasses) one needs to locate the element containing the given point before the field can be evaluated, a process that may be very time consuming. In FE methods, field evaluation usually take place within integration loops where the relevant el-

[2] The class `Ptv(real)` is designated for short vectors typically used to represent spatial points. Similarly, the class `Ptm(real)` is designated for small matrices and is used to represent tensorial quantities such as spatial transformations and vector field gradients.

ement information is already known[3]. In such cases the alternative function `valueFEM` is used instead. Finally, a third evaluation function `valueNode` is provided that accepts a global node number as argument. This function is convenient if we want to output the field values at all nodal points to a file for visualisation, etc.

The class `Fields` has a corresponding set of functions (`gradient...`) that return the field derivatives $v_{i,j}$ with respect to the spatial coordinates through a `Ptm(real)` object. Finally, the `divergence...` functions return the sum $v_{i,i}$, i.e., the trace of the gradient matrix $[v_{i,j}]$.

Each subclass of `Fields` has its own implementation of each evaluation function which is optimized for that particular representation. Some functions may however be undefined for some subclasses (`valueNode` has no meaning for `FieldsPiWisConst` whereas `derivativeNode` has a meaning for neither `FieldsPiWisConst` nor `FieldsFE`, as these values typically are undefined at nodal points. Further details on the field class hierarchies may be found in [3] and in the Diffpack reference manuals [4].

2.4 An Object-Oriented Implementation of SPR

Our implementation of SPR is based on the C++ classes shown in Figure 4. The classes are in the following described in a bottom-up sequence.

Classes for Representation of the Recovery Equations. First, we have the following two classes representing the equations involved in the recovery operation:

[3] In Diffpack, the information connected to a particular element, such as element type, local coordinates and weights of current integration point, etc., is stored in an object of the class `FiniteElement`.

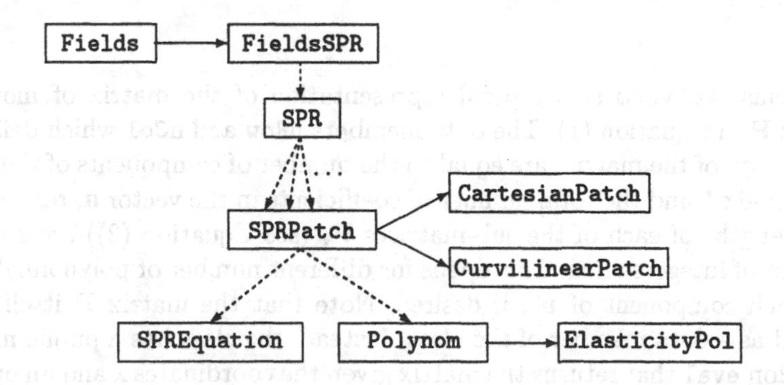

Fig. 4. Basic classes of the SPR procedure. The solid arrows indicate class inheritance ("is a"-relationship) whereas the dashed arrows indicate pointers/references to internal data objects ("has a"-relationship).

```
class Polynom
{
protected:
  int              nRow, nCol; // Dimension of the P-matrix
  VecSimplest(int) nMon; // No. of monomials for each comp.

public:
  Polynom (const VecSimplest(int)& nmon) { redim(nmon); };
  Polynom (const int m, const int n = 1) { redim(m,n); };
  Polynom () { nCol = nRow = 0; };
 ~Polynom () {};

  virtual bool redim (const int m, const int n = 1);
          bool redim (const VecSimplest(int)& nmon);

  virtual void eval (Mat(real)& P, const Ptv(real)& x, real w) const;
  virtual void eval (Mat(real)& P, const Ptv(real)&, Ptv(real)& w) const;
  ...
};

class SPREquation
{
  Mat(real)              A; // Coefficient matrix
  VecSimplest(Vec(real)) R; // Right-hand-side vectors

public:
  SPREquation (const int neq, const int nrhs) { redim(neq,nrhs); };
  SPREquation () {};
 ~SPREquation () {};

  void redim    (const int m, const int n);
  void assemble (Mat(real)& P, const Ptv(real)& v);
  void solve    (VecSimplest(Vec(real))& a);
  ...
};
```

The class Polynom is a general representation of the matrix of monomial terms $\mathbf{P}$ in Equation (1). The data members nRow and nCol which define the dimension of the matrix, are equal to the number of components of the recovered field $\boldsymbol{v}^*$ and the total number of coefficients in the vector $\mathbf{a}$, respectively. The lengths of each of the sub-matrices $\mathbf{P}_i$ (see Equation (2)) are stored in the list of integers nMon. This opens for different number of polynomial terms for each component of $\boldsymbol{v}^*$, if desired. Note that the matrix $\mathbf{P}$ itself is not stored as a data member of the class. Instead, the class has a public member function eval that returns the matrix given the coordinates x and an optional weighting factor w as arguments. The matrix returned contains the complete polynomials up to the order defined by nMon for each field component. An overloaded version of eval takes a vector of individual weighting factors for each field component as argument.

The member functions **eval** are here declared **virtual** such that subclasses of **Polynom** may be derived for other polynomials. For instance, we could derive the subclass **ElasticityPol**, as indicated in Figure 4, representing the statically admissible polynomial (4) for 2D elasticity. Since such polynomials are problem dependent, we would need one subclass for each problem type.

The second class defined above, **SPREquation**, represents the linear system of equations (5) and has both the coefficient matrix **A** and the right-hand-side vector **R** as members. The assembly of the equation system is handled by the member function **assemble** which takes the matrix **P** and the values of the FE field $\boldsymbol{v}^h$ as arguments, whereas **solve** computes and returns the solution vector.

If the same set of monomials is used for each component of $\boldsymbol{v}^*$, the system of equations (5) will decouple into **nRow** equal sub-systems, only with different right-hand sides. In such cases, the equations are assembled and solved as a multiple right-hand-side system, but with a much smaller coefficient matrix. The solution is then returned by member function **solve** as a list of **Vec(real)** objects (one for each field component).

If the SPR procedure is enhanced with equilibrium constraints, either as proposed by Wiberg *et al.* [13], or by using statically admissible polynomials [6,8], a coupling is introduced between the different field components in the coefficient matrix and decoupling is thus not possible. The solution **a** is then returned in a single **Vec(real)** object. The classes **Polynom** and **SPREquation** are designed such that they can handle such cases with only minor further modifications.

Classes for Representation of Nodal Patches. To administer the operations performed on a single patch (see Figure 1), we define the following base class:

```
class SPRPatch
{
protected:
  int               patchNode;
  VecSimplest(int) elements;
  Handle(GridFE)        mesh; // Pointer to the FE mesh

  Handle(Polynom)           P; // Pointer to patch polynomial
  VecSimplest(Vec(real))  a; // Coefficients of the polynomial
  SPREquation             eqs; // The patch equation system

  virtual Ptv(real) global2local (Ptv(real)& X, int e = 0) { return X; };
          Ptv(real) evaluate (Ptv(real)& Xloc, const real w);
  ...

public:
```

```
  SPRPatch (const int node, const PchPolType type, GridFE& mesh);
  SPRPatch () { patchNode = 0; }
  ~SPRPatch () {};

  void init (const int node, const PchPolType type, GridFE& mesh);

  void assembleAndSolve (Fields& v, bool deriv);

  virtual Ptv(real) valueFEM (FiniteElement& fe, const real w);
  virtual Ptv(real) valuePt  (Ptv(real)& Xglob, const real w);

  ...
};
```

The class contains the node number identifying patch; `patchNode`, the list of `elements` that form it and a pointer[4] to the FE mesh, represented by the Diffpack class `GridFE`, on which the patch is defined. The object also has the patch polynomial and an associated list of coefficient vectors and the patch equation system as data members. The polynomial is represented by a pointer to a `Polynom`-object, which may be assigned to any subclass of `Polynom`, if desired.

The data members `patchNode`, `elements`, `mesh` and `P` are initialized by the member function `init` which takes the node number, the polynomial type and the FE mesh as arguments. The list of elements that form the patch are found from the node-to-element connectivity table which is stored within the `GridFE`-object. The patch equation system (5) is assembled and solved by the member function `assembleAndSolve`, which takes any vector field as argument and a Boolean variable indicating whether field derivatives or the field itself are going to be recovered. If this parameter is `TRUE`, the member function `v.gradientFEM` of the given `Fields` object is used in the assembly of the right-hand side. Otherwise, the function `v.valueFEM` is used. The latter option is suitable if we have a piece-wise constant field (class `FieldsPiWisConst`), e.g. the computed velocity field of a finite volume CFD-simulation, from which we want to obtain a corresponding continuous field.

The recovered field over the patch (see Equation (1)) is evaluated by the member functions `valueFEM` or `valuePt`, which both use the protected member function `evaluate`. The latter function takes the local patch coordinates of the point in question as arguments, which are obtained from the global coordinates via the virtual member function `global2local`. In the base class `SPRPatch`, this function is just an identity mapping (i.e. the global coordinates are used as patch coordinates). However, two subclasses have been derived in the present study implementing different approaches for computing the local coordinate systems:

[4] Pointers to objects of type `classType` are in Diffpack represented by another class `Handle(classType)`, which works as ordinary C-pointers, but have some additional features such as reference counting, automatical deallocation, etc.

`CartesianPatch` : Implements the local Cartesian system proposed by Kvamsdal and Okstad [6].
`CurvilinearPatch` : Implements the curvilinear coordinate system proposed by Ramsay and Sbresny [10].

All data members and member functions of `SPRPatch`, except for the function `global2local`, are inherited by the two subclasses. The subclasses also contain some additional data members holding the information needed to perform the coordinate transformation.

Class for Administration of the SPR Procedure. The next class administers all the patches created for a given FE mesh:

```
class SPR
{
  VecSimplest(Handle(SPRPatch)) patches;

  VecSimplest(int) patchPtr; // Index for the patches array
  VecSimplest(int) nPchConn; // No. of patches for each node
  ...

public:
  SPR () {};
  ~SPR () {};

  void init (GridFE& g,...);
  void recover (Fields& vh, bool deriv, Handle(Fields)& vs);

  void valuePt (Ptv(real)& v, const Ptv(real)& x);
  void valueFEM (Ptv(real)& v, FiniteElement& fe);
  void valueNode (Ptv(real)& v, const int node);
  ...
};
```

The patches are here stored as a list of pointers to `SPRPatch`-objects. Thus, each pointer may be assigned to a dynamically allocated `SPRPatch`-object, or to an object of any of its subclasses. If every node in the FE mesh was associated with one and only one patch each, the list `patches` could contain one unique entry for each node and the global node number could be used directly as the index into the list. However, as discussed in Section 2.2, some nodes along the exterior boundary may be associated with two or more patches (see Figure 2c). On the other hand, two or more nodes may also refer to the same patch; for the corner element shown in Figure 2a all the four element nodes would refer to the same shaded patch. To handle these situations, the class `SPR` has two additional lists of integers as data members. The first list, `patchPtr`, maps a node number to an index into the list `patches` giving the first patch associated with that node. The other list gives the number of patches that are associated with each node. Thus, for a given node n, the indices of the patches associated with that node are given by patchPtr(n)...patchPtr(n)+nPchConn(n)-1.

The member functions of the SPR class consist of an initialization function that constructs all the patches for a given FE mesh, and one recovery function that performs the recovery operation for a given vector field vh. The function recover loops over all the patches and invokes the assembleAndSolve function for each patch. It then returns a pointer to the recovered field vs. This object will be of class FieldsSPR which is described below. Notice the general interface of the recover function; it takes any Fields as input and returns a pointer to another Fields. The class SPR is therefore easily fitted into a hierarchy of recovery procedures with recover as a virtual function. Other members of this hierarchy could then be direct nodal averaging and global least-squares fit, etc.

The class SPR has three evaluation functions which correspond to those of the Fields base class described in Section 2.3. The function valueFEM evaluates the recovered field at a specific point within an element by combining the related patch polynomials according to Equation (6). At a nodal point, only the patch that is associated with that node will contribute to the global field defined by (6), since the weight N^a will be zero for all other patches. This is utilized by the function valueNode which accesses only one patch[5] and no weighting is needed. Although the function valueFEM will work also at nodal points the function valueNode will thus be far more efficient. The third evaluation function, valuePt, is now defined in terms of the other two functions; it first performs a search for the element or the nodal point matching the given spatial point, and then calls valueFEM or valueNode accordingly.

Representation of the Recovered Field. The class SPR is not declared as a subclass of Fields although the evaluation functions would fit into that base class. Instead, we define a separate Fields subclass which has a reference to a SPR-object as its major data member:

```
class FieldsSPR : public Fields
{
  SPR& spr; // Reference to the SPR-object

  int  ncomp, nsd;  // Dimension of the gradient tensor
  bool derivatives; // Are derivatives recovered ?
  ...

public:
  FieldsSPR (SPR&, const char* name = NULL);
```

[5] Unless we have a similar situation as illustrated by Figure 2c). We then need to access two patches and the field values obtained from the two patches are averaged.

```
 ~FieldsSPR ();

  virtual void valuePt (Ptv(real)& v, const Ptv(real)& x);
  virtual void valueFEM (Ptv(real)& v, FiniteElement& fe);
  virtual void valueNode (Ptv(real)& v, const int node);

  virtual void gradientPt (Ptm(real)& g, const Ptv(real)& x);
  virtual void gradientFEM (Ptm(real)& g, FiniteElement& fe);
  virtual void gradientNode (Ptm(real)& g, const int node);

  virtual real divergencePt (const Ptv(real)& x);
  virtual real divergenceFEM (FiniteElement& fe);
  virtual real divergenceNode (const int node);
  ...
};
```

The main motivation for this is the increased flexibility. If more than one field is to be recovered for a given FE discretization there is no need to construct a separate patch structure for each field to recover. Instead, we may re-use the same SPR-object for each recovered field. Note, however, that the present implementation does not allow for two `FieldsSPR`-objects to refer to the same SPR-object concurrently, since the coefficients of the patch polynomials are stored within the SPR-object and not within `FieldsSPR`.

The class `FieldsSPR` re-implements all the virtual evaluation functions defined for the base class `Fields` (see Section 2.3) using the corresponding member functions of the given SPR-object. The data member `derivatives` corresponds here to the second argument to the `SPR::recover` function, and is used to determine whether the field derivatives or the field values themselves are available, as follows:

```
void FieldsSPR::valuePt (Ptv(real)& val, const Ptv(real)& x)
{
  if (derivatives)
    error("FieldsSPR::valuePt","No field values");
  else
    spr.valuePt (val,x);
}

void FieldsSPR::gradientPt (Ptm(real)& grad, const Ptv(real)& x)
{
  if (derivatives)
    convert2Ptm (grad,spr.valuePt(x),ncomp,nsd);
  else
    error("FieldsSPR::gradientPt","No derivatives");
}
```

The other `FieldsSPR` evaluation functions are implemented similarly. The global function `convert2Ptm` used above converts a given `Ptv(real)`-object into a corresponding $\texttt{ncomp} \times \texttt{nsd}$ `Ptm(real)`-object.

An object of class `FieldsSPR` may now be treated as any other vector field object in the user's application program (provided he/she knows whether derivatives have been recovered or the field values themselves). For instance, if the user has some class for evaluating a norm of the solution fields of some problem, the user may provide a `FieldsSPR` object as argument to the norm-evaluating function as well as objects of any other `Fields` subclass. Such use of the `FieldsSPR` class will now be discussed further in the subsequent Section.

3 Recovery-Based Error Estimation

3.1 Introduction

In recovery-based error estimation, the error in the FE solution is assessed through the difference between the computed FE solution and a corresponding higher-order solution, which has been recovered from the FE solution by means of some post-processing technique, such as SPR. This results in a pointwise error field

$$\boldsymbol{e} = \boldsymbol{v}^* - \boldsymbol{v}^h. \tag{7}$$

To obtain a quantitative measure of this error various types of norms are used. Any norm can be considered as a functional taking one or more fields as arguments and returning a real number. They can be of a rather general type, such as the classical L_2-norm

$$\|\boldsymbol{e}\|_{L_2(\Omega)} := \sqrt{\int_\Omega \boldsymbol{e} \cdot \boldsymbol{e} \, \mathrm{d}V}, \tag{8}$$

or more problem-specific as the often-used energy norm

$$\|\boldsymbol{e}\|_{E(\Omega)} := \sqrt{a(\boldsymbol{e}, \boldsymbol{e})_\Omega} \tag{9}$$

where $a(\cdot, \cdot)_\Omega$ denotes a problem-dependent bi-linear form, which usually corresponds to some integrated form of the governing differential equation of the problem at hand. For linear elasticity, for instance, we have for two displacement fields $\boldsymbol{u}$ and $\boldsymbol{v}$

$$a(\boldsymbol{u}, \boldsymbol{v})_\Omega = \int_\Omega \boldsymbol{\nabla} \boldsymbol{u} : \boldsymbol{C} : \boldsymbol{\nabla} \boldsymbol{v} \, \mathrm{d}V \tag{10}$$

where $\boldsymbol{\nabla}$ denotes the gradient operator and $\boldsymbol{C}$ represents the fourth-order constitutive tensor.

Error estimators based on the energy norm are often referred to as *asymptotically exact*, which means that the estimated error approaches the exact error as the FE mesh is refined, provided the mesh is refined adaptively. For a deeper discussion on the theoretical aspects of error estimation and adaptive refinement, we refer to [6] and references therein.

3.2 Class Hierarchy for Norms

In the design of a class hierarchy for representing norms of different type, one should aim to identify the data and operations that are common for all kinds of norms and put those into a general base class. The problem-specific norms should then be implemented as derived subclasses that inherit the common features from the base class. The common feature may be that all norms are evaluated by numerical integration over a given FE discretization of the domain by means of some given quadrature rule. The differences between the various norms are only the integrands. An implementation that reflects this philosophy is presented below. The main classes involved are depicted in Figure 5.

Base Classes. The base class for all norms is defined as follows:

```
class Norm
{
protected:

  class Integrand
  {
  public:
    Integrand () {};
   ~Integrand () {};

    virtual real eval (FiniteElement& fe);
  };

  GridFE& Omega;

  real numItgOverOmega (Vec(real)& enor, Integrand& the_integr);
  ...

public:
  Norm (GridFE& g) : Omega(g) {};
 ~Norm () {};

  virtual real a (Vec(real)& enor, Fields& v1, Fields& v2) = 0;
  virtual real e (Vec(real)& enor, Fields& v1, Fields& v2);
  ...
};
```

The class `Norm` has as its only data member a reference to a `GridFE` object, `Omega`, which represents the FE mesh over which the numerical integration involved in the norm evaluation is performed. The class has three important member functions; the protected function `numItgOverOmega` and two public virtual functions `a` and `e`.

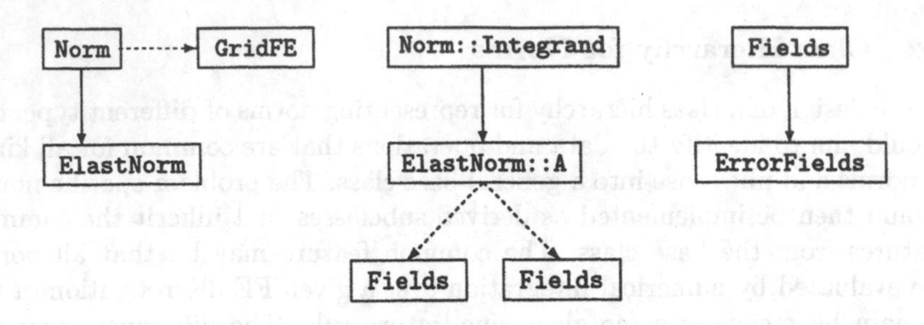

Fig. 5. Some classes for error estimation. The solid arrows indicate class inheritance ("is a"-relationship) whereas the dashed arrows indicate references to internal data objects ("has a"-relationship).

The member function **numItgOverOmega** implements the numerical integration procedure used to evaluate the norm. This involves looping over all elements in the mesh Omega; and for each element, a loop over appropriate quadrature points within the element. At each quadrature point the norm integrand must be evaluated, multiplied with the corresponding integration point weight and Jacobian, and accumulated. The return value of the function is the accumulated norm over all elements. In error estimation and adaptive procedures, we are also interested in the contributions from each element to the global norms. The individual element norms are thus stored in an array (class Vec(real)) which is provided as the first argument to numItgOverOmega.

The function Norm::e evaluates the error norm expression (9), or equivalently, using the other virtual member function Norm::a:

```
real Norm::e (Vec(real)& enor, Fields& u, Fields& v)
{
  ErrorFields e_u(v,u); // The point-wise error  e_u = v - u

  real global = sqrt(a(enor,e_u,e_u));
  nor.apply(sqrt);
  return global;
}
```

The function Norm::e uses another Fields subclass, ErrorFields, which defines the point-wise error field (7). The error norm is then evaluated by providing the error field object as argument to the function Norm::a. Note that Norm::a is declared purely virtual; it has to be defined in a derived subclass (see below).

Since the norm integrand may involve a varying number of fields depending on the type of norm, and perhaps also a set of other parameters as well, we use another *nested* base class, Norm::Integrand, for representing such integrands. This class has a virtual member function eval that returns the integrand value at a given point. An object of this class is then supplied as argument to numItgOverOmega which then invokes the associated eval member function within the integration loop.

Subclasses for Isotropic Linear Elasticity. Different types of norms may now be implemented by deriving subclasses of Norm and Norm::Integrand. Let us consider the energy norm in linear elasticity (10). This norm is now implemented by the subclass ElastNorm and the associated Integrand subclass ElastNorm::A as follows:

```
class ElastNorm : public Norm
{
protected:
  real E, nu; // Material parameters

  class A : public Integrand
  {
    real   E, nu; // The material parameters
    Fields& u, v; // Two displacement fields

    real evaluateIt (Ptm(real)& gradU, Ptm(real)& gradV);

  public:
    A (real _E, real _nu, Fields& _u, Fields& _v) : u(_u), v(_v)
      { E = _E; nu = _nu; };
    ~A () {};

    virtual real eval (FiniteElement& fe);
  };

public:
  ElastNorm (GridFE& g, real _E, real _nu) : Norm(g)
    { E = _E; nu = _nu };
 ~ElastNorm () {};

  virtual real a (Vec(real)& enor, Fields& v1, Fields& v2);
  ...
};
```

We here consider an *isotropic* elastic problem where the constitutive tensor C is uniquely defined by two material parameters E and ν (Young's modulus and Poisson's ratio). Since these material parameters are considered to be known for a given problem[6], they are defined as data members of the class ElastNorm and are initialized by the constructor.

[6] Note that the two material parameters also may be represented by two scalar fields, since they in general are functions of space. However, here we assume they are constant throughout the domain for simplicity.

On the other hand, several displacement fields may be considered for a given problem; fields obtained from the FE analysis, from derivative recovery, or from some analytical solution, etc. The two fields that should be considered are therefore provided as arguments to the norm evaluation function `ElastNorm::a`, which evaluates the integral (10) for the given two fields. The implementation of this function is now particularly simple; it consists of a singe call to the inherited function **numItgOverOmega** with an instance of the nested class `ElastNorm::A` as argument:

```
real ElastNorm::a (Vec(real)& enor, Fields& u, Fields& v)
{
  return numItgOverOmega (enor,A(E,nu,u,v));
}
```

All the necessary data for the norm evaluation is provided as arguments to the constructor of the latter class, such that the interface to the member function **numItgOverOmega** can be retained on a very generalized level.

Implementation of Integrand Subclasses. When implementing the evaluation function of an integrand subclass, one should keep in mind that these functions are called within the integration loop, and therefore need to be efficient. It is therefore important to use the 'correct' field-evaluation functions for the fields involved. For instance, if we know that a field is defined on the same FE grid that we are integrating over, we can utilize that to speed up the norm function. However, we also want the function to be able to handle fields defined on other grids, or maybe on no grid at all (analytical fields, etc.).

To illustrate these issues, we consider the implementation of the evaluation function for the nested class `ElastNorm::A` defined above[7]:

```
real ElastNorm::A::eval (FiniteElement& fe)
{
  Ptm(real) gradU, gradV;

  const GridFE& Omega = fe.grid();
  const Ptv(real)& Xg = fe.getGlobalEvalPt();

  if (u.getGridBase() == &Omega)
    gradU = u.gradientFEM (fe);
  else // Different grid
    gradU = u.gradientPt (Xg);
  if (&v == &u)
```

[7] The private member function **evaluateIt** of `ElastNorm::A` evaluates the triple tensor product $\nabla \boldsymbol{u} : \boldsymbol{C} : \nabla \boldsymbol{v}$, given the two gradients $\nabla \boldsymbol{u}$ and $\nabla \boldsymbol{v}$. In an actual implementation, the code for this function is in-lined in the **eval** function.

```
    gradV = gradU; // Identical fields
  else if (v.getGridBase() == &Omega)
    gradV = v.gradientFEM (fe);
  else
    gradV = v.gradientPt (Xg);

  // Evaluate the integrand grad{U}:[C]:grad{V}
  return evaluateIt (gradU,gradV);
}
```

We here use the **gradientFEM** function for fields that are attached to the grid **Omega** used in the numerical integration[8]. As discussed earlier (see Section 2.3), this is the most efficient evaluation function for fields defined by point-wise values, such as **FieldsFE**, etc. This is true also for the class **FieldsSPR**, as discussed in Section 2.4. For fields that are *not* attached to **Omega**, we must use the more general evaluation function **gradientPt**. This function takes the global coordinates of the current integration point as argument, which may be obtained through the Diffpack function **FiniteElement::getGlobalEvalPt**. If the field is attached to an other grid, a search procedure is then invoked within the **gradientPt** function to find the corresponding element or node within that grid. This will of course slow down the norm evaluation considerably in such cases.

The two fields involved in the norm expression (10) are often identical, as in the error norm (9). It is then not necessary to evaluate more than one field in the **eval** function, and so we may save computational time. To check if two field objects are identical we can simply compare their pointers; if the pointers are equal, they refer to the same object.

We may alternatively define the class **ElastNorm::A** in terms of *stress* fields $\boldsymbol{\sigma}$ $(= \{\sigma_{xx}, \sigma_{yy}, \sigma_{xy}\}^{\mathrm{T}}$ in 2D) or *strain* fields $\boldsymbol{\varepsilon}$ $(= \{\varepsilon_{xx}, \varepsilon_{yy}, \varepsilon_{xy}\}^{\mathrm{T}})$ directly, instead of displacement gradients. We would then use the functions **valueFEM** and **valuePt** within the member function **ElastNorm::A::eval**, and the function **evaluateIt** must be modified accordingly such that it evaluates the matrix product $\boldsymbol{\sigma}_1 \mathbf{C}^{-1} \boldsymbol{\sigma}_2$ or $\boldsymbol{\varepsilon}_1 \mathbf{C}\, \boldsymbol{\varepsilon}_2$, respectively, where $\mathbf{C}$ now denotes the constitutive *matrix* for isotropic linear elasticity.

3.3 An Error Estimation Example

To demonstrate the use of the classes discusses above, we consider a linear elastic problem with known analytical solution. Such problems may be used to test the quality of the error estimation procedures, as the analytical solution enables us to compute also the *exact* error in the FE solution.

[8] The Diffpack function **FiniteElement::grid** returns a reference to the grid that the numerical integration is performed over, whereas the function **Fields::getGridBase** returns access to the attached grid in the form of a base class for all types of grids.

For this example, we assume the existence of the following two additional vector field classes:

`FieldsEX` Represents the exact displacement field for a given test problem through some known function. It can be defined as a subclass of `FieldsFunc`

`StressField` A subclass of `Fields` that defines the stress field for a given displacement field and material data. Its evaluation functions then use the `gradient...` functions of the given displacement field to set up the stress vector.

Error Estimation based on Recovered Gradients. The code segment needed for computing discretization errors based on displacement derivatives is now simply as follows:

```
GridFE   Omega; // The FE mesh
real     E,nu;  // Material parameters
FieldsFE u_h;   // The FE displacement field
FieldsEX u;     // The exact displacement field
...

// 0. Setup and solve the FE problem ...
...

// 1. Define the error norm
ElastNorm errest(Omega,E,nu);
Vec(real) elnorms;

// 2. Initialize the patch recovery procedure
SPR recovery;
recovery.init(Omega,...)

// 3. Recover C0-continuous displacement gradients
Handle(Fields) u_r;
recovery.recover (u_h,TRUE,u_r);

// 4. Estimated error in u_h
error  = errest.e(elnorms,u_h,u_r);
norm_u = sqrt(errest.a(elnorms,u_h,u_h) + error*error);
relerr = 100 * error / norm_u;

// 5. Exact error in u_h
exact1 = errest.e(elnorms,u_h,u);
norm_u = sqrt(errest.a(elnorms,u,u));
relex1 = 100 * exact1 / norm_u;

// 6. Exact error in u_r
exact2 = errest.e(elnorms,u_r,u);
```

```
relex2 = 100 * exact2 / norm_u;

// 7. Effectivity index
effind = error / exact1;
...
```

Here we specify TRUE as the second argument to the function SPR::recover in step 3 to indicate that the gradients of the given field u_h should be recovered. The output argument u_r, representing the recovered displacement field $\boldsymbol{u}^*$, will then refer to a FieldsSPR object where only the gradient.. member functions are defined (see Section 2.4). Consequently, it is possible to only evaluate the gradient of the field $\boldsymbol{u}^*$ and not the field values themselves. However, the field values are not needed when measured in the energy norm.

The steps 4–6 above evaluate the estimated error in the FE solution $\|\boldsymbol{u}^* - \boldsymbol{u}^h\|_E$, the exact error in the FE solution $\|\boldsymbol{u} - \boldsymbol{u}^h\|_E$, and the exact error in the recovered solution $\|\boldsymbol{u} - \boldsymbol{u}^*\|_E$, respectively, in terms of the energy norm (9). We also evaluate the corresponding *relative* percentage errors which may be easier to interpret by the user than the absolute errors. Finally, we compute the effectivity index as the ratio between the estimated and exact error.

Error Estimation Based on Recovered Stress Components. Alternatively, the error estimation may be based on recovered stresses instead of displacement gradients. The steps 3–6 would then be replaced by

```
// 3a. Define the FE and exact stress fields
StressField s_h(u_h,E,nu);
StressField s(u,E,nu);

// 3. Recover a C0-continuous stress field
Handle(Fields) s_r;
recovery.recover (s_h,FALSE,s_r);

// 4. Estimated error in s_h
error  = errest.e(elnorms,s_h,s_r);
norm_u = sqrt(errest.a(elnorms,s_h,s_h) + error*error);
relerr = 100 * error / norm_u;

// 5. Exact error in s_h
exact1 = errest.e(elnorms,s_h,s);
norm_u = sqrt(errest.a(elnorms,s,s));
relex1 = 100 * exact1 / norm_u;

// 6. Exact error in s_r
exact2 = errest.e(elnorms,s_r,s);
relex2 = 100 * exact2 / norm_u;
```

We now specify FALSE as the second argument to SPR::recover in step 3 to indicate that field values, and not derivatives, should be recovered. The output argument `s_r`, now representing the recovered stress field σ^*, will then refer to a `FieldsSPR` object where only the `value..` member functions are defined (see Section 2.4).

Each stress component may be considered as a linear combination of the displacement gradient components. In the former approach, the linear combination is performed *after* the recovery step, whereas now it is performed *before* the recovery. The two approaches will therefore yield slightly different results for the estimated errors $\|u^* - u^h\|_E$ and $\|\sigma^* - \sigma^h\|_E$. The exact errors $\|u - u^h\|_E$ and $\|\sigma - \sigma^h\|_E$, however, will be identical. Notice also that the recovery step in the latter approach may be more computationally efficient than in the former due to the lower number of (symmetric) stress components compared to the number of (unsymmetric) displacement gradient components.

3.4 Error Estimation for the Navier–Stokes Equations

We now consider error estimators for the Navier–Stokes equations. Two popular error norms for this type of problems are (see also [9,1])

$$\|e\|_E = \sqrt{a(e,e)_\Omega} \;; \qquad a(u,v)_\Omega := \int_\Omega \nu\, \nabla u : \nabla v \, \mathrm{d}V \tag{11}$$

$$\|e\|_* = \sqrt{a(e,e)_\Omega + d(u^h)_\Omega} \;; \; d(u)_\Omega := \int_\Omega \nu\, (\nabla \cdot u)^2 \, \mathrm{d}V \tag{12}$$

where $e = u^* - u^h$ now represents the estimated point-wise error in the FE *velocity* field u^h, and ν is the dynamic viscosity parameter. The first norm (Equation (11)) is equivalent to the energy norm of linear elasticity (see Equation (9). The second norm (Equation (11)) is due to Ainsworth [1], and accounts also for the error in the divergence of the FE velocity field (which should be zero).

An implementation of these two norms in the framework presented above may now be as follows:

```
class NavierNorm : public Norm
{
protected:

  real   nu;    // Dynamic viscosity parameter
  String type; // Which norm to evaluate, (1.11) or (1.12)

  class A : public Integrand
  {
    real nu;
    Fields& u, v;

  public:
```

```
    A (real _nu, Fields& _u, Fields& _v) : u(_u), v(_v) { nu = _nu; };
   ~A () {};

    virtual real eval (FiniteElement& fe);
  };

  class D : public Integrand
  {
    real nu;
    Fields& u;

  public:
    D (real _nu, Fields& _u) : u(_u) { nu = _nu; };
   ~D () {};

    virtual real eval (FiniteElement& fe);
  };

public:
  NavierNorm (GridFE& g, real _nu) : Norm(g) { nu = _nu; };
 ~NavierNorm () {};

  void setType (const String& _type) { type = _type; };

  virtual real a (Vec(real)& enor, Fields& u, Fields& v);
          real d (Vec(real)& enor, Fields& u);
  virtual real e (Vec(real)& enor, Fields& u, Fields& v);

  ...
};
```

We now have two `Norm::Integrand` subclasses, namely `NavierNorm::A` and `NavierNorm::D`, which correspond to the two terms of Equation (12). We also have two associated norm evaluation functions, `NavierNorm::a` and `NavierNorm::d`, respectively, with the following simple definitions:

```
real NavierNorm::a (Vec(real)& enor, Fields& u, Fields& v)
{
  return numItgOverOmega (enor,A(nu,u,v));
}

real NavierNorm::d (Vec(real)& enor, Fields& u)
{
  return numItgOverOmega (enor,D(nu,u));
}
```

For this class (in contrast to the class `ElastNorm` discussed in Section 3.2), we also re-implement the virtual function `Norm::e` as follows:

```
real NavierNorm::e (Vec(real)& enor, Fields& u, Fields& v)
{
  ErrorFields e_u(v,u); // Point-wise error e_u = v - u

  real global = a(enor,e_u,e_u);  // Energy norm

  if (type == "Ainsworth's norm") // Add the div(u) term
  {
    Vec(real) dnor;
    global += d(dnor,u);
    enor.add(dnor,1,1.0);
  }
  enor.apply(sqrt);
  return sqrt(global);
}
```

This function may be used to evaluate both of the two norms given above, depending on the value of the type-variable which is assigned using the setType member function. Alternatively, we could have splitted the class NavierNorm as defined above into two different classes with separate implementations of the e member-function.

Finally, the implementations of the evaluation functions for the integrand classes NavierNorm::A and NavierNorm::D are given below. The structure of these functions is very similar to the corresponding function ElastNorm::A::eval discussed above. The function Ptm(real)::inner used in NavierNorm::A::eval computes the inner-product (also known as the double-dot product) between *this and another Ptm(real) object, which represent second-order tensors.

```
real NavierNorm::A::eval (FiniteElement& fe)
{
  Ptm(real) gradU, gradV;

  const GridFE& Omega = fe.grid();
  const Ptv(real)& Xg = fe.getGlobalEvalPt();

  if (u.getGridBase() == &Omega)
    u.gradientFEM (gradU,fe);
  else // Different grid
    u.gradientPt (gradU,Xg);

  if (&v == &u) // Identical fields
    return nu * gradU.inner(gradU);

  if (v.getGridBase() == &Omega)
    v.gradientFEM (gradV,fe);
  else // Different grid
    v.gradientPt (gradV,Xg);
```

```
  return nu * gradU.inner(gradV);
}

real NavierNorm::D::eval (FiniteElement& fe)
{
  real divU;

  const GridFE& Omega = fe.grid();
  const Ptv(real)& Xg = fe.getGlobalEvalPt();

  if (u.getGridBase() == &Omega)
    divU = u.divergenceFEM (fe);
  else // Different grid
    divU = u.divergencePt (Xg);

  return nu * divU*divU;
}
```

The above demonstrates clearly how compact a new type of norm for error estimation may be implemented when using the general base class `Norm`. Basically, the main effort is the programming of the norm integrand through a `Norm::Integrand` subclass. This is somewhat similar to the philosophy of the `FEM` base class in Diffpack [3,4,7]; subclasses of `FEM` basically need only to implement the integrand of the weak form of the PDE to be solved. The `FEM` class then contains member functions performing the numerical integration and assembly of element matrices, operations that are independent of the actual PDE.

4 Numerical Example: Linear Elasticity

4.1 Infinite Plate with a Circular Hole

Figure 6 shows a finite quarter portion of an infinite isotropic linear-elastic plate with a circular hole subjected to unidirectional unit tension. The analytical solution to this problem is known (e.g. see [14]) and it is a popular benchmark problem among error estimation developers [14,6,2,13]. The analytical stress distribution is applied as Neumann boundary conditions on the boundaries $x = 4.0$ and $y = 4.0$, whereas symmetry conditions are applied on the boundaries $x = 0$ and $y = 0$, as indicated in Figure 6. Plane strain conditions are assumed.

The Hole problem is analysed with Diffpack using the two mesh sequences shown in Figure 7. The first sequence consists of a series of uniformly refined, structured meshes. The second mesh sequence contains a series of unstructured meshes based on adaptive refinement, where the estimated local error in one mesh is used to indicate the desired element size in the subsequent mesh [6].

4.2 Numerical Results

In the present simulations we use the energy norm error estimator based on recovered *stresses*, i.e. the second alternative discussed in Section 3.3. To assess the quality of the error estimator, we study the convergence of the estimated relative global error and the associated effectivity index, i.e.

$$\eta^* := \frac{\|\boldsymbol{\sigma}^* - \boldsymbol{\sigma}^h\|_E}{\sqrt{\|\boldsymbol{\sigma}^h\|_E^2 + \|\boldsymbol{\sigma}^* - \boldsymbol{\sigma}^h\|_E^2}} \times 100\% \tag{13}$$

$$\Theta := \|\boldsymbol{\sigma}^* - \boldsymbol{\sigma}^h\|_E \;/\; \|\boldsymbol{\sigma} - \boldsymbol{\sigma}^h\|_E \tag{14}$$

which are computed in steps 4 and 7 of Section 3.3 (`relerr` and `effind`, respectively). The denominator of Equation (13) represents an estimate for $\|\boldsymbol{\sigma}\|_E$; the energy norm of the exact stress field. We here assumed that the FE solution always underestimates the exact solution in terms of total strain energy. This is true for problems with only homogeneous Dirichlet boundary conditions and when standard displacement-based finite elements are used. In our study, we use the standard bi-linear quadrilateral element with full 2×2 Gauss integration of all stiffness terms.

The convergence of the indicators (13) and (14) is plotted in Figure 8 for the two mesh sequences. We observe that the convergence rate of the estimated error is considerably higher for the adapted mesh sequence than

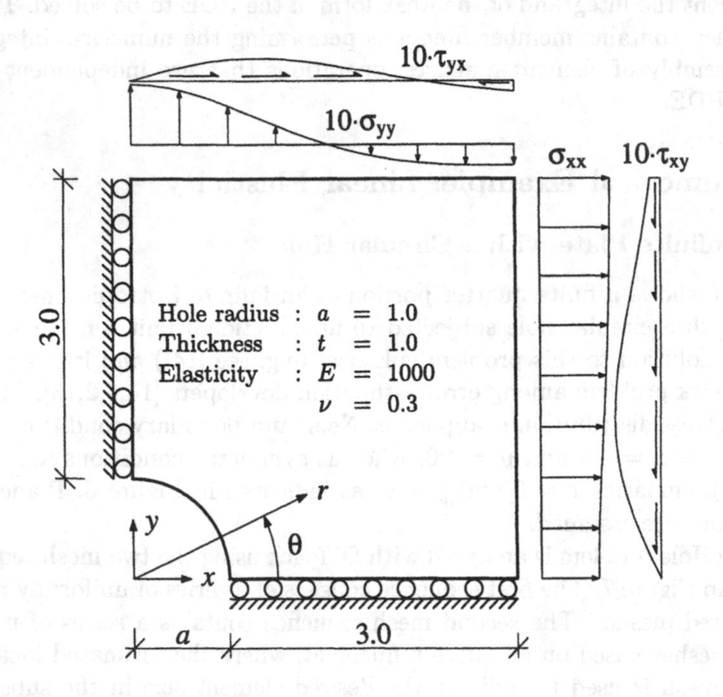

Fig. 6. The Hole problem: Geometry and properties.

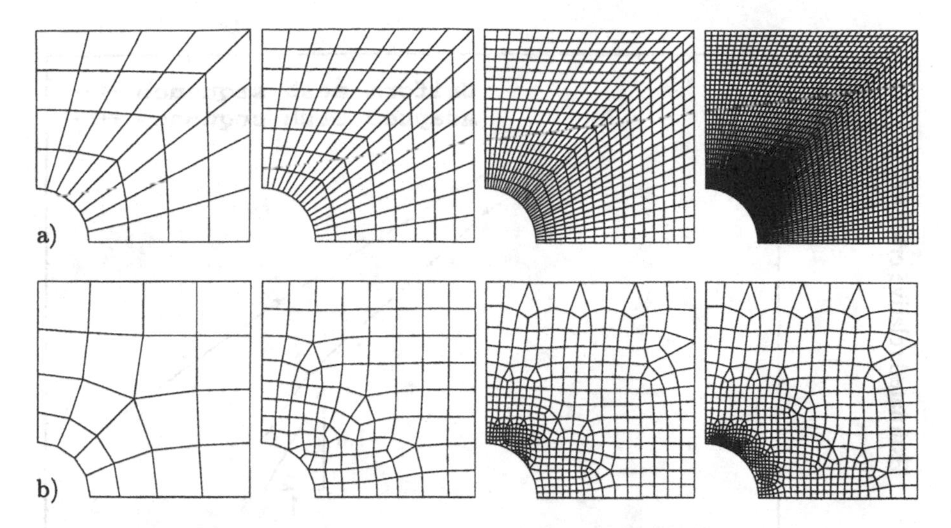

Fig. 7. The finite element meshes used in the present study. a) Uniform regular mesh sequence. b) Adapted irregular mesh sequence.

for the uniform sequence. This is also reflected by the associated effectivity indices, which both approach unity as the mesh is refined, but much faster for the adapted mesh sequence than for the uniform sequence. This is due to the more uniform error distribution obtained for the adapted mesh sequence, which implicitly fulfil the superconvergence properties better.

4.3 Computational Cost

In addition to the numerical results, we would also like to know the computational cost of the various parts of the post-processing module, both compared to the cost of the Finite Element Analysis (FEA) itself and compared to a similar program written in FORTRAN[9].

In Figure 9a) we plot as functions of mesh size the consumed CPU-time for Step 2 (Initialization), Step 3 (Field recovery) and Step 4 (Error estimation) of Section 3.3. The total time of the three steps (Total) and the time consumed by the FE analysis itself (FEA; Step 0) is also plotted.

Comparing the Total time and the FEA time, we see that the FEA has a considerably higher growth rate than the post-processing steps and for the finest mesh the post-processing consumes less than 10% of the time spent by the FEA. The reason for this difference is that the FEA involves the solution

[9] The simulations reported here are performed on a SGI Origin 200 computer running the IRIX 6.4 operation system. The computer has two R10000 processors (180MHz), 512 Mbytes main memory, 32 Kbytes instruction cache and 32 Kbytes data cache.

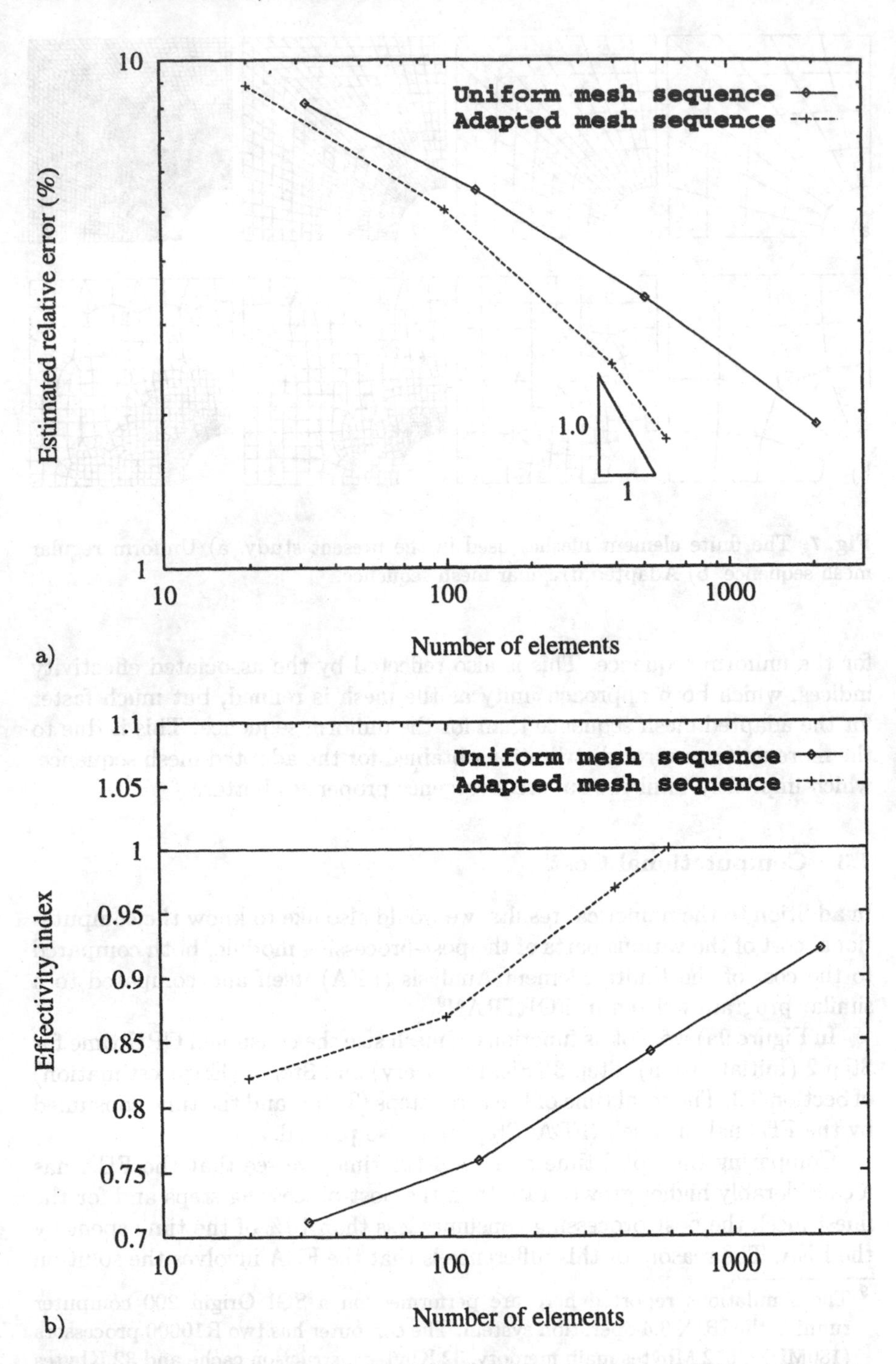

Fig. 8. Some error estimation results for the Hole problem. a) Estimated relative global error. b) Associated effectivity index.

of a global system of equations[10], whereas the post-processing steps consists of local computations only (linear growth rate).

Of the three post-processing steps, the error estimation (Step 4) consumes most CPU-time. This is mainly due to the large amount of field evaluations involved in this step. Note also that the initialization step (Step 2) is more costly than the field recovery itself for all meshes, and its growth rate appears also to be slightly higher. The reason for this is that this step involves the construction of a set of global connectivity arrays relating nodes, elements and element boundaries, which are used by the patch recovery procedure. However, in time-dependent problems where the FE solution is computed in increments, the initialization time is less important, since it is done only once whereas the field recovery and error estimation is performed for several increments.

In Figure 9b) we plot the ratio of the CPU-time consumed by the present C++ program and by a similar program written in FORTRAN. The FORTRAN program performs the same three steps as the C++ program, but the implementation is somewhat different. The main difference is that the evaluation of the recovered field now is performed during the recovery step, and the field values are stored in a global array which is used in the error estimation. This is why the error estimation ratio in Figure 9b) is so high (it is approximately 7.5 for the finest mesh); it corresponds to a very low time consumption of the error estimation step of the FORTRAN program.

However, studying the curve for the total time (which is the most important one) we see that this ratio decreases as the mesh size increases, starting from ≈ 2 for the coarse mesh, and from ≈ 500 elements the C++ program uses *less* time than the FORTRAN implementation.

5 Numerical Example: Navier–Stokes

5.1 The Jeffery–Hamel Flow Problem

The Jeffery–Hamel flow is the exact solution to the incompressible Navier–Stokes equations on a wedge-shaped domain with a source or sink at the origin (see Figure 10a). The analytical solution to this problem has been derived in [12] (pp. 184–189) and gives the radial velocity u_r (the angular velocity u_θ being zero) and the pressure p as

$$u_r = \frac{U_0}{r} f(\eta) \; ; \; \eta = \frac{\theta}{\alpha} \tag{15}$$

$$p = p_0 - \frac{1}{2}\rho\left(\frac{U_0}{r/r_0}\right)^2 \left[f^2 \pm \frac{1}{Re\,\alpha} f''\right] \tag{16}$$

where U_0 is the radial velocity at a given point $r = r_0$ and $\theta = 0$, ρ is the mass density of the fluid and p_0 is an arbitrary constant, here selected such

[10] A direct sparse matrix solver is used to solve the global linear system of equations.

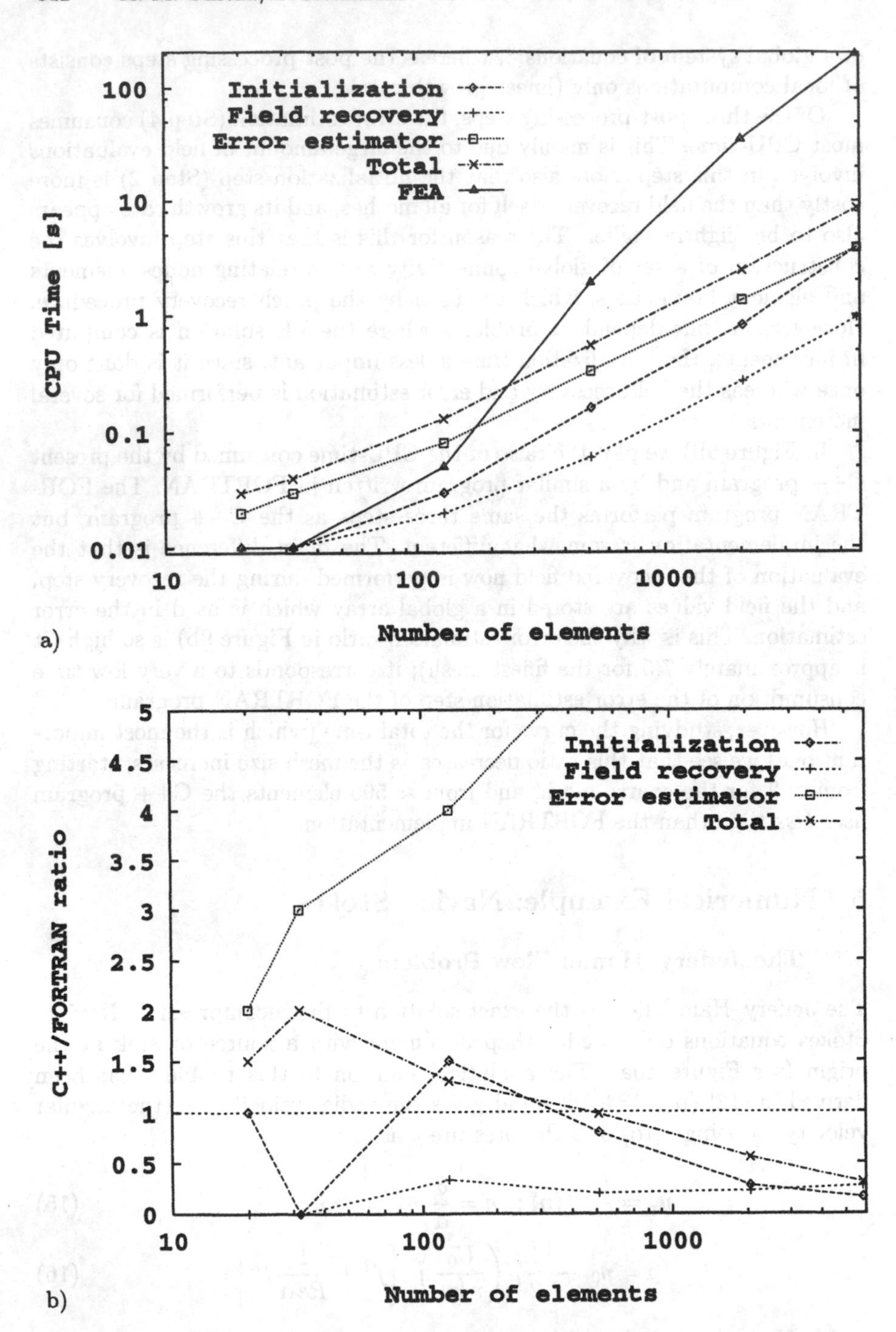

Fig. 9. CPU time consumption in various parts of the error estimation module for the uniform meshes (computations only, file-I/O is not included). a) Absolute CPU time. b) Relative CPU time compared with the FORTRAN module.

that $p = 0$ for $r = r_0$ and $\theta = 0$. The Reynolds number Re is given as $Re = U_0 r_0/\alpha$. The + sign in front of the last term in (16) is for flow in the positive x-direction and the − sign for flow in the negative x-direction.

The function $f(\eta)$ is determined through the differential equation

$$f''' + 2\,Re\,\alpha\, f f' + 4\alpha^2 f' = 0 \tag{17}$$

with boundary conditions $f(0) = 1$, $f(1) = 0$ and $f'(0) = 0$. This third order, non-linear equation has been solved numerically using MATLAB and the solution for our choice of parameters, $Re = 10$ and $\alpha = \pi/4$ is plotted in Figure 10b). The associated first and second derivatives are plotted in Figures 10c) and 10d), respectively.

The Cartesian components of the velocity field are $u_x = u_r \cos\theta$ and $u_y = u_r \sin\theta$, respectively. With $x = r\cos\theta$ and $y = r\sin\theta$, we may write the velocity field in terms of Cartesian coordinates as

$$\boldsymbol{u}(x,y) = \begin{Bmatrix} u_x \\ u_y \end{Bmatrix} = \frac{U_0}{r^2} f(\eta) \begin{Bmatrix} x \\ y \end{Bmatrix}. \tag{18}$$

The gradient of $\boldsymbol{u}$ may be obtained through partial differentiation using the substitution of variables $r^2 = x^2 + y^2$ and $\eta = \alpha \tan^{-1} \frac{y}{x}$. This yields

$$\boldsymbol{\nabla u} = \frac{U_0}{\alpha r^4} f'(\eta) \begin{bmatrix} -xy & x^2 \\ -y^2 & xy \end{bmatrix} + \frac{U_0}{r^4} f(\eta) \begin{bmatrix} y^2 - x^2 & -2xy \\ -2xy & x^2 - y^2 \end{bmatrix}. \tag{19}$$

A class `JHvelocity` may now be defined as a `FieldsFunc` subclass, where the member functions `valuePt` and `gradientPt` are defined using the expressions (18) and (19), respectively. This class may then be used to assess the exact error of the computed solution through the `NavierNorm` class defined in Section 3.4.

5.2 Numerical Results

The shaded quadrilateral section shown in Figure 10a) is analysed with the flow in the positive x-direction. The exact velocity (18) is imposed as Dirichlet conditions along both the inflow and outflow boundaries. The pressure is then determined up to a constant and here it is prescribed equal to zero at the point $\{x, y\} = \{1.0, 0.0\}$. It should be noted that this point-wise condition is not well-posed[11]. However, as long as we do not use adaptive refinement it does not cause a severe problem.

The simulations are performed with the FE codes NAVSIM [5], which is a 2D code based on linear triangles, and SPECTRUM [11], which is a commercially available 3D code. The SPECTRUM simulations are performed on a 3D

[11] A better constraint would be to claim that $\int_\Omega p^h \, \mathrm{d}V$ should be equal to a given constant. Such a condition is not applicable, however, without modifying the FE code.

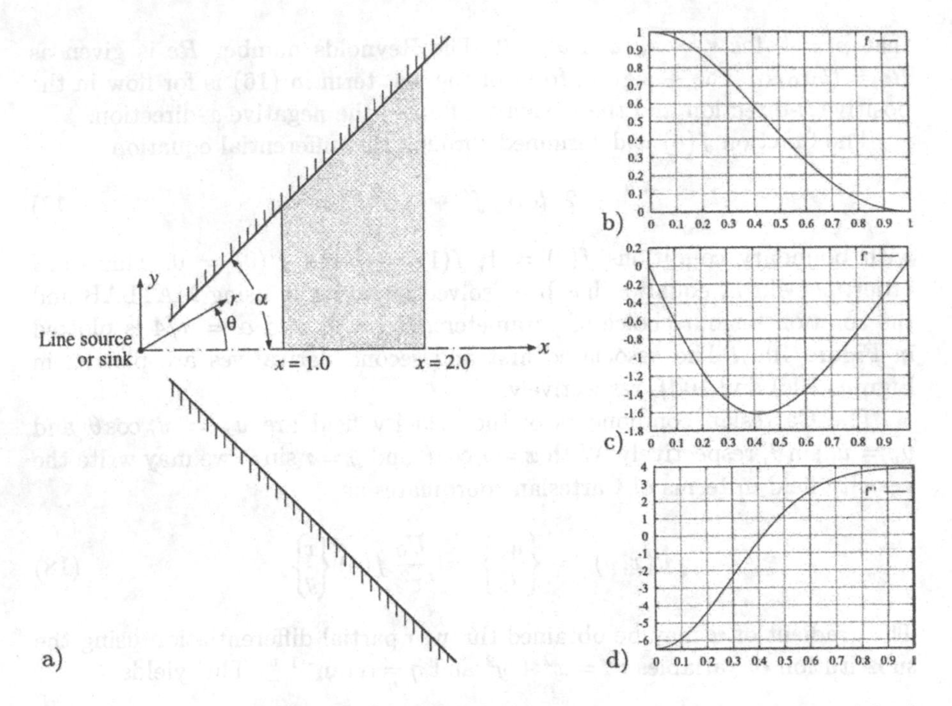

Fig. 10. The Jeffery–Hamel flow problem: a) Geometry. b) The function $f(\eta)$. c) First derivative $f'(\eta)$. d) Second derivative $f''(\eta)$.

slice discretized with one layer of 8-noded hexahedrons. The post-processing, however, is performed on an equivalent 2D model also for these simulations. A series of four uniform meshes are used for each code (respectively, 10×10, 20×20, 40×40 and 80×80 elements[12]).

In Figure 11a) the estimated errors (11) and (12) are plotted as functions of mesh size. Both norms have been normalized with the quantity $\|\boldsymbol{u}\|_E = \sqrt{a(\boldsymbol{u}, \boldsymbol{u})}$, where $\boldsymbol{u}$ is the analytical velocity field (18), to obtain the relative measures. We observe that all four curves have approximately the same rate of convergence. As in the previous example (see Section 4.2), we would have obtained a higher rate with adaptive mesh refinement.

The associated effectivity indices are plotted in Figure 11b). We notice here that both error estimators work better for the SPECTRUM simulations (bi-linear elements) compared with the NAVSIM simulations (linear triangles). We also observe that the error norm (12) has slightly better effectivity than the energy norm (11).

[12] These numbers represent the size of the SPECTRUM meshes. Of course, the NAVSIM meshes contain twice as many elements since it uses triangular elements. However, the number of nodes (and degrees of freedom) is equal for the two codes.

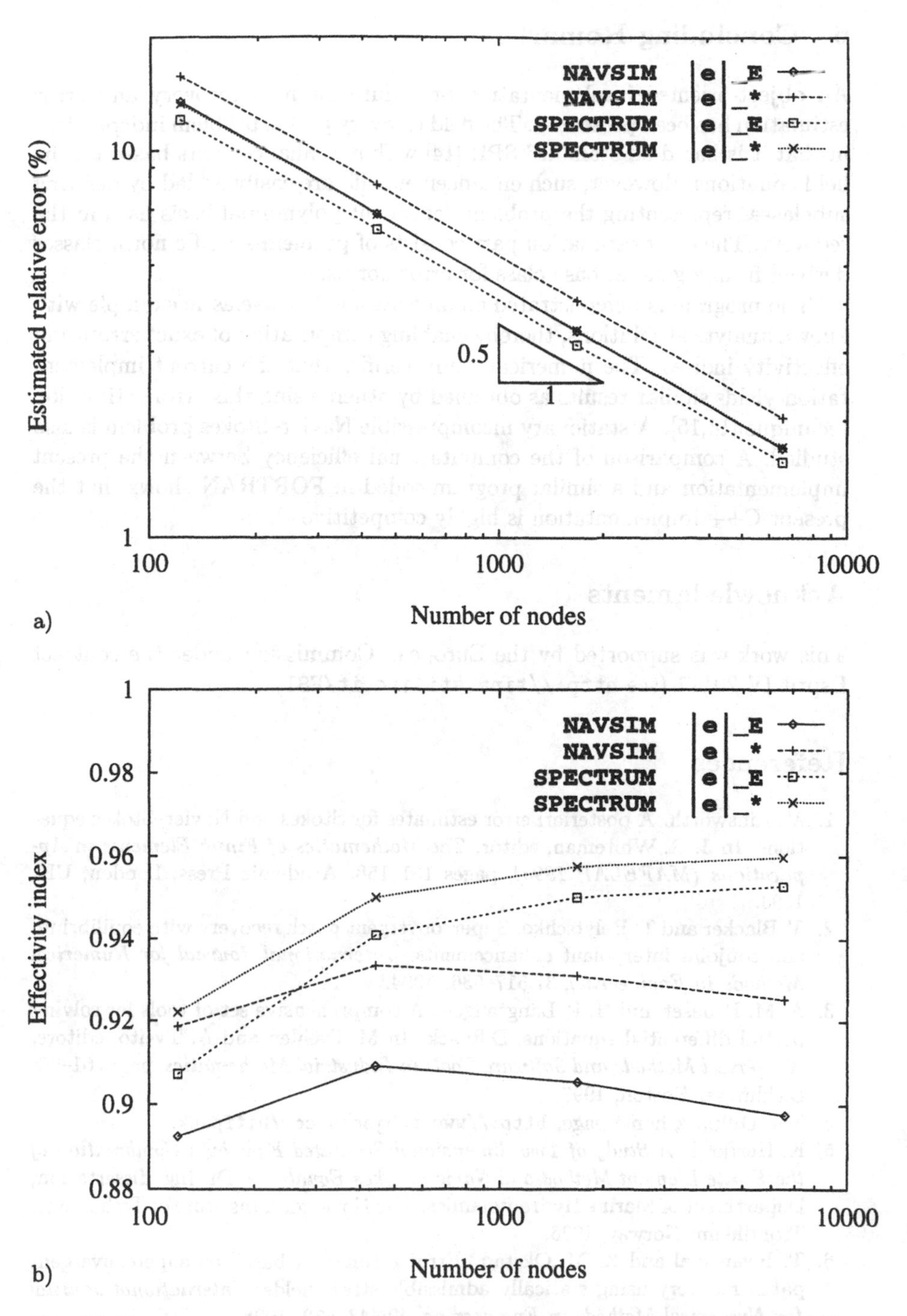

Fig. 11. Some error estimation results for the Jeffery–Hamel flow problem. a) Estimated relative global errors. b) Associated effectivity indices.

6 Concluding Remarks

An object-oriented implementation procedure for field recovery and error estimation has been presented. The field recovery part is problem independent in that it is based on standard SPR [14] without enhancements based on the field equations. However, such enhancements [6] are easily added by deriving subclasses representing the problem-dependent polynomial basis used in the recovery. The error estimation part consists of problem-specific norm classes derived from a general base class for error norms.

The program is demonstrated on an isotropic linear-elastic example with known analytical solutions, thereby enabling computation of exact errors and effectivity indices. The numerical study verifies that the current implementation yields similar results as obtained by others using this error estimation technique [14,15]. A stationary incompressible Navier–Stokes problem is also studied. A comparison of the computational efficiency between the present implementation and a similar program coded in FORTRAN shows that the present C++ implementation is highly competitive.

Acknowledgments

This work was supported by the European Commission under the contract Esprit IV 20111 (see `http://tina.sti.jrc.it/FSI`).

References

1. M. Ainsworth. A posteriori error estimates for Stokes and Navier–Stokes equations. In J. R. Whiteman, editor, *The Mathematics of Finite Elements in Applications (MAFELAP 1993)*, pages 151–158. Academic Press, London, UK, 1993.
2. T. Blacker and T. Belytschko. Superconvergent patch recovery with equilibrium and conjoint interpolant enhancements. *International Journal for Numerical Methods in Engineering*, 37:517–536, 1994.
3. A. M. Bruaset and H. P. Langtangen. A comprehensive set of tools for solving partial differential equations; Diffpack. In M. Dæhlen and A. Tveito, editors, *Numerical Methods and Software Tools in Industrial Mathematics*, pages 61–90. Birkhäuser Boston, 1997.
4. The Diffpack home page. `http://www.nobjects.com/Diffpack`.
5. K. Herfjord. *A Study of Two-dimensional Separated Flow by a Combination of the Finite Element Method and Navier–Stokes Equations.* Dr. Ing. dissertation, Department of Marine Hydrodynamics, The Norwegian Institute of Technology, Trondheim, Norway, 1996.
6. T. Kvamsdal and K. M. Okstad. Error estimation based on superconvergent patch recovery using statically admissible stress fields. *International Journal for Numerical Methods in Engineering*, 42:443–472, 1998.
7. H. P. Langtangen. *Computational Partial Differential Equations – Numerical Methods and Diffpack Programming.* Springer-Verlag, 1999.

8. K. M. Okstad, T. Kvamsdal, and K. M. Mathisen. Superconvergent patch recovery for plate problems using statically admissible stress resultant fields. *International Journal for Numerical Methods in Engineering*, 44:697–727, 1999.
9. K. M. Okstad and T. Kvamsdal. Error estimation of Navier–Stokes computations based on superconvergent patch recovery. In *Proceedings of the 10th Nordic Seminar on Computational Mechanics*, pages 155–158. Tallinn, Estonia, October 1997.
10. A. C. A. Ramsay and H. Sbresny. Evaluation of some error estimators for the four-noded Lagrangian quadrilateral. *Communications in Numerical Methods in Engineering*, 11:497–506, 1995.
11. Centric Engineering Systems Inc., Palo Alto, California. *SPECTRUM Solver (ver. 2.0) Command Reference and Theory Manual*, May 1993.
12. F. M. White. *Viscous Fluid Flow*. McGraw–Hill, 1974.
13. N.-E. Wiberg, F. Abdulwahab, and S. Ziukas. Enhanced superconvergent patch recovery incorporating equilibrium and boundary conditions. *International Journal for Numerical Methods in Engineering*, 37:3417–3440, 1994.
14. O. C. Zienkiewicz and J. Z. Zhu. The superconvergent patch recovery and *a posteriori* error estimates. Part 1: The recovery technique. *International Journal for Numerical Methods in Engineering*, 33:1331–1364, 1992.
15. O. C. Zienkiewicz and J. Z. Zhu. The superconvergent patch recovery and *a posteriori* error estimates. Part 2: Error estimates and adaptivity. *International Journal for Numerical Methods in Engineering*, 33:1365–1382, 1992.

Designing an ODE Solving Environment

Dana Petcu and Mircea Drăgan

Department of Mathematics, Western University of Timişoara, Romania
Email: {petcu, dragan}@info.uvt.ro

Abstract. The volume and diversity of the available numerical software for initial value ordinary differential equations have become a problem for the common users. Designing a new solving environment is justified only if it answers some critical user needs, like that of deciding the type of problems to which a specific software can be optimally applied (thus enabling the choice of appropriate software for a specific problem). Such critical user needs will be discussed here, and some ideas will be suggested. Special attention is paid to the class of parallel numerical methods for ordinary differential equations. A proposal for a dedicated solving environment is described, and the facilities of a prototype are presented.

1 Introduction

The numerical solution of initial value problems (IVPs) for ordinary differential equations (ODEs) is an active field of research, motivated by challenging real-life applications. Therefore, a very wide variety of numerical methods and codes have been developed. Many of them were constructed to help users solve their real problems without consulting a mathematician. Unfortunately, the number of such efficient software packages remains relatively small. There are also several user needs not met by currently highly-used codes. Among these, we mention: friendly software interface, help assistance, guidance in the solving process and for the interpretation of the results, facilities for solving special problems, fast solution methods for large-scale problems. For these user needs some methods have been proposed, but the corresponding software is only in an experimental stage.

In this chapter we discuss the approach to solving differential equations of the form:

$$y'(t) = f(t, y(t)),\ t \in [t_0, t_0 + T], \qquad y(t_0) = y_0,$$

where $f : [t_0, t_0 + T] \times \mathrm{R}^d \to \mathrm{R}^d$, and y has values in some convex domain $\Omega \subset \mathrm{R}^D$ such that $\sup_{y \in \Omega} \|f'_y(t, y)\| \leq L$, L being the Lipschitz constant.

The chapter is organized as follows. Section 2 describes briefly the problems faced by non-experts when solving an ODE for which an exact solution cannot be easily found. Section 3 discusses the options for exploiting parallel computers, and outlines the difficulties involved. Section 4 describes the prototype expert system EpODE, representing an attempt to endow both experts and non-experts with the full power of a wide range of numerical ODE solvers. Section 5 contains some concluding remarks. Several examples of EpODE facilities are presented in Appendix A.

2 Solving ODEs with Available Numerical Software

We discuss in this section the problems faced by non-experts when confronted with having to solve non-trivial ODEs.

2.1 Solving ODEs with a General Problem Solver

Computer Algebra Systems (CAS) like Mathematica or Maple are general problem solvers which allow both symbolic computation and numerical computations of the solutions. Designing a CAS that can be applied to a wide range of mathematical problems is a very difficult task. The basic idea is to construct a kernel able to solve any simple problem, and to use special library functions and procedures for more complicated problems. These packages of dedicated functions can be thought of as problem-dedicated software and the library containing them can be easily extended.

Unfortunately, combining in the same tool symbolic and numerical computations can be a drawback especially for large-scale problems. If the ODE system consists of hundreds of equations, the only reasonable solution is to integrate it numerically. In order to describe a new numerical method which fits a specific problem, the user must know how to handle the tool language (which might become a problem in itself).

The information about an iterative method designed for solving IVP for ODEs does not always include specification of details, such as the stepsize to be chosen for obtaining an approximate solution within a given error range. The user must therefore not only be a person accustomed with the tool language, but also an expert in numerical analysis, to be able to supply the missing information. Moreover, the selection of the method is user dependent, and not problem dependent (as it is normal), thus the user must know the method which is appropriate for his problem; a wrong selection implies wrong results.

2.2 Solving ODEs with a Dedicated Software Package

Selecting the solving package, from the variety of available (free or commercial) tools, is a great problem for the common user. Many dedicated software packages have been constructed for ODEs. There are at least two kinds of such packages: those with a friendly user interface for describing the problem and with build-in solution methods (like the program ODE [3]), and those represented by packages of programs in a standard programming language (like ODEPACK [8]).

The main idea for designing the first category of packages is to allow the user to concentrate on the problem, not on the solution procedure. In general, such packages are designed in a black-box style: a fixed number of methods (the most frequently used ones) are implemented, and the user must select one of them. Depending on the method and the help providing facilities, the solving task can have different degrees of difficulty.

We emphasize that situations in which all methods fail to solve the problem can be easily encountered (for example, if the tool has only explicit methods, and the problem is stiff). In such a situation, the user has no other solution than trying an alternative tool.

The dimension of the problem can be another restriction, if the package allows only a limited number of equations.

The above-mentioned restrictions can be partially eliminated if the user takes advantage of a package of procedures and functions written in a high level programming language which describe some numerical methods for solving the given problem (such codes dominate the field of scientific computing). Modular design and code reuse are the main points in analyzing the success of such a type of dedicated software (see more details in [1]). The problem is usually described by the user as a function. The problem dimension depends on the information storage scheme. The main difficulty consists in the correct selection of the procedures depending on the problem to be solved, a task which is by no means simple. The GAMS [2] (Guide for Available Mathematical Software programs) provides on-line facilities for selecting an adequate solution procedure, useful for stiff/nonstiff initial value ODEs as well.

2.3 Solving ODEs with a Problem Solving Environment

Critical user needs, such as having a broad knowledge of numerical software, and knowing where this software can be optimally applied, can be addressed by using a problem solving environment (PSE) software system which supports the solving of problems from a certain class using a high level user interface.

When designing a PSE, the user's comfort and the usefulness of the output are the primary goals. A PSE must efficiently carry out the routine parts of the solving process without interference from the user, it should help the user specify the problem, it should select algorithms, and it should determine the problem oriented parameters of the algorithms.

The selection mechanism, by which a subsystem or a subprogram will be used to solve the problem, ranges from simple decision trees to expert systems whose knowledge base is provided by experts in the field.

Scientific projects have been started recently for constructing expert systems that provide on-line consultation for mathematical software, for the evaluation and the debugging of the software solving the ODEs, and for solver selection. Successful prototypes like SAIVS and ODEXPERT [12] have been constructed. SAIVS (Selection Advisor for Initial Value Software) is a prototype automated consultant for evaluating software for solving initial value ODEs. ODEXPERT is a prototype knowledge-based expert system for initial value ODE solver selection which has the capability to investigate certain properties of the given problem (like linearity or stiffness).

2.4 Solving Stiff ODE Systems is Time Expensive

Stiffness is very difficult to characterize explicitly. In physics and chemistry, a real system which is modeled by an ODE system with components supposing some evolution in different dials of time is treated as a stiff system. The components of practical interest are those with a slow variation. A large variation of a solution component relative to the value $1/T$ determines the stiffness character.

The essence of the stiffness phenomenon consists of the fact that the exact solution includes components with some transitory components that are hard to follow by the numerical solution, which arises in a step by step iterative process.

The numerical methods based on polynomial interpolation are successful in the integration of a stiff system only if they are implicit, and, consequently, assume certain procedures for solving nonlinear difference equations. In order to numerically estimate the solution of implicit nonlinear equations associated to stiff problems, the method of simple iterations is out of question, since the convergence condition cannot be satisfied for reasonable step size values (relative to the integration interval). A Newton-like iterative process must be used instead, meaning that the Jacobian matrix must be computed at some iterative steps, which is expensive in terms of computational effort.

More information about stiff problems and numerical solutions for them can be found in [7] or [15].

2.5 Solving Large ODE Systems on a Common Computer

Very large systems of ODEs arise in solution methods for time dependent partial differential equations (PDEs). If the method of lines is used for the transformation between PDEs and ODEs, the number of the ODEs depends on the number of spatial grid points considered for the approximation, by finite differences, of the partial derivatives. In particular, the method of lines applied to parabolic PDEs usually results in a stiff system, provided that the spatial grid is moderately fine-grained. If the problem is stiff and nonlinear, explicit methods are not suited and implicit methods must be used instead. Because of this, such a problem is often insolvable on a common computer.

Unfortunately, a robust software for parallel computers or distributed systems is not yet available for solving very large ODE systems.

2.6 Solving ODEs Using Different Numerical Methods

Some additional remarks must be made concerning the available software packages for ODEs.

We must emphasise the lack of environments which would allow the researcher to make convenient and rapid comparisons of alternative methods and strategies. Different codes or procedures are normally used for different solving methods and this approach can sometimes influence the test results (concerning the execution times for solving the same problem).

The results about the properties of the problems or solution methods must be checked in practice. By changing, for example, the implicit equation solver associated with an implicit method, the method stability properties can be modified. Finally, the convergence of the approximate solution will be put in question. Adopting a strategy for studying and controlling such changes from theory to practice, the IVP solution procedure will be simplified.

An efficient way must be found to systematically uncover methods appropriate to particular problems, to create and assess new methods, and to supply some missing information.

3 Survey of Parallel Solvers for ODEs

We discuss in this section the options for exploiting parallel computers and we outline the difficulties involved. We must emphasize the lack of environments which would allow the researcher to make convenient and rapid comparisons of alternative parallel integration schemes. Several parallel codes for ODEs (like PSODE [18]) have been developed to demonstrate the possible improvements relative to most known sequential codes (like LSODE [8]), without claiming to be available to any user, to be computer platform independent, or to be good solvers for a wide range of ODEs.

3.1 Motivation

A natural approach for answering the need for faster solvers consists in the use of parallel computers. The main problem concerns the effective use of the huge computational power provided by a parallel computer. In order to be efficient, the software designed for such a machine should be based on algorithms constructed for the multi-processor architecture.

A parallel solution method can be used with success especially if the ODE system is very large, if the solutions are needed in real time, if there is a large period of integration, if repeated integration is required, or if function evaluations are expensive (such problems are not solvable, at this moment, in a reasonable time on a serial machine).

3.2 Designing Parallel Algorithms and Codes for Solving ODEs

Great chances lie in the development of essentially new methods whose analytic and numerical structure take into account both the deep nature of the particular underlying problem and the characteristics of the target parallel architecture.

The implementation questions that need to be addressed for developing a parallel code include the avoidance of deadlocks and non-determinacies, balancing of the load of work among the processors, the choice of communication models (synchronous or asynchronous), and the avoidance of communication bottlenecks.

Comparing the algorithms by measuring their performances on parameters like the speed-up can be a very difficult task, because a comparison has to be made theoretically between the execution time of the parallel algorithms on a parallel computer, and the execution time of the fastest existent sequential algorithm running on a same kind of processor. Unfortunately, in the case of the IVP for ODEs we do not know which is the fastest sequential algorithm, since algorithm performances are problem dependent.

3.3 Classification of Parallel Methods for ODEs

Parallelism in the field of ODEs requires the knowledge of a special structure of the IVP, or a great number of redundant calculations.

The ways of achieving parallelism in solving ODEs can be classified by the following categories [18]: parallelism across the system (across space), parallelism across the method, and parallelism across the steps (across time). More details about this subject can be found in [17].

Parallelism Across the System. In the equation segmentation method [13], the various components of the system of ODEs are distributed amongst available processors. This technique is especially effective in explicit solution methods. Massive parallelism can be obtained for very large systems of ODEs. If the system is irregular (the differential equations are not similar), load balancing between processors can pose a significant problem. The parallelism across the system is problem and user dependent.

Parallelism Across the Method. Each processor executes a different part of a method. This approach has the double advantage of being application independent (it does not require user intervention or special properties of the given ODE), and of avoiding load balancing problems due to the IVP. Moreover, a small number of processors is requested which can be an advantage when we discuss a modest parallel or distributed computational system. Unfortunately, many existing algorithms that perform well on a sequential computer hardly profit from a parallel configuration. The main disadvantage is the limited speed-up.

Explicit methods for ODEs do not stand much chance for massive parallelism unless they can be used for large systems of equations. In some cases, two or more stages can be performed in parallel (some examples in [9]).

Implicit one-step methods. The implicit Runge-Kutta methods have the drawback of requiring at each step the solution to an *sd*-dimensional system of equations, where s and d are the number of stages respectively the dimension of the IVP. By contrast, linear multistep methods require the solution of a d-dimensional system of equations. The introduction of parallel computers has changed the scene. Solving with Runge-Kutta methods in a suitable parallel iteration process leads to integration methods that are more efficient and much robust than the best sequential methods based on multistep methods.

The inherent potential for parallelism of implicit Runge-Kutta methods can be investigated by stressing the important role of a sparse stage system [10]. The main idea is to construct a new Runge-Kutta method that consists essentially of a number of disjoint Runge-Kutta methods implemented on different processors, which together combine to produce the approximation. In particular, parallelism in diagonally implicit Runge-Kutta methods comes about because the stages within a block are shared amongst the processors.

One possible way to construct Runge-Kutta methods which are suitable for implementation in a parallel environment is by requiring that the method is of multi-implicit type, i.e. the eigenvalues of the associated matrix are all real and distinct. Then the system in the s variables of the intermediate stages is decoupled into s systems of equations with the dimension equal to the one of the ODE system, using a simple transformation (the s systems can be solved independently on different processors). A more complicated transformation can also be applied to an arbitrary fully implicit Runge-Kutta method, at the level of the stage system, in order to decouple some independent subsystems [11].

Extrapolation methods possess a high degree of parallelism and offer an extremely simple technique for generating high-order methods [7]. However, a detailed analysis shows that a sophisticated load balancing scheme (like that proposed in [15]) is required in a parallel system to achieve a good speed-up.

Block methods are easily adapted to a parallel mode with little apparent alteration of the solution [13]. The obvious drawback of these schemes is that their parallel performance is inherently limited (the performance is dependent on the number of nodes in the scheme).

Predictor-corrector methods. The main idea is to compute the predictor and correctors simultaneously, but for different solution times [9]. When using multistep methods as basic formulae for a predictor-corrector scheme, the following drawbacks of the parallel versions can be identified: data communications are required at each step (reducing the method performance), problems with a smaller region of method stability, dependence of the predictor equations on the predicted values, multi-evaluations of the system function at the same time point. Of particular interest is the case where the corrector method is of Runge-Kutta type (PIRK methods [18]), when one uses an approach similar to a diagonally implicit Runge-Kutta method, but with better convergence properties.

Parallelism Across the Steps. It seems to be the only possibility for using large-scale parallelism on small problems. Contrary to the step-by-step idea, several steps are performed simultaneously, yielding numerical approximations in many points of the independent variable axis (the time). Some continuous time iteration methods, waveform relaxations, must be used to decouple the ODE system, and henceforth to discretize the resulting subsystems, by solving them concurrently. The number of discrete points handled simultaneously is the degree of parallelism of the method. The main weakness of waveform relaxation is that the iteration process may suffer from slow convergence or even divergence.

While waveform methods offer many potential benefits for parallelism, extracting those benefits complicates the code considerably. Several parallel codes for solving large ODE systems, resulting from parabolic equations, have been implemented with success; moreover, in [19] it is shown that the proposed multigrid waveform relaxation method outperforms the standard sequential solvers for some parabolic PDEs.

4 Prototype of an ODE Solving Environment

We implemented a prototype called EpODE (ExPert system for Ordinary Differential Equations),in order to address some of the user problems mentioned in Section 2.

4.1 Main Goals

EpODE's main goals are the following:

1. to be a tool for solving nonstiff, stiff, or large systems of ODEs;
2. to provide an automatic identification of problem properties and method properties, in order to choose automatically an adequate solution method for a problem;
3. to be a tool for describing, analyzing, and testing new types of iterative methods for ODEs, including those proposed for parallel or distributed implementation (using real or simulated parallel machines).

4.2 Defining and Checking the Problem

The user problem can be easily specified. The requested information consists of the ordinary differential equations, described in a natural (mathematical) form, the initial values, and the integration interval.

The information provided to the user consists of the properties of the problem, like linearity or non-linearity, the existence of independent subsystems (information used in parallelism across the system), the classification as nonstiff or stiff problem, or the impact on the computation time of a system function evaluation. At least one hundred (real or test) classical problems can be loaded, solved, modified and re-saved (including stiff problems and ODEs obtained from PDEs).

4.3 Defining and Asking for a Numerical Method

In order to provide good solvers, the most widely used numerical methods for stiff and nonstiff problems were incorporated as basic tools. The current set of methods (about one hundred) includes Runge-Kutta methods (explicit and implicit ones), multistep methods (like Adams-Bashforth-Moulton or Gear's backward differentiation formulae, different predictor-corrector schemes), multi-derivative multistep methods, block methods, hybrid methods, nonlinear multistep methods, some general linear methods, and special methods designed for stiff methods (like those presented in [15]).

The user interface is designed to allow the description of an arbitrary iterative method for solving ODEs. Exceptions to this rule are the extrapolation methods (which must be rewritten as Runge-Kutta schemes in order to be recognized as iterative methods for solving ODEs), the integration formulae with free parameters others than the stepsize [15], the linear multistep formulae with matrix coefficients, or the W-methods [7]. The requested information consists of the equations (described in mathematical form), the starting procedure in the case of a multistep method, and the implicit equation solver in the case of an implicit method (an item from a fixed list). Methods can be added or removed according to the user's current fields of interest.

The report about the method properties specifies: the method classification as explicit or implicit, one- or multi-step, one- or multi-derivative, one- or multi-stage, the method order and the error constant, the method stability characteristics for the most common case of linear problems with the characteristic matrix eigenvalues on the real negative half-axis, the impact on the overall computation time of one method step.

4.4 Computation Results

The properties of the problem and the method are used in the computation process in order to establish a starting stepsize, the estimated number of steps, or the estimated computation time. Variable stepsize schemes can also be applied.

The method and the computation parameters can be given by human expert, or can be the task of the automatic selector. Therefore, the tool can be used to underline the effects of overpassing the stepsize restrictions imposed by accuracy or stability. The quantity of explanation messages can be controlled by the user. The automatic method selection is based on a simple decision tree, and depends on the type of problem which will be solved, the admitted global error, and a maximum for the computation time. In the actual version of EpODE, only stiff and nonstiff problems are treated differently.

The approximate values of the solution can be visualized using some graphic facilities in two- and three-dimensional space or using some tabular form. The numerical results can also be saved in order to be interpreted within other tools.

Classical performance measurements, like computation time, number of function evaluations, or the estimated error, are provided after each solving process. These measurements can be used for comparing distinct methods applied to the same problem.

EpODE allows a researcher to compare efficiently the performance of ten or twenty methods for a given problem.

4.5 Prototype Particularities

The main characteristics of EpODE, which distinguish it from other ODE solving environments, are the following:

(a) interpreter mode for describing problems and solution methods;
(b) the solvers are implemented in a uniform way: all solvers behave in a coherent way and have the same calling sequence;
(c) the tool is independent from other software packages, with the exception of PVM[1] [6] used for parallel or distributed computations;
(d) one has portability not only for sequential codes, but also for parallel ones (except for the graphical interface, currently designed for X-Windows and Windows'95), and the codes can be easily recompiled for any computer architecture (the first version of EpODE was written in C++);
(e) friendly user interface for describing not only new problems, but also new iterative solving methods;
(f) broad functionality required by modern modelling environments and simulation tools;
(g) a dynamic allocation strategy is used in order to optimize the storage requirements of any number of differential equations, or of any number of method equations;
(h) special codes are constructed to support parallelism across system, method and steps, for the case of a distributed memory parallel architecture (in the message-passing model of intercommunication between processing elements – couples processor-memory).

4.6 Implementation Details

EpODE has five major components:

1. a user interface with help facilities, the front end of which permits the description of an IVP for ODEs or an iterative method for ODEs, the control of the solution computation process, and the interpretation of the

[1] PVM (Parallel Virtual Machine) is a software system that enables a collection of heterogeneous computers to be used as a coherent and flexible concurrent computational resource. The individual computers may be shared- or local-memory multiprocessors, vector supercomputers, specialized graphics engines, or scalar workstations, that may be interconnected by a variety of networks.

results of the computation; help facilities are provided in order to assist the user in using the software;
2. a property detection mechanism containing the procedures for establishing some properties of an IVP for ODEs or those of an iterative method;
3. a mechanism for selecting the solution procedure, implementing the decision tree for the selection of the class of iterative methods according to the properties of the IVP for ODEs, and for the selection of one method from this class according to the solution accuracy requirements and time restrictions;
4. a sequential computing procedure: a generic solution procedure whose parameters are selected according to the current problem and the serial method;
5. a parallel computing procedure, a generic solution procedure with message-passing facilities for intercommunication of more than one computation process.

The Solution Procedure. When constructing a unique solution procedure for any type of iterative method for ODEs we took into account the variety of mathematical forms that a solving method can have (one- or multi-stage, one- or multi-step, one- or multi-derivative, explicit or implicit methods). A unique solution procedure was also used in the Godess project [14], but this is designed only for Runge-Kutta methods.

EpODE allows to define customized difference methods for ODEs (which can be included in the class of A-methods) of the form:

$$Y_{n+1} = AY_n + h_n\Phi(t_n, Y_n, Y_{n+1}, h), \quad n \geq 0, \quad Y_0 = \Psi(h).$$

We distinguish: the iterative formula, the starting procedure, the implicit equation solver, the error control procedure. Theoretical studies have been concentrated especially on the iterative formula. Numerical tests done with EpODE confirm that the other components have also a great influence on the properties of the method.

For each method part a special interpreter for the user commands was build. For example, in order to define a new iterative formula, the user must specify the variables, the right side of the formula, the variables whose values will be stored, and the link between the old and the new values of the variables.

Uniformity in defining the difference methods in EpODE allows to construct a unique procedure for interpreting the data about an arbitrary method. The parameters of such a procedure are the outputs of the interpreters for the iterative formula, the starting procedure, the implicit solver, the error control procedure. The outputs are provided in a condensed form, like evaluation trees of some arithmetic expression.

A unique solution procedure with different parameters, depending on the current solution method, guarantees a better comparison (relative to a given

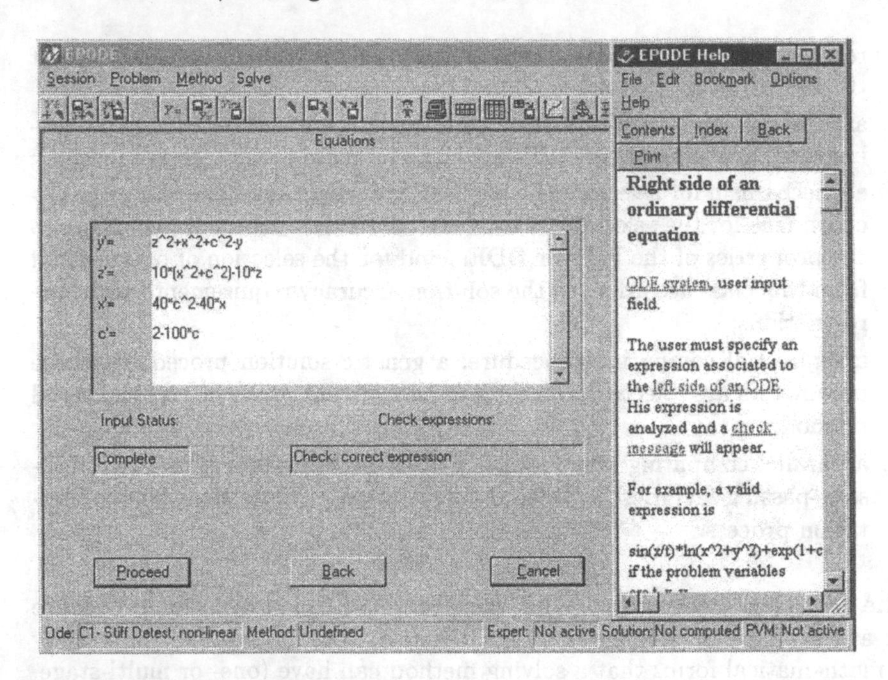

Fig. 1. Describing an ODE system from the STIFF DETEST package [4]: $y' = z^2 + x^2 + c^2 - y$, $z' = 10(x^2 + c^2) - 10z$, $x' = 40c^2 - 40x$, $c' = 2 - 100c$, $y(0) = z(0) = x(0) = c(0) = 1$, $T = 20$.

problem) between two different methods, than one between the results of distinct implementations of the two solution methods.

Detecting the Properties of the Method. The most important properties of the method are automatically identified, so that the ones which were theoretically founded can be checked, and, moreover, the influence of different selection criteria (like the solution procedure for implicit equations) can be outlined (for example, the fact that the implicit Euler method coupled with a fixed iteration procedure loses the properties of A-stability). The class of detected properties is not restricted only to those which are deciding the convergence of the approximate solution to the exact one (method order, for example), or those which decide the applicability to a special class of problems (like the stiff ones – depending on the bounds of the method stability region), but also include those which decide the applicability of the method on a parallel or distributed computing machine (depending on the method's data-flow graph, if some computations can be done simultaneously).

We take here, for example, the delicate problem of finding the order of the method. Formulae were provided for each type of the classical iterative formula, but the user does not always have the possibility of knowing it in

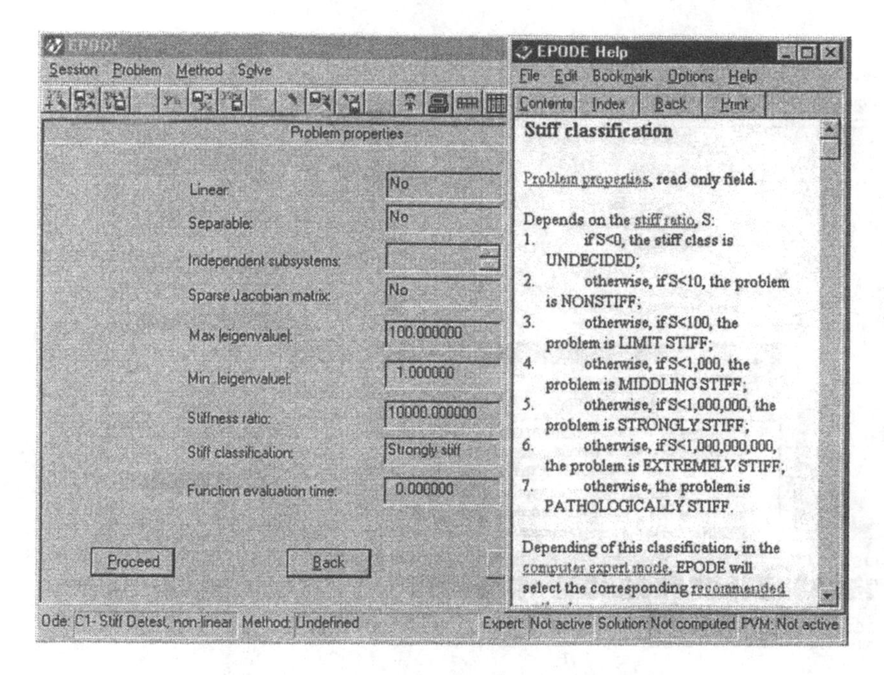

Fig. 2. Properties detected for the above IVP: stiffness and nonlinearity.

a particular case. For detecting the optimal stepsize for the computational process, at least an estimation of the order of the method must be provided.

In EpODE the task of detecting the order of the method is done automatically, depending on the input data received from the user about the iterative formula, and the starting and error control procedures. To do that, the method is applied many times to a test equation for a finite number of steps. The differences between the exact solutions and the approximate solutions are used in order to find the coefficients c_i from the following equality:

$$y(t_n) - y_n = c_0 h f(t_n, y(t_n)) + c_1 \frac{h^2}{2} f'(t_n, y(t_n)) + \cdots$$

Note that the iterative formula order does not decide alone the order of the solution scheme, as the order of the method can be changed drastically by the starting procedure, if this is not accurate. The implicit equation solver can also influence the order of the method in the case where it is designed to be applied with a fixed number of steps.

Parallel and Distributed Codes. Our tool combines both sequential codes and parallel/distributed codes. The two kinds of codes receive the same input data and present the output data in a similar fashion. Execution times of the sequential, concurrent, distributed and parallel implementation of the

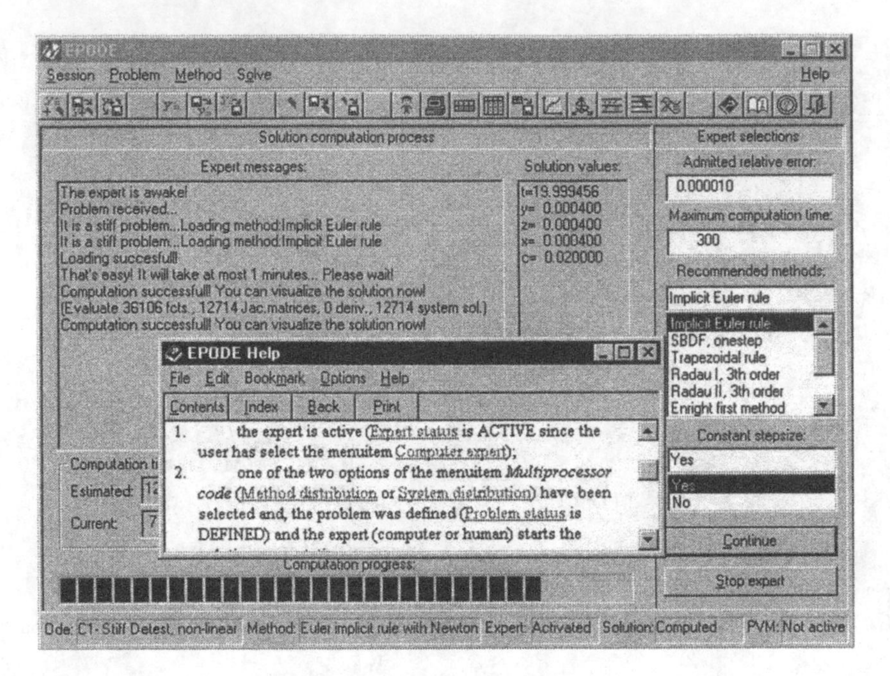

Fig. 3. Expert in action: the implicit Euler rule can offer an approximate solution, at the requested accuracy, for the IVP from Figure 1.

same algorithm can be measured and correctly compared since the codes are running under similar conditions. The user has the freedom of choosing between the three types of parallelism.

Special codes were designed in EpODE, using the facilities of PVM [6] in order to allow concurrent computations on some local workstation network or on a parallel machine with distributed memory (in the message passing model for process inter-communication). Therefore, EpODE can be seen as an environment for testing parallel and distributed solution methods.

The basic members of the classes of parallel solvers can be found in this tool. Parallelism across the system can be exploited by (automatic) assignment of each independent block of differential equations to a different processor (equation segmentation method). Parallelism across the method can be exploited by the (automatic) distribution of the computational effort from each single integration step among a number of processors; to do that a data-flow graph and a partition of this graph on distinct processes and stages are established. Parallelism across the steps is represented by a waveform relaxation method of Jacobi type.

More details about parallel computations with EpODE can be found in [16]. Test results and comparisons with other similar tools are provided in [17].

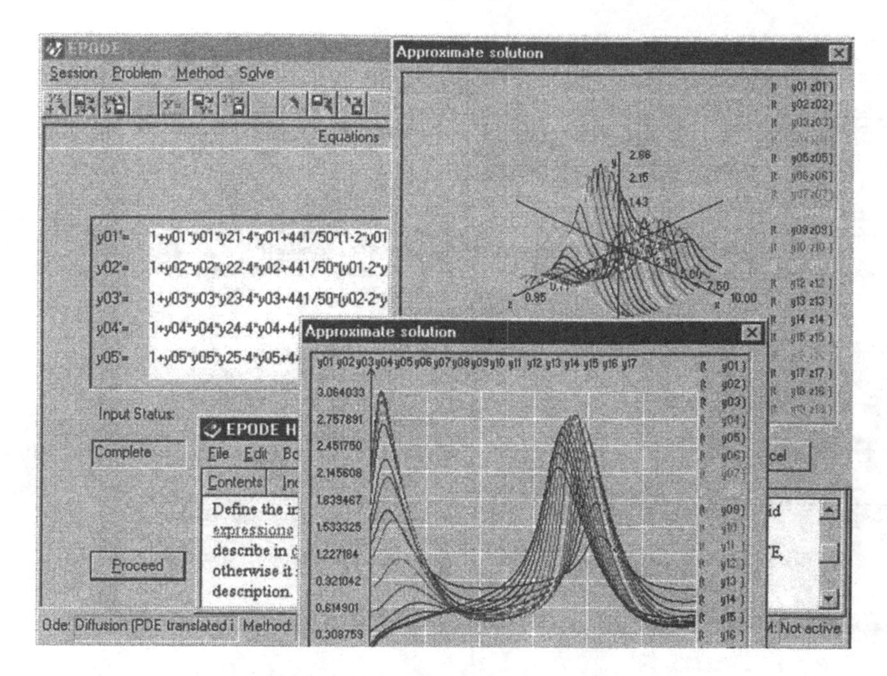

Fig. 4. Graphic representations of an approximate solution of a large system of ODEs obtained from the Brusselator diffusion equation. [7]

5 Lesson Learned

5.1 Motivation

A system of ODEs can easily fall outside the boundaries within which symbolic or numerical computer systems can be expected to produce reliable results. In large-scale problems, where the ODE may be a system of hundreds of stiff differential equations, any general-purpose tool will be overwhelmed. Even for those packages that can accommodate such a system, determining the suitability of the methods employed remains a task for mathematicians with numerical expertise. Comparison with a range of alternative schemes is unlikely to be an option.

5.2 Proposed ODE Solver

EpODE offers the benefit of being able to run a large set of alternative schemes for each particular problem, and to have tabulated convergence data and error estimates for each method. The appropriate methods and steppers would suggest themselves.

The main idea behind the prototype construction was that all the solvers behave in a consistent way and have the same calling sequence, so the under-

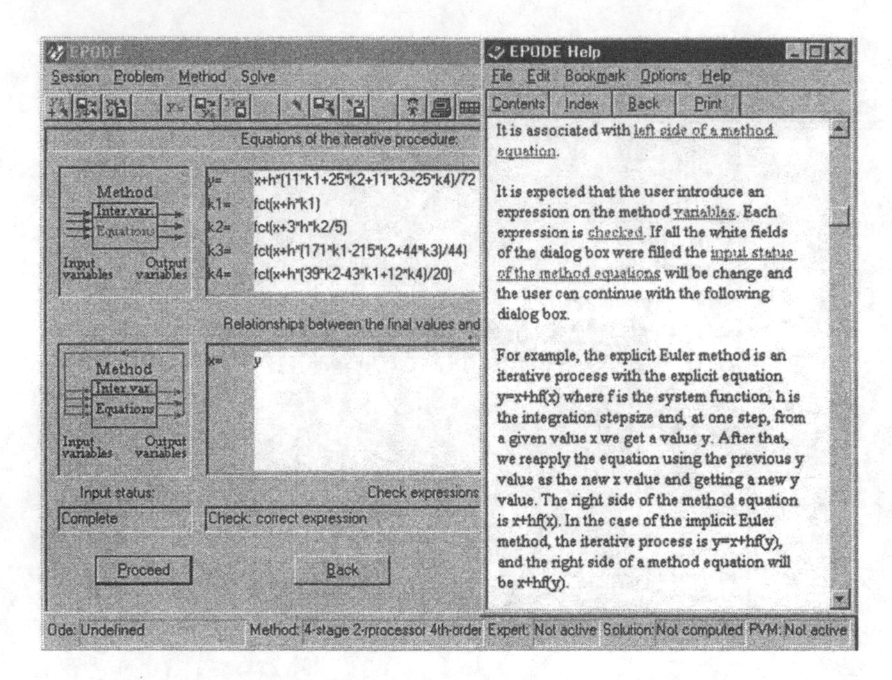

Fig. 5. Describing the Runge-Kutta method $y_{n+1} = y_n + h[11(k_1 + k_3) + 25(k_2 + k_4)]/72$, where $k_1 = f(y_n + hk_1)$, $k_2 = f(y_n + 3hk_2)$, $k_3 = f(y_n + h(171k_1 - 215k_2 + 44k_3)/44)$, $k_4 = f(y_n + h(39k_2 - 43k_1 + 12k_4)/20)$.

lying libraries can easily be fitted to modeling environments and simulation tools.

The practical solution for solving large systems of ODEs which was adopted is the use of parallel and distributed codes.

5.3 Future Developments

In EpODE 1.0 the method and the parameter selection mechanism are based on a simple decision tree. In a future version, this tree will be refined in order to include other categories of IVP than nonstiff or stiff problems.

In order to transform the tool into a veritable expert system, the learning component must be improved, e. g. by "learning from errors". The current knowledge-acquisition mechanism must be improved by retaining the decisions taken in the solution procedure for future usage. To this end, a future version of the expert part will be written in Clips, not in C++, like the actual version.

The idea of independence from other tools must be revised in a particular way. Our system must provide an independent way of solving problems, but it must also have the possibility to use some facilities of other tools which seem to be more performant than the ones proposed by the current tool. In

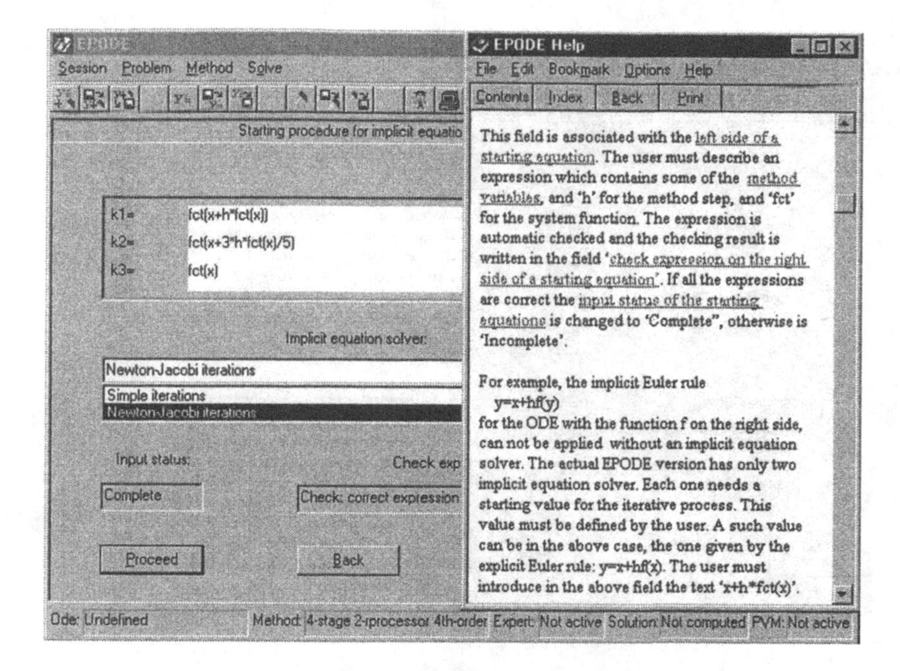

Fig. 6. Implicit solver for the Runge-Kutta method described in Figure 5.

order to fulfil such a request, the facility of a solution procedure with modules from distinct applications will be designed and implemented. Technically this can be done using OpenMath [5] which is a standard for communicating mathematical objects between computer programs.

From the computational point of view better schemes for estimating the global error will be implemented.

The condition of a unique solution procedure has imposed some restrictions on the form of the iterative method which the user can describe. For example, the equations of an extrapolation method must be rewritten like a Runge-Kutta method before using EpODE. A family of integration formulae with some free parameters, others than the stepsize (used, for example, for exponential fitting) cannot be described in the first version of EpODE. Therefore, the interpreter will be improved to give more freedom not only in the method description, but also for the problem description.

An automatic translator from time dependent PDEs to ODEs must also be added. The stability tests must be enlarged (from the stability on the real negative to for the full region of absolute stability, for example). Stiff separability must be also exploited, i.e. it must be allowed to apply different stepsizes in different components of the ODE system. Facilities for using schemes with variable formulae must be provided.

The computer platform independence is actually achieved only for the computing procedures. The user interface will be rewritten in Java in order

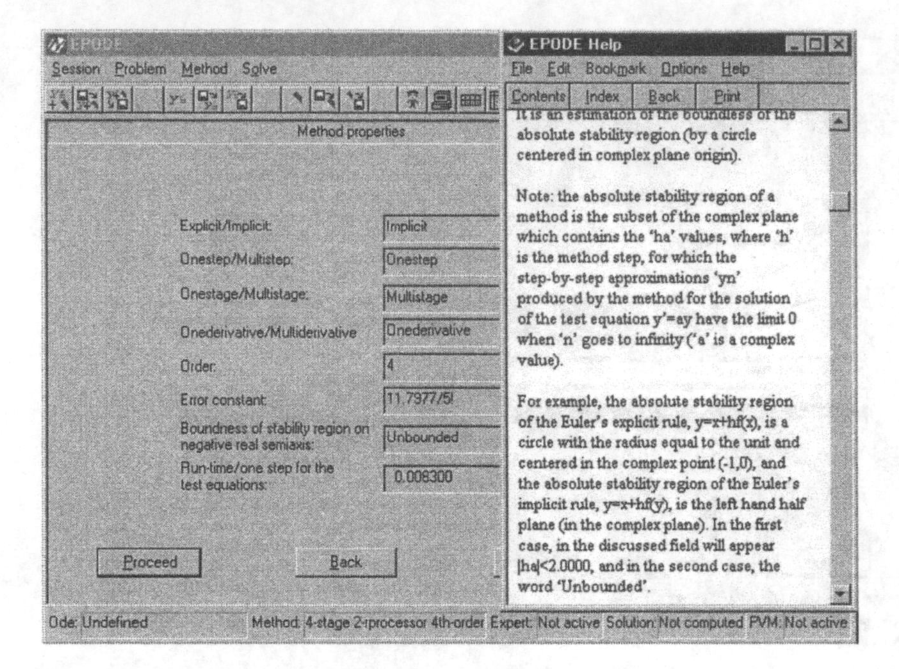

Fig. 7. Properties detected for the solution scheme based on the Runge-Kutta method from Figure 5: an implicit method of 4th order and at least A_0-stable.

to achieve independency also for the graphical interface, and to offer free access to EpODE for any Web user.

Improving the parallel/distributed codes is also on our agenda, especially by implementing schemes for decoupling the ODE system, parallel iterative solvers for implicit equations, or general waveform relaxation methods.

Acknowledgments

We express our sincere gratitude to the reviewers of this chapter who have made invaluable suggestions for improvement. We wish also to thank Dr. Tudor Jebelean and Dr. Razvan Gelca for reading it carefully and making innumerable corrections and remarks.

A EpODE Facilities by Examples

We present in Figures 1-8 some user interface facilities. Supplementary material (theoretical aspects, examples, figures, and test results) about the subject discussed in this chapter can be found at
http://www.info.uvt.ro/~petcu/epode. A free version of EpODE is also available at this address, for Windows'95, Linux, and Sun Solaris.

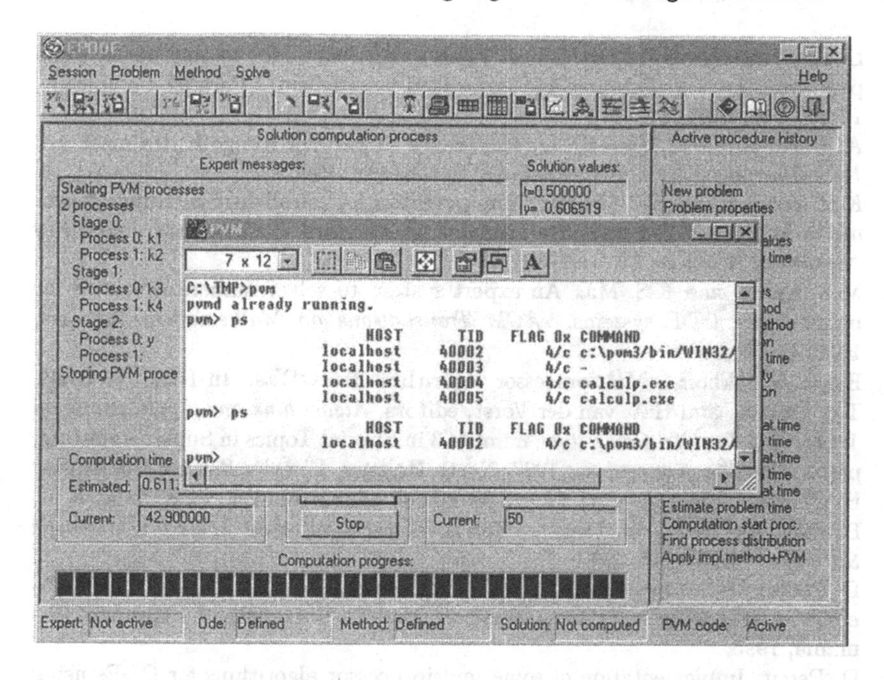

Fig. 8. Applying parallelism across the Runge-Kutta method from Figure 5: k_1 and k_2, respectively k_3 and k_4 can be computed in two distinct processes.

References

1. R.C. Aiken. Stiff packages. In R.C. Aiken, editor, *Stiff Computations*, pages 155–170, Oxford, 1985. Oxford University Press.
2. R.F. Boisvert, S.E. Howe, and D.J. Kahaner. GAMS: A framework for the management of scientific software. *ACM Transactions on Mathematical Software*, 11(4):313–356, 1985.
3. K. Briggs. ODE – a program for interactive solution of systems of ordinary differential equations. http://archives.math.utk.edu/topics/ordinaryDiffEq.html.
4. W.H. Enright, T.E. Hull, and B. Lindberg. Comparing numerical methods for stiff systems of ODEs. *BIT*, 15:10–48, 1975.
5. M. Gaetano, C. Huchet, and W. Neun. The realization of an OpenMath server for Reduce. In V. Gerdt and E. Mayr, editors, *Computer Algebra in Scientific Computing*, pages 48–55, St. Petersburg, 1998. Euler International Mathematical Institute.
6. A. Geist, A. Beguelin, J. Dongarra, W. Jiang, R. Manchek, and V. Sunderam. *PVM: Parallel Virtual Machine. A users' guide and tutorial for networked parallel computing.* MIT Press, Cambridge, 1994.
7. E. Hairer, S.P. Nørsett, and G. Wanner. *Solving Ordinary Differential Equations II. Stiff and Differential-Algebraic Problems*, volume 14 of *Springer Series in Computational Mathematics*. Springer-Verlag, Berlin, Heidelberg, New York, 1991.
8. A.C. Hindmarsh. Brief description of ODEPACK - a systemized collection of ODE solvers. http://netlib.org/odepack.

9. D. Hutchinson and B.M.S. Khalaf. Parallel algorithms for solving initial value problems: front broadening and embedded parallelism. *Parallel Computing*, 17:957–968, 1991.
10. A. Iserles and S.P. Nørsett. On the theory of parallel Runge-Kutta methods. *IMA Journal of Numerical Analysis*, 10:463–488, 1990.
11. K.R. Jackson and S.P. Nørsett. The potential for parallelism in Runge-Kutta methods. Part I: Runge-Kutta formulas in standard form. *SIAM Journal of Numerical Analysis*, 32(1):49–82, 1995.
12. M.S. Kamel and K.S. Ma. An expert system to select numerical solvers for initial value ODE systems. *ACM Transactions on Mathematical Software*, 19(1):44–61, 1993.
13. E.J.H. Kerckhoffs. Multiprocessor algorithms for ODEs. In H.J.J. te Riele, T.J. Dekker, and H.A. van der Vorst, editors, *Algorithms and Applications on Vector and Parallel Computers*, number 3 in Special Topics in Supercomputing, pages 325–346, Amsterdam, 1987. North-Holland, Elsevier Science Publishers.
14. H. Olsson. Object-oriented solvers for initial value problems. In E. Arge, A.M. Bruaset, and H.P. Langtangen, editors, *Modern Software Tools for Scientific Software*. Birkhäuser, 1997.
15. D. Petcu. *Multistep methods for stiff initial value problems*, volume 50 of *Mathematical Monographs*. Printing House of Western University of Timişoara, Romania, 1995.
16. D. Petcu. Implementation of some multiprocessor algorithms for ODEs using PVM. In M. Bubak, J. Dongarra, and J. Wasniewski, editors, *Recent Advances in Parallel Virtual Machine and Message Passing Interface*, number 1332 in Lectures Notes in Computer Science, pages 375–383, Berlin, Heidelberg, New York, 1997. Springer-Verlag.
17. D. Petcu. *Parallelism in solving ordinary differential equations*, volume 64 of *Mathematical Monographs*. Printing House of Western University of Timişoara, Romania, 1998.
18. B.P. Sommeijer. Parallel iterated Runge-Kutta methods for stiff ordinary differential equations. *Journal of Computational and Applied Mathematics*, 45:151–168, 1993.
19. S. Vanderwalle and R. Piessens. Numerical experiments with nonlinear multigrid waveform relaxation on a parallel processor. *Applied Numerical Mathematics*, 8:149–161, 1991.

Generalized Maps in Geological Modeling: Object-Oriented Design of Topological Kernels

Yvon Halbwachs and Øyvind Hjelle

SINTEF Applied Mathematics, Oslo, Norway
Email: {Yvon.Halbwachs, Oyvind.Hjelle}@math.sintef.no

Abstract. A boundary based topological representation technique known as generalized maps is presented and applied to modeling the topology of complex geological structures. Extensions of this concept for capturing geologists' needs for specifying topological relationship information between topological cells are suggested. Object-oriented design of topological kernels based on generalized maps is given and issues of implementation in C++ are discussed. Examples of applications are presented for modeling the topology of surface triangulations and geological fault networks.

1 Introduction

The geometric and topological characterization of a geological model is crucial in various application domains such as geophysics and reservoir modeling. Building boundary topology representations of geological models is extremely difficult since the spatial geometry of the surface boundary elements delimiting volumes can generate very complex structures.

In geometric modeling, the boundary representation and manipulation of objects is based on elementary operations such as boolean operations. Although these tools are adequate for the CAD industry, they need to be extended in the realm of geological modeling because they were not designed with the same issues in mind. For instance, during the process of modeling geological structures various intermediate types of non-manifold geometry, such as hanging edges and surface patches, may occur which cannot be efficiently handled by many traditional boundary representation techniques used in CAD systems.

Volumes in geological structures are delimited by geological faults and horizons. Faults represent cracks or fractures in the earth's crust in which there has been movement of one or both sides relative to the other. Faults are important regarding oil occurrences, so correct representation of regions with faults is vital for geologists. The topology and geometry of geological fault systems may take very complex forms and has been a challenge to software engineers developing geological modeling systems.

We present a robust and flexible boundary topology representation model based on so-called *generalized maps*, or G-maps, which have extensive capabilities for modeling geological objects with complex topology. First, the theory of G-maps is presented and compared to other existing boundary representation techniques. Then, object-oriented design of topological kernels and implementation in the C++ programming language are proposed. We proceed with an example of modeling the topology of a surface triangulation, and finally, the construction of geological fault networks based on a topological kernel is discussed.

2 Topological Modeling

Boundary based topological models (B-reps) have become the standard way to represent the topology of 3D geometric objects on computers, e.g. in CAD systems. The principle is to represent objects by their boundaries, for example, a 3D cube by its six faces where the faces are tied together with some relationship information. Usually an object is subdivided further such that both volumes, faces, edges and vertices are represented explicitly together with the positional relationships between these entities. The different B-rep based topological representations used by today's geometric modeling systems differ in their level of subdivision, the relationships established between the topological elements, and how they distinguish between the topological model and the embedding model, e.g. geometry.

A topological framework introduced by Weiler [17,18] has been widely used in many modeling systems, for example in the 3D geological modeling system GOCAD [7]. Weiler improved and extended the *winged-edge* structure [2], where the topology was defined using the *edge* as the basic element, and introduced the notion of *half-edge* as the basic element of the topology. Moreover, the *radial edge* structure was introduced for non-manifold boundary based geometric modeling representation [18].

Recently, more robust and rigorously defined approaches to boundary based topological representation of complex objects have been developed based on *combinatorial maps* [16]. The n-dimensional hypermap, or *n-h-map* [5], is a formal algebraic approach to topological modeling. The n-dimensional generalized map, or n-G-map [10], that has proved to be well adapted to modeling complex geological structures [8,15], is a special case of hypermaps and derived by introducing constraints. Many other models can also be derived from hypermaps [3].

These general combinatorial models have many advantages over traditional models. They are easy to implement since they are based on a small number of concepts only. They have extensive capabilities for non-manifold topology representation, they are independent of a special data structure, and it is easy to distinguish between topology and embedding. Since the distinction between topology and embedding is clear, a generic topological kernel independent of a special data structure can be implemented.

2.1 Generalized Maps

In this section we give the algebraic definition of an n-G-map which will serve as the topological engine of the geological modeler specified in Section 3. We also briefly introduce some theory of n-G-maps.

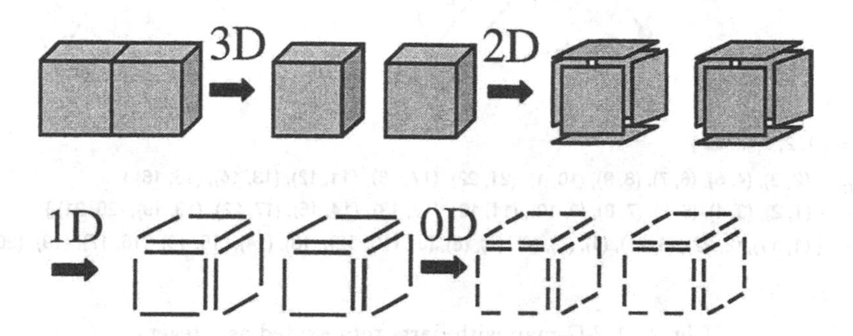

Fig. 1. Subdivision of a model into volumes (3-cells), faces (2-cells), edges (1-cells) and vertices (0-cells).

Consider the subdivision of a simple 3D model shown in Figure 1. The model is first subdivided into two volumes, then each volume is subdivided into faces by pulling the faces apart, and each face is subdivided into edges. Finally each edge is subdivided into two half-segments, each of which we call a *dart*. The dart is thus the lowest level of the topological representation and can be thought of as a half-segment of a half edge. We observe that each dart is incident to exactly one vertex, edge, face and volume which we call *0-cells*, *1-cells*, *2-cells* and *3-cells* respectively.

With this simple exercise we have already entered the very core of G-maps: A G-map, like other so-called *cellular* topological B-rep models [16,12], is a graph that consists of a set of such unique basic dart elements for representing the topological model, and a set of functions operating on the darts for exploring the topology.

Before giving a precise algebraic definition of an n-G-map, we give some basic notions.

Let d be a single member of a set D. A *permutation* $\alpha(d)$ is a function that maps d onto an element in D, $\alpha : D \rightarrow D$. If the *composition* $\alpha \circ \alpha(d) = \alpha(\alpha(d)) = d$ for all $d \in D$, we call α an *involution*. An element $d \in D$ is called a *fixed point* of a permutation α if $\alpha(d) = d$.

Definition 1 (Generalized map). A generalized map of dimension $n \geq -1$, or an n-G-map, $G = (D, \alpha_0, \alpha_1, \ldots, \alpha_n)$ is a graph that consists of a non-empty finite set D of darts, and $n + 1$ involutions α_i on D with the

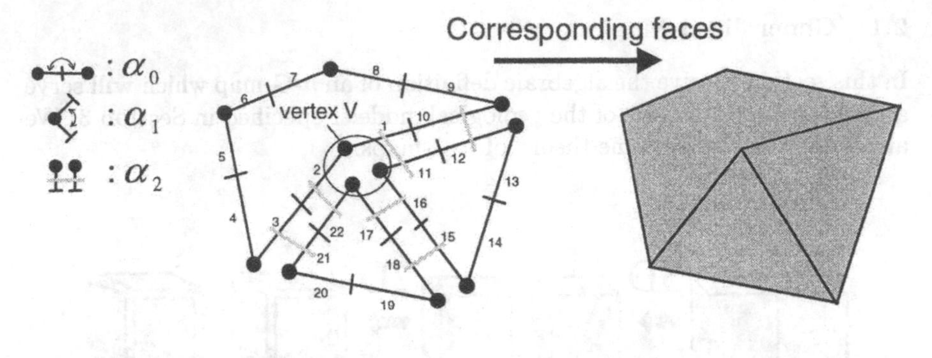

D = { 1, 2, 3, ..., 22 }

α_0 = { (2, 3), (4, 5), (6, 7), (8, 9), (10, 1), (21, 22), (17, 18), (11, 12), (13, 14), (15, 16) }

α_1 = { (1, 2), (3, 4), (5, 6), (7, 8), (9, 10), (11, 16), (12, 13), (14, 15), (17, 22), (18, 19), (20, 21) }

α_2 = { (1, 11), (2, 22), (3, 21), (4), (5), (6), (7), (8), (9), (10, 12), (13), (14), (15, 18), (16, 17), (19), (20) }

Fig. 2. A 2-G-map with darts represented as integers.

condition that $\alpha_i \circ \alpha_j$ is an involution for $j \geq i+2$, and α_k is without fixed points for $k < n$.

If $n = -1$, there are no involutions and thus G consists only of a set of darts D. Lienhardt [11] describes the involutions α_i, $i = 0, \ldots, n$ of an n-G-map as links between darts and defines the involutions constructively by recursion on the dimension n starting from $n = 0$. With the above definition of an involution in mind and referring to Figure 2 which shows a 2-G-map, $G = (D, \alpha_0, \alpha_1, \alpha_2)$, embedded in the plane with darts represented as integers, we summarize the definitions thus:

$\alpha_0(d)$ links dart d to the opposite dart at the same side of an edge; $\alpha_1(d)$ links d to the dart incident to the same vertex and face as d but incident to another edge; $\alpha_2(d)$ links d to the dart at the opposite side of the edge and incident to the same vertex as d. Likewise, for a 3-G-map, $\alpha_3(d)$ links d to a dart in another volume at the opposite side of a face and incident to the same vertex. In general, for $0 \leq k \leq n$, $\alpha_k(d)$ links d to a dart of another k-cell but incident to all other cells as d. Note, however, that all fixed points satisfy $\alpha_n(d) = d$, for example, $\alpha_2(6) = 6$ in Figure 2.

An involution can be represented as a set of couples and singles as indicated in Figure 2, where singles correspond to fixed points which, c.f. the definition above, can only be present for α_n. Thus, a couple (d_i, d_j) in the set of α_k involutions is equivalent to $\alpha_k(d_i) = d_j$ (and $\alpha_k(d_j) = d_i$), and a single (d) is equivalent to $\alpha_k(d) = d$. If $\alpha_n(d)$ is without fixed points for all $d \in D$, then the graph G is without boundaries. For example, a 2-G-map, $G = (D, \alpha_0, \alpha_1, \alpha_2)$, representing the topology of a subdivision of a sphere into connected triangles, has no fixed points for α_2.

The process of establishing the involutions in a G-map, i.e. the couples and singles in Figure 2, is called *sewing.* We say that two darts d_i and d_j are *k-sewed,* or α_k*-sewed,* in a G-map if $\alpha_k(d_i) = d_j$ (and $\alpha_k(d_j) = d_i$).

From an implementation point of view, darts and involutions need not be represented explicitly as they might be inherent in an underlying data structure or mapped from another topology representation [15].

Definition 2 (Orbit and k-orbit). Let Γ be a subset of the involutions $\alpha_0, \dots, \alpha_n$ in an n-G-map and $d \in D$, then an orbit $\langle \Gamma \rangle (d)$ of a dart d is the set of all darts in D that can be 'reached' by successively applying the involutions $\alpha_k \in \Gamma$ in any order starting from d. The k-orbit of a dart d in an n-G-map is defined as the orbit $\langle \{\alpha_i,\ i \neq k,\ 0 \leq i \leq n\} \rangle (d)$.

For 2-G-maps, the 0-orbit, 1-orbit and the 2-orbit is the set of all darts incident to a vertex, edge and face respectively. For example, the 2-orbit $\langle \alpha_0, \alpha_1 \rangle (1)$ of the dart $d = 1$ in Figure 2, is the set of all darts reached by applying the composition $\alpha_0 \circ \alpha_1(d)$ repeatedly starting with $d = 1$, i.e. all darts incident to the vertex V. Similarly, for a 3-G-map the 3-orbit $\langle \alpha_0, \alpha_1, \alpha_2 \rangle (d)$ is the set of all darts incident to a volume. Thus, a k-orbit, $\langle \Gamma \rangle (d)$ of a dart d is always *associated* with a specific k-cell of the G-map. We observe that the orbit $\langle \alpha_0, \alpha_1, \alpha_2 \rangle (d)$, d being any dart of D in Figure 2, is the set of all darts in D, i.e. the 2-G-map is a *connected graph.*

2.2 The Embedding Model

So far, we have focused on the topological model only. The *embedding model* is the information we associate with the topological elements in the model. For a 2-G-map we will typically associate geometric embeddings such as points (0-embeddings), curves (1-embeddings) and surfaces (2-embeddings) to vertices, edges and faces respectively. Thus, a k-embedding is associated with a k-cell in the G-map. Likewise, since a k-orbit $\langle \Gamma \rangle (d)$ of a dart d is associated with a specific k-cell, a k-embedding is also associated with a k-orbit.

2.3 An Extended 3-G-map for Geological Modelling

In the geological modeler we describe later, we use a 3-G-map and extensions derived from it, for the boundary topology representation of geological structures. The 3-G-map is thus embedded in the 3-dimensional Euclidean space where 2-cells represent the topology of geological faults and horizons. The geometric embeddings for 2-cells will typically be represented as surface triangulations and spline surfaces. Extensions of the standard G-map representation defined in Section 2.1 are necessary since the topology of geological structures are more complex than what can be represented by the standard G-map definition. In addition, geologists require topological relationship information that is based on geological knowledge in addition to positional geometric relationship information.

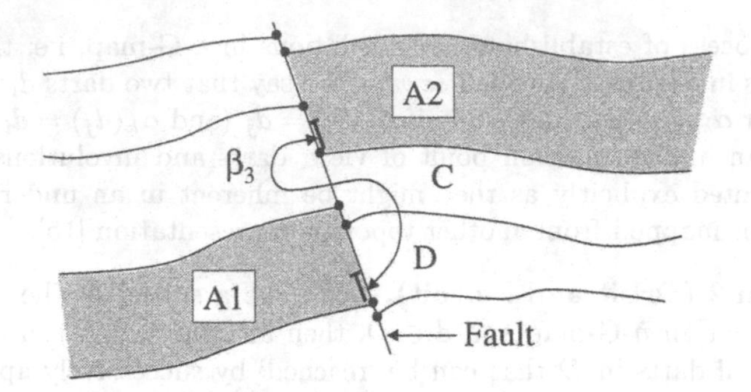

Fig. 3. Cross-section through a 3D geological structure with a geological fault. A β_3 involution is introduced in the topological model for linking two geological layers.

Figure 3 shows a cross-section through a geological structure where geological layers are displaced relative to a fault. A topology representation deduced from the geometry only will typically link the volume $A1$ to the volumes C and D through α_3 involutions as described previously. Geologists may also require relationship information between the geological structures $A1$ and $A2$ as indicated in the figure, based on the knowledge that they have formerly belonged to the same geological layer. This can be obtained by the standard G-map definition by defining the involutions accordingly, i.e. the α_3 involutions link darts in $A1$ incident to the fault to darts in $A2$, but then we lose important geometric positional relationship information between $A1$, C and D. We solve this by introducing new involutions $\beta_3 : D \rightarrow D$ in the 3-G-map that link darts between volumes based on the geological relationship information as shown in the figure.

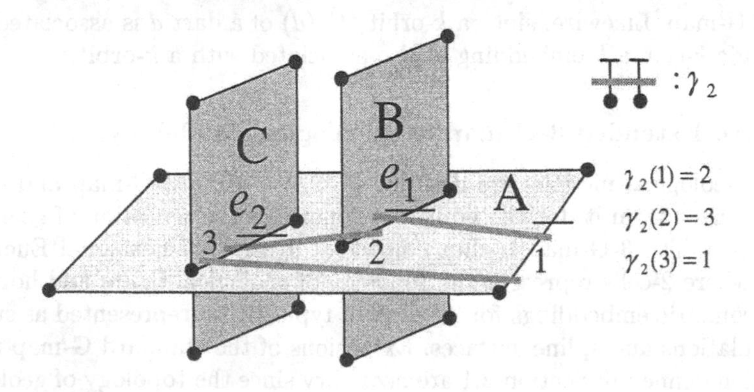

Fig. 4. γ_2 involutions for linking a disconnected 2-G-map graph.

Another example requiring extension of the standard G-map representation is shown in Figure 4. The face A is "intersected" by two faces B and C to form new internal edges e_1 and e_2 in face A. The boundary topology model can be represented by a standard 2-G-map $G = (D, \alpha_0, \alpha_1, \alpha_2)$, but since the vertices of the new internal edges are not on the external boundary of face A, there is no "path" from the boundary of face A to the internal edges through involutions. Thus, G will be a disconnected graph which makes traversal of the topology difficult to implement. We introduce a new function $\gamma_2 : D \to D$ for connecting such disconnected components in the graph as depicted in the figure. Holes in a face can be modeled similarly with γ_2 involutions, and volumes inside volumes in a 3D model can be linked topologically using γ_3 functions. Note that γ_2 and γ_3 are permutations and not involutions since, in general, $\gamma \circ \gamma(d) \neq d$ as exemplified in the figure.

Our extended G-map with the new involutions and permutations used in the geological modeler specified later now reads

$$G' = (D, \alpha_0, \alpha_1, \alpha_2, \alpha_3, \beta_3, \gamma_2, \gamma_3). \tag{1}$$

Note that we have not introduced new darts in this modified G-map definition - only new involutions β and permutations γ operating on the darts.

3 Object-Oriented Implementation of n-G-maps

Our goal is to design a topological kernel for n-dimensional generalized maps for arbitrary dimensions n, that can handle both micro and macro topology and provide functionality to store any kind of user defined embeddings. An object-oriented approach is used to avoid code duplication and to specialize gradually the theoretical concepts into a robust implementation.

3.1 A Simple 2-G-map Kernel

We first consider a simple case of implementing a 2-G-map $G = (D, \alpha_0, \alpha_1, \alpha_2)$ without extensions and with geometric embeddings only, i.e. the k-embeddings for $k = 0, 1, 2$ are points, curves and surface patches respectively. With these restrictions we can model a collection of connected 2-cells (faces) and their boundaries embedded in the plane or in the 3-dimensional Euclidean space such that the embedded 2-G-map graph represents a subdivided 2-manifold (a surface geometry).

Figure 5 shows, somewhat simplified, a class diagram representing a possible implementation of a 2-G-map kernel with embeddings. There are three main classes for representing topology and embedding, and for building the topology structure.

- `2-G-map` contains a list of darts and functions for building the topology, i.e. the sewing operations operating on darts for establishing the involutions as defined in Section 2.1. In addition, these functions also establish links between topology and embedding through the darts of the 2-G-map.
- `Dart`: Implements a dart of a 2-G-map with links to other darts through involutions α_0, α_1 and α_2 (c.f. Figure 2). If a dart represents a fixed point of an involution, it points to the dart itself. In addition the dart class has references to the various embeddings. To make updating of the model easier, we follow this rule: *only one dart of a specific k-orbit has a reference to the corresponding k-embedding.* For example, in Figure 2 the 1-orbit $\langle\alpha_0, \alpha_2\rangle$ (2) which is the set of darts $(2, 3, 21, 22)$, has only one of its darts carrying the information of the associated 1-embedding (which is a geometric curve).
- `Embedding`: Base class for geometric embeddings. The specific implementation for each k-embedding, $k = 0, 1, 2$, i.e. point, curve and surface, is derived from this class. The class has also a reference to one of the darts of the associated k-orbit and a reference to the `2-G-map` object.

In addition to the basic classes described above, two classes provide functionality for traversing the topology structure:

- `G-mapIterator`: Iterate through the k-orbits of the 2-G-map for a fixed k. For example, all edges of the graph can be visited when the class is initialized with a parameter $k = 1$.
- `OrbitIterator`: Iterate through the k-orbits linked to a specific p-orbit given a dart of the p-orbit. For example, all the edges attached to a specific vertex can be explored when $k = 1$ and $p = 0$. In Figure 2, one could explore the edges incident to the vertex V by starting from dart 1 and calling a function `OrbitIterator::next` twice retrieving the darts 16 and 2.

Many other algorithms exploring graphs in algorithmic graph theory [13] can also be implemented in addition to the iteration methods above.

The proposed 2-G-map kernel provides a sound basis for modeling subdivision of 2-manifolds (surfaces). It makes a clear distinction between topology and geometry, as embeddings are only referenced through one single topological element, the dart. Implementation in C++ is more or less straightforward, and the amount of code is limited. A 3-G-map kernel for modelling subdivision of 3D objects can be designed similar to the 2-G-map kernel.

Nevertheless, there are some restrictions. First of all, the implementation will be fixed for a specific dimension. The limitation to standard k-orbits makes it impossible to introduce new topological entities such as the β and γ involutions suggested in the extended G-map in Section 2.3. In addition, the topological part of the model is not extendible from an application programmer's point of view since the classes have a fixed predefined implementation without virtual functions.

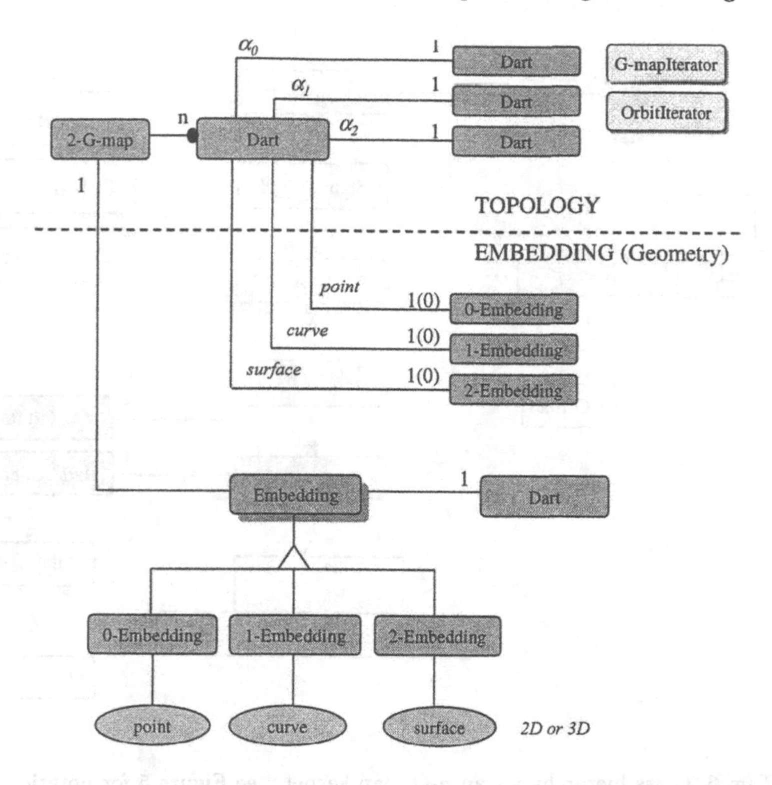

Fig. 5. Class diagram for a 2-G-map kernel. The line with a bullet at the end indicates a list and the lines with a "1" at the end indicate a pointer. The shaded box indicates an abstract C++ class, and the diamond symbol indicates class inheritence.

3.2 A Generic n-G-map Kernel

In this section we elaborate further on the basic ideas given above for the 2-G-map kernel and present a generic topology kernel for n-G-maps without the restrictions mentioned above. The main philosophy behind the design is that an arbitrary data structure can be accessed through the topology kernel. In addition, it provides flexible means for associating embeddings to the topological cells, not necessarily having a geometric interpretation. For example, graphic attributes can be specified for presentation purposes.

Figure 6 shows the global class hierarchy for the topological kernel with embeddings. As indicated by the shaded abstract C++ classes , the library is generic in the sense that it allows an application programmer to add derived classes at various locations in the class hierarchy. The classes `n-G-map`, `Dart` and `Embedding` are similar to those of Section 3.1 except that they can handle any dimension n, and they can deal with any kind of involutions and orbits. The classes `G-mapIterator` and `OrbitIterator` are also basically as

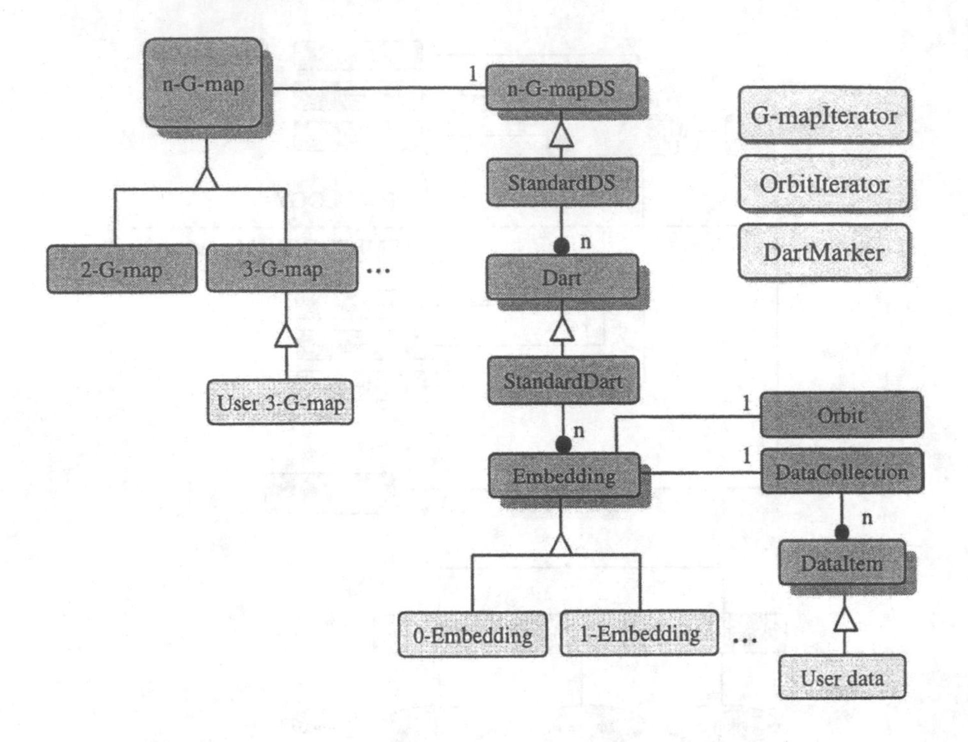

Fig. 6. Class hierarchy for an n-G-map kernel. See Figure 5 for notation.

defined previously, but extended in such a way that they can handle arbitrary involutions and orbits, such as β and γ involutions defined in the extended G-map definition in Section 2.3.

A hierarchy of operations for generalized maps embedded in 3-dimensional Euclidean space, which detailed the high level operations involved when building 3D geometric objects, was specified algebraically by Bertrand & Dufourd [3]. We have used these ideas for establishing the interface between arbitrary data structures and the basic operations in the n-G-map kernel.

The classes `n-G-mapDS` *(n-dimensional generalized map data structure)* and `Dart` declare these basic operations through pure virtual functions in C++. Their actual implementation in derived classes establishes the link between the topology kernel and a specific underlying data structure. Thus, building an interface to a data structure consists in deriving a class from `n-G-mapDS`, a class from `Dart`, and implementing functions which are virtual in the base classes.

Figure 7 shows two different implementations. The first, which is provided by the kernel library and denoted the "standard implementation", implements the topology representation as a list of darts similar to the 2-G-map kernel in Section 3.1 (classes `StandardDS` and `StandardDart`). The other is

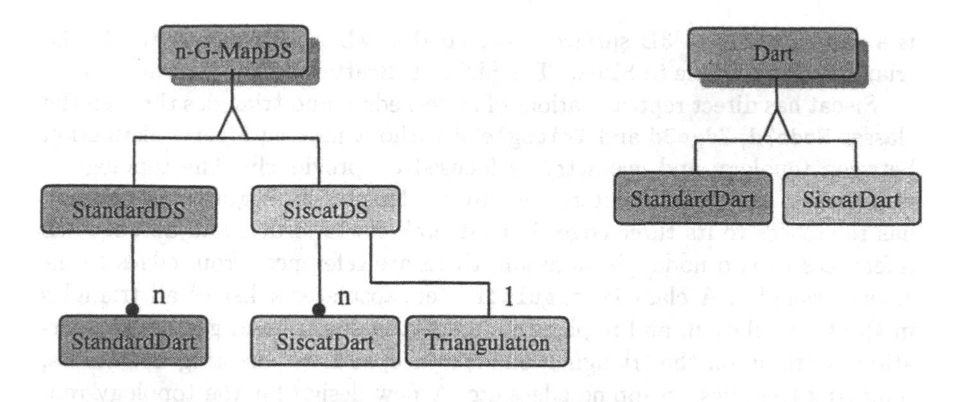

Fig. 7. **Interface to data structures for the generic n-G-map kernel.**

an interface to the data structure for surface triangulations in Siscat, The SINTEF Scattered Data Library [14], which will be outlined in more detail in Section 3.3.

An application programmer can define new topological entities such as β and γ involutions in the extended n-G-map definition in Section 2.3. Since the n-G-map kernel should deal with any kind of involutions defined by an application programmer, we suggest that involutions be described by enumerations in the standard implementation. An orbit, which is implemented in a class `Orbit`, is thus created from a set of enumerations and initialized with a dart, c.f. Definition 2, and attached to an embedding as indicated in Figure 6. This standard implementation is similar to the 2-G-map kernel in Section 3.1, except that references from topology to embedding are managed by functions in the dart classes instead of by pointers.

The classes `DataCollection` and `DataItem` provide mechanisms for assigning properties to embeddings, for example information that do not necessarily have a geometric meaning, such as colors for drawing objects on a graphic display. An in-depth description of these mechanisms can be found in [9].

Many exploration algorithms require a tool for associating one or more boolean marks with a dart. The class `DartMarker` provides functionality to set and remove marks, and to check whether a dart of the G-map is marked or not. The current implementation stores references to marked darts in an array, but a hash table would be more efficient when checking whether a dart is marked or not.

3.3 A Micro Topology Example

We give an example of how to interface an existing topology structure with the n-G-map kernel described in the previous section. The topology structure is a subdivision of a 3D surface into triangles, which is implemented in the triangulation module in Siscat, The SINTEF Scattered Data Library [14].

Siscat has direct representations of nodes, edges and triangles through the classes `Node3d`, `Edge3d` and `Triangle3d` without making a clear distinction between topology and geometry as focused on previously. The topology is represented as a fixed pointer structure as depicted in Figure 8: A triangle has references to its three edges in counterclockwise order, and an edge has references to two nodes. In addition, there are references from edges to incident triangles. A class `Triangulation` encapsulates a list of all triangles in the triangulation, and implements many topological and geometric operations working on the triangulation structure, such as inserting new nodes, removing triangles, swapping edges etc. A new design for the topology representations based on the concepts of darts and involutions, similar to that of G-maps, was suggested in [1].

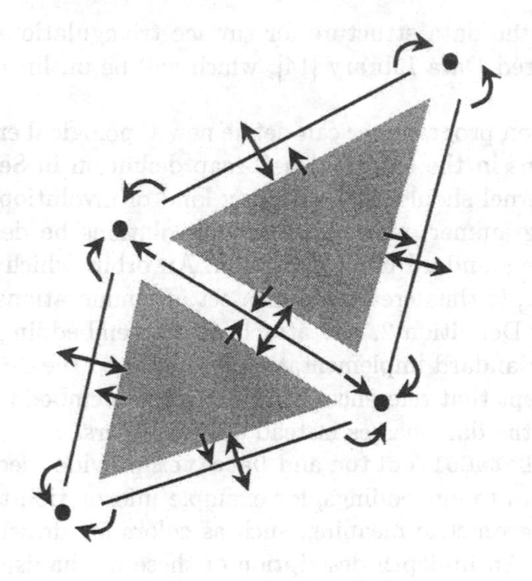

Fig. 8. Pointer structure for surface triangulations in Siscat.

The aim is to attach the data structure for triangulations in Siscat to the n-G-map topology kernel by deriving classes from `n-G-mapDS` and `Dart` as described in Section 3.2. Rather than allocating memory for each dart of the equivalent 2-G-map of the triangulation, we use an idea from [1] where a dart is a dynamic element which changes its content and position in the topological model through α involutions.

Consider the triple (*node, edge, triangle*) drawn as shaded elements in Figure 9a. Figure 9b shows the equivalent 2-G-map graph of the same triangulation. The following table shows what elements of the triple are kept fixed under involutions α_i, $i = 0, 1, 2$:

	node	**edge**	**triangle**
α_0		fixed	fixed
α_1	fixed		fixed
α_2	fixed	fixed	

We see that an α_i involution changes only the ith element of the triple. This property reflects exactly the definition of a dart in a 2-G-map. In other words, a dart of a Siscat triangulation is equivalent to a consistent triple (*node, edge, triangle*). In this context, a dart can be implemented as a class derived from `Dart` with three pointers to: `Node3d`, `Edge3d`, and `Triangle3d` objects.

Implementation of the involutions α_i, $i = 0, 1, 2$, is straightforward by changing node, edge or triangle respectively in the triple, see Figure 9c.

- α_0: The edge in the triple has pointers to two nodes. Replace the node in the triple with the other node referenced by the edge.
- α_1: Replace the edge in the triple with the other edge incident to the node and the triangle in the triple.
- α_2: If the edge in the triple is an internal edge in the triangulation, it has pointers to two triangles. Replace the triangle with the other triangle referenced by the edge. If the edge is on a boundary, the dart (triple) represents a fixed point of the α_2 involution and the triple is not changed.

Some operations must be implemented in order to initialize a dart with a (*node, edge, triangle*) triple. The 2-G-map class should provide mechanisms for creating a dart from different types of information, for example, building a dart from a node, from an edge, or from a triangle. With the pointer structure described above, building a dart from a node would be of order $O(n)$, n being the number of triangles. The two other operations would be $O(1)$.

4 A 3D Geological Modeler Based on 3-G-maps

A 3D geological modeler based on a 3-G-map kernel was developed in the ESPRIT project *OMEGA* (Object Oriented Methods Environment for Geoscience Applications). One of the goals of the project was to develop a set of tools for generating geological models from interpreted seismic data. The general approach was to base the modeler on CAD (Computer Aided Design) technology, and add geometric and topological functionality required for modeling 3D geological objects. The resulting software, which we call the OMEGA platform, is based on the CAD system *CAS.CADE* [4] from Matra Datavision, *Siscat*, The SINTEF Scattered Data library [14], and a topological G-map kernel similar to the kernel proposed in Section 3.2. A detailed presentation of the work done on geometry and topology construction was presented in [6].

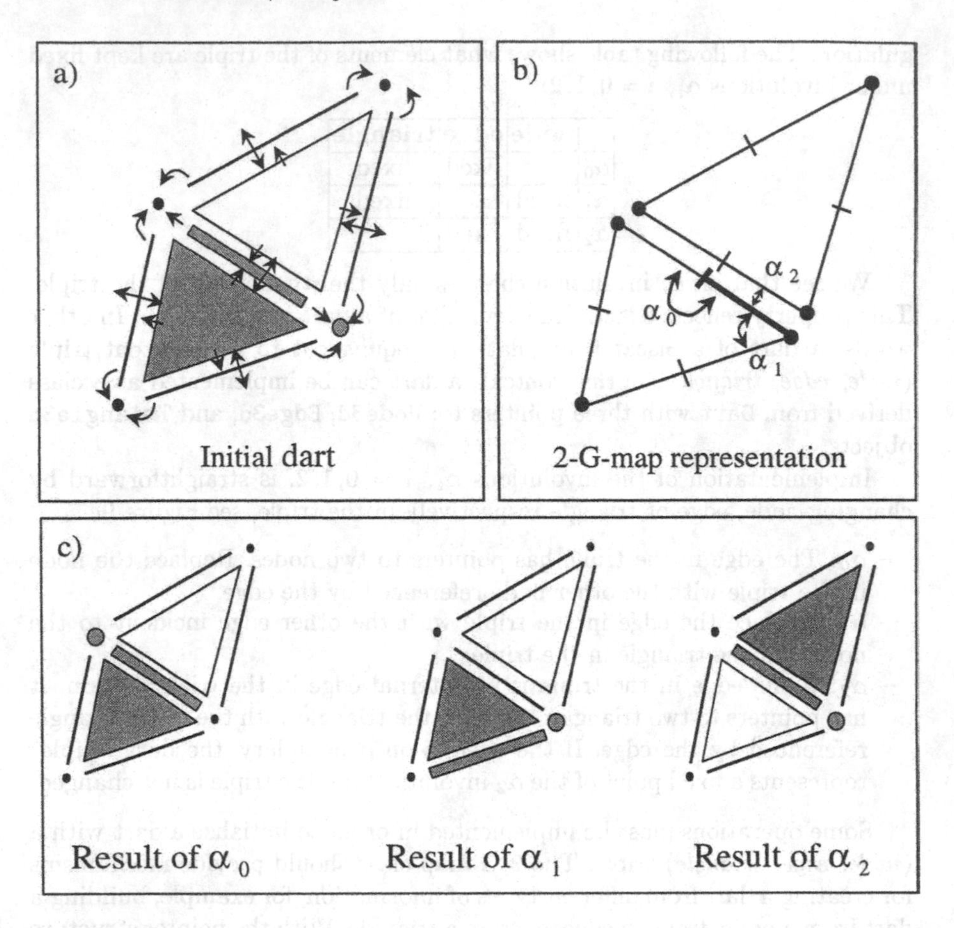

Fig. 9. Involutions on a triangulation structure.

4.1 Fault Network Modeling

The main problem studied in OMEGA concerns solid modeling of fault networks, i.e. the topology construction of subdivision of volumes delimited by 3D surfaces being geological faults. This can also be the starting point for constructing other topological models for geological modeling since it often involves the same methodology. The general approach is to describe a fault network by:

- *a set of geological faults*: a fault is represented as a 3D parametric surface
- *a working volume*: a closed bounding box covering the volume of interest. This is the volume to be subdivided by the surfaces representing faults
- *a set of relations between faults*: so far, only the relations *"Fault A stops on Fault B"* have been used, see Figure 10.

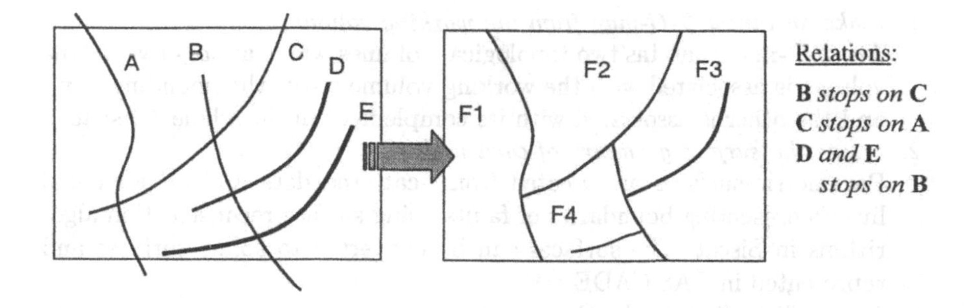

Fig. 10. Geological fault network shown in two dimensions.

Figure 10 shows a 2-dimensional equivalent to a fault network where the working volume is a rectangle and geological faults are curves. The working volume and an initial set of five faults (A, B, C, D and E) are shown left, and a resulting topological model embedded in the plane is shown right. Four distinct faces are created, where two of them, $F3$ and $F4$, have non-manifold "hanging" edges.

The construction process inserts each fault of the network individually into the working volume. As each fault is inserted, it creates a new subdivision of the working volume which is used in the next insertion step. Geometric information, such as the intersection curve between two surfaces or connectivity relations between faults, specifies which topological operations are needed in the subdivision process.

4.2 Implementation

The topology representation in CAS.CADE is based on incident graphs where k-cells have no direct references to topological cells of higher dimension, for example, there are no references from edges to faces, see Figure 11. This is not sufficient for a geological modeler that must provide fast iteration algorithms working on huge geological models with complex topology. Thus, an n-G-map kernel similar to the topological kernel in Section 3.2, but restricted to $n = 3$, was implemented.

The topology structure is a macro topology, that is, 3D points, curves and parametric surfaces are associated with topological vertices, edges and faces respectively. A topological volume is represented by its boundary faces only and has no associated geometry.

The main topology construction software implemented concerns construction of fault networks as described in the previous section. Basically, the steps are as follows:

1. *make an initial 3-G-map from the working volume*
   The 3-G-map contains two topological volumes which are α_3-sewed. One volume is associated with the working volume itself (the bounding box), and the other is associated with its complement in the whole 3D space.
2. *create the surface geometry of each fault*
   Parametric surfaces are created from scattered data and/or polygonal lines representing boundaries of faults using surface reconstruction algorithms in Siscat. The surfaces can be converted to spline surfaces and represented in CAS.CADE.
3. *insert all faults into the G-map*
   The geometry of each fault is associated with a topological face of a 3-G-map. When this step is accomplished, the G-map is composed of the initial working volume and the faces associated to faults. Still, topological relations for carrying out involutions and permutations are not established.
4. *compute intersection curves between faults and update the G-map*
   For each pair of faults:
   (a) Compute the intersection curve, if any, using surface-surface intersection algorithms in CAS.CADE.
   (b) The topological faces in the G-map associated with the faults are subdivided accordingly, and topological relations are updated (including β-involutions and γ-permutations of the extended G-map definition in Section 2.3).
   (c) Remove geometry outside the working volume by clipping faults against the bounding box. Update the G-map accordingly.

When the last step is accomplished, the fault network is represented as a 3-G-map embedded in the 3-dimensional Euclidean space. Several tools allow us to find shortest paths between vertices, display the topological model or the geometric model, iterate on the topological cells, etc.

It should be mentioned that there are weak points in the current implementation and work that remains. The most critical point is step 4 above. Intersecting surfaces relies on the surface-surface intersection algorithms in CAS.CADE which were originally designed for CAD models. Although they seem to work for simple geological surfaces with C^1 continuity, more complex and irregular surfaces are not handled correctly. Also, surfaces with only C^0 continuity, like surface triangulations embedded with planar patches, cannot be handled by the current intersection algorithms.

5 Conclusion and Further Work

Generalized maps provide a sound basis for implementing the topological representations of geometric models on computers. The great advantage of G-maps over many other topological models is the ease and efficiency with which one can navigate through the model. The few and clear basic concepts

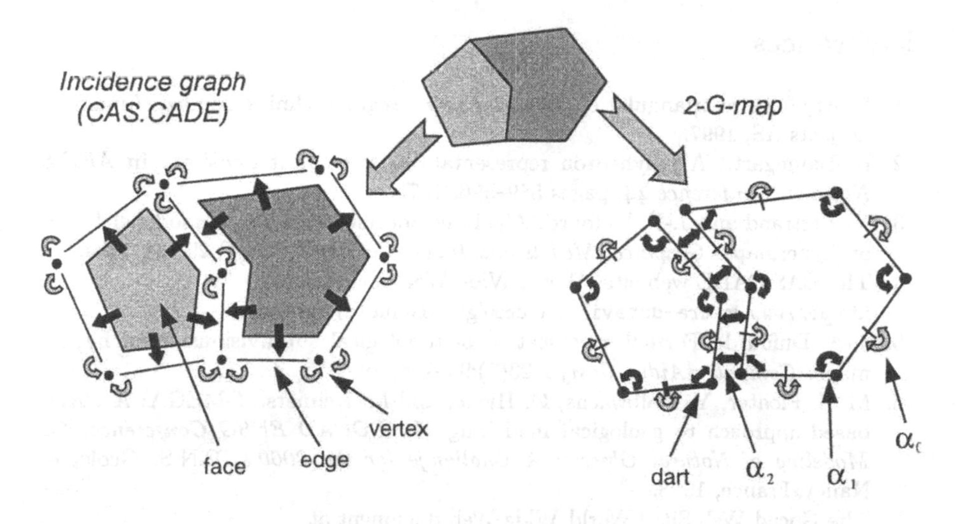

Fig. 11. Incident graph representation in CAS.CADE (left) and the corresponding 2-G-map graph $G = (D, \alpha_0, \alpha_1, \alpha_2)$ (right).

that G-maps are based on are intuitive and easy to deal with for application programmers and end users. Topological kernels are easy to implement, and moreover, the G-map concept can easily be extended with new functionality to meet the needs of a particular application.

On the other hand, implementing the topology construction operations might be cumbersome. In the aforementioned OMEGA project, we rely heavily on surface-surface intersection algorithms for finding the 3D curve embeddings in the topological model, and the topology construction process is more or less automatic. Many difficulties have been encountered such as finding the correct intersection curves between surfaces, determining the orientation of curves and surfaces, and sewing the topological elements together into a topological model without anomalies.

An alternative would be to choose a less ambitious approach by implementing more interactive facilities allowing the user to guide the topology construction process in more detail. A careful design of such manipulation and decision tools would greatly improve the topology construction process.

Acknowledgements

This work was supported partly by the European ESPRIT Project 23245 - OMEGA, and partly by the Research Council of Norway under the research program 110673/420 (Numerical Computations in Applied Mathematics). The authors also thank Professor Michel Perrin at Ecole des Mines de Paris for fruitful discussions on fault network modeling.

References

1. E. Arge. Siscat triangulation, ideas for new design. Technical report, Numerical Objects AS, 1997.
2. B. Baumgart. A polyhedron representation for computer vision. In *AFIPS National Conference 44*, pages 589–596, 1975.
3. Y. Bertrand and J.-F. Dufourd. Algebraic specification of a 3d-modeller based on hypermaps. *Graphical Models and Image Processing*, 56(1):29–60, 1994.
4. The CAS.CADE web site. World Wide Web document at `http://www.matra-datavision.com/gb/products/cascade/`.
5. J.-F. Dufourd. Formal specification of topological subdivisions using hypermaps. *Computer Aided Design*, 23(2):99–116, 1991.
6. M. S. Floater, Y. Halbwachs, Ø. Hjelle, and M. Reimers. OMEGA: A CAD-based approach to geological modelling. In *GOCAD ENSG Conference, 3D Modeling of Natural Objects: A Challenge for the 2000's*, E.N.S. Géologie, Nancy, France, 1998.
7. The Gocad Web Site. World Wide Web document at `http://www.ensg.u-nancy.fr/GOCAD/gocad.html`.
8. Y. Halbwachs, G. Courrioux, X. Renaud, and P. Repusseau. Topological and geometric characterization of fault networks using 3-dimensional generalized maps. *Mathematical Geology*, 28(5):625–656, 1996.
9. Ø. Hjelle. Data structures in siscat. Report STF42 A96007, SINTEF Applied Mathematics, 1996.
10. P. Lienhardt. Subdivision of n-dimensional spaces and n-dimensional generalized maps. In *5th ACM Symposium on Computational Geometry*, pages 228–236, Saarbrucken, Germany, 1989.
11. P. Lienhardt. Topological models for boundary representation: A comparison with n-dinensional generalized maps. *Computer-Aided Design*, 23(1):59–82, 1991.
12. P. Lienhardt. N-dimensional generalized maps and cellular quasi-manifolds. *Intern. Jour. Computational Geometry and Applications*, 4(3):275–324, 1994.
13. J. A. Mchugh. *Algorithmic Graph Theory.* Prentice-Hall Inc., 1990.
14. The Siscat Home Page. World Wide Web document at `http://www.oslo.sintef.no/siscat`.
15. B. L. Stephane Contreaux and J.-L. Mallet. A celluar topological model based on generalized maps: The GOCAD approach. In *GOCAD ENSG Conference. 3D Modelling of Natural Objects: A Challenge for the 2000's*, Nancy, 1998.
16. W. T. Tutte. *Graph Theory*, volume 21 of *Encyclopedia of Mathematics and its Applications.* Addison-Wesley, Reading, MA., 1984.
17. K. Weiler. Edge based data structures for solid modeling in curved-surface environments. *IEEE Computer Graphics and Applications*, 5(1):21–40, 1985.
18. K. Weiler. *Topological Structures for Geometric Modeling.* PhD thesis, Rensselaer Polytechnic Institute, 1986.

Author Index

Editorial Policy

§1. Volumes in the following four categories will be published in LNCSE:

i) Research monographs
ii) Lecture and seminar notes
iii) Conference proceedings
iv) Textbooks

Those considering a book which might be suitable for the series are strongly advised to contact the publisher or the series editors at an early stage.

§2. Categories i) and ii). These categories will be emphasized by Lecture Notes in Computational Science and Engineering. **Submissions by interdisciplinary teams of authors are encouraged.** The goal is to report new developments – quickly, informally, and in a way that will make them accessible to non-specialists. In the evaluation of submissions timeliness of the work is an important criterion. Texts should be well-rounded, well-written and reasonably self-contained. In most cases the work will contain results of others as well as those of the author(s). In each case the author(s) should provide sufficient motivation, examples, and applications. In this respect, Ph.D. theses will usually be deemed unsuitable for the Lecture Notes series. Proposals for volumes in these categories should be submitted either to one of the series editors or to Springer-Verlag, Heidelberg, and will be refereed. A provisional judgment on the acceptability of a project can be based on partial information about the work: a detailed outline describing the contents of each chapter, the estimated length, a bibliography, and one or two sample chapters – or a first draft. A final decision whether to accept will rest on an evaluation of the completed work which should include

- at least 100 pages of text;
- a table of contents;
- an informative introduction perhaps with some historical remarks which should be accessible to readers unfamiliar with the topic treated;
- a subject index.

§3. Category iii). Conference proceedings will be considered for publication provided that they are both of exceptional interest and devoted to a single topic. One (or more) expert participants will act as the scientific editor(s) of the volume. They select the papers which are suitable for inclusion and have them individually refereed as for a journal. Papers not closely related to the central topic are to be excluded. Organizers should contact Lecture Notes in Computational Science and Engineering at the planning stage.
In exceptional cases some other multi-author-volumes may be considered in this category.

§4. Category iv) Textbooks on topics in the field of computational science and engineering will be considered. They should be written for courses in CSE education. Both graduate and undergraduate level are appropriate. Multidisciplinary topics are especially welcome.

§5. Format. Only works in English are considered. They should be submitted in camera-ready form according to Springer-Verlag's specifications. Electronic material can be included if appropriate. Please contact the publisher. Technical instructions and/or TEX macros are available via http://www.springer.de/author/tex/help-tex.html; the name of the macro package is "LNCSE – LaTEX2e class for Lecture Notes in Computational Science and Engineering". The macros can also be sent on request.

General Remarks

Lecture Notes are printed by photo-offset from the master-copy delivered in camera-ready form by the authors. For this purpose Springer-Verlag provides technical instructions for the preparation of manuscripts. See also *Editorial Policy*.

Careful preparation of manuscripts will help keep production time short and ensure a satisfactory appearance of the finished book. The actual production of a Lecture Notes volume normally takes approximately 12 weeks.

The following terms and conditions hold:

Categories i), ii), and iii):
Authors receive 50 free copies of their book. No royalty is paid. Commitment to publish is made by letter of intent rather than by signing a formal contract. Springer-Verlag secures the copyright for each volume.

For conference proceedings, editors receive a total of 50 free copies of their volume for distribution to the contributing authors.

Category iv):
Regarding free copies and royalties, the standard terms for Springer mathematics monographs and textbooks hold. Please write to Peters@springer.de for details. The standard contracts are used for publishing agreements.

All categories:
Authors are entitled to purchase further copies of their book and other Springer mathematics books for their personal use, at a discount of 33,3 % directly from Springer-Verlag.

Addresses:

Professor M. Griebel
Institut für Angewandte Mathematik
der Universität Bonn
Wegelerstr. 6
D-53115 Bonn, Germany
e-mail: griebel@iam.uni-bonn.de

Professor D. E. Keyes
Computer Science Department
Old Dominion University
Norfolk, VA 23529–0162, USA
e-mail: keyes@cs.odu.edu

Professor R. M. Nieminen
Laboratory of Physics
Helsinki University of Technology
02150 Espoo, Finland
e-mail: rni@fyslab.hut.fi

Professor D. Roose
Department of Computer Science
Katholieke Universiteit Leuven
Celestijnenlaan 200A
3001 Leuven-Heverlee, Belgium
e-mail: dirk.roose@cs.kuleuven.ac.be

Professor T. Schlick
Department of Chemistry and
Courant Institute of Mathematical
Sciences
New York University
and Howard Hughes Medical Institute
251 Mercer Street, Rm 509
New York, NY 10012-1548, USA
e-mail: schlick@nyu.edu

Springer-Verlag, Mathematics Editorial
Tiergartenstrasse 17
D-69121 Heidelberg, Germany
Tel.: *49 (6221) 487-185
e-mail: peters@springer.de
http://www.springer.de/math/
peters.html

Lecture Notes in Computational Science and Engineering

Vol. 1 D. Funaro, *Spectral Elements for Transport-Dominated Equations.* 1997. X, 211 pp. Softcover. ISBN 3-540-62649-2

Vol. 2 H. P. Langtangen, *Computational Partial Differential Equations.* Numerical Methods and Diffpack Programming. 1999. XXIII, 682 pp. Hardcover. ISBN 3-540-65274-4

Vol. 3 W. Hackbusch, G. Wittum (eds.), *Multigrid Methods V.* Proceedings of the Fifth European Multigrid Conference held in Stuttgart, Germany, October 1-4, 1996. 1998. VIII, 334 pp. Softcover. ISBN 3-540-63133-X

Vol. 4 P. Deuflhard, J. Hermans, B. Leimkuhler, A. E. Mark, S. Reich, R. D. Skeel (eds.), *Computational Molecular Dynamics: Challenges, Methods, Ideas.* Proceedings of the 2nd International Symposium on Algorithms for Macromolecular Modelling, Berlin, May 21-24, 1997. 1998. XI, 489 pp. Softcover. ISBN 3-540-63242-5

Vol. 5 D. Kröner, M. Ohlberger, C. Rohde (eds.), *An Introduction to Recent Developments in Theory and Numerics for Conservation Laws.* Proceedings of the International School on Theory and Numerics for Conservation Laws, Freiburg / Littenweiler, October 20-24, 1997. 1998. VII, 285 pp. Softcover. ISBN 3-540-65081-4

Vol. 6 S. Turek, *Efficient Solvers for Incompressible Flow Problems.* An Algorithmic and Computational Approach. 1999. XVII, 352 pp, with CD-ROM. Hardcover. ISBN 3-540-65433-X

Vol. 7 R. von Schwerin, *Multi Body System SIMulation.* Numerical Methods, Algorithms, and Software. 1999. XX, 338 pp. Softcover. ISBN 3-540-65662-6

Vol. 8 H.-J. Bungartz, F. Durst, C. Zenger (eds.), *High Performance Scientific and Engineering Computing.* Proceedings of the International FORTWIHR Conference on HPSEC, Munich, March 16-18, 1998. 1999. X, 471 pp. Softcover. 3-540-65730-4

Vol. 9 T. J. Barth, H. Deconinck (eds.), *High-Order Methods for Computational Physics.* 1999. VII, 582 pp. Hardcover. 3-540-65893-9

Vol. 10 H. P. Langtangen, A. M. Bruaset, E. Quak (eds.), *Advances in Software Tools for Scientific Computing.* 2000. X, 357 pp. Softcover. 3-540-66557-9

Vol. 11 B. Cockburn, G. Karniadakis, C.-W. Shu (eds.), *Discontinuous Galerkin Methods: Theory, Computation and Applications.* 2000. X, 452 pp. Hardcover. 3-540-66787-3

For further information on these books please have a look at our mathematics catalogue at the following URL: `http://www.springer.de/math/index.html`

Druck: Strauss Offsetdruck, Mörlenbach
Verarbeitung: Schäffer, Grünstadt

